Models for Investors in
Real World Markets

Models for Investors in Real World Markets

JAMES R. THOMPSON
Noah Harding Professor of Statistics
Rice University

EDWARD E. WILLIAMS
Henry Gardiner Symonds Professor of Management
Rice University

M. CHAPMAN FINDLAY, III
Principal: Findlay, Phillips and Associates

WILEY-
INTERSCIENCE

A JOHN WILEY & SONS, INC., PUBLICATION

Library of Congress Cataloging-in-Publication Data:

Thompson, James R
  Models for investors in real world markets / James R. Thompson, Edward E Williams, M. Chapman Findlay, III
     p cm — (Wiley series in probability and statistics)
  Includes bibliographical references and index.
     ISBN 0-471-35628-X (cloth)
  1 Investments—Statistical methods. 2 Securities—Statistical methods. 3. Stocks—Statistical methods I. Williams, Edward E. II Findlay, M. Chapman. III. Title. IV. Series
     HG4515.2 T48 2002
     332 63'22—dc21   2002009949

To my wife, Ewa Majewska Thompson
James R. Thompson

To Susan, Laura, David, Garrett, Morgan, and Jacklyn
Edward E. Williams

To my wife, Beatrice Findlay
M. Chapman Findlay, III

Contents

Preface

The bloom is off investment strategies based on *risk neutral probabilities*. Operating from their famed trading room, the Black Box, the "wise men" at Enron acted like holders of higher truth derived from sacred scrolls. Lesser beings who could not comprehend the soundness of Enron's strategy simply assumed they were missing something and went along. The result was the largest bankruptcy in American history, with tens of thousands thrown out of work and wondering what happened to their Enron-based 401k plans. The Big Five accounting firm of Andersen had some explaining to do as to how it had failed to question six straight years of highly imaginative Enron bookkeeping. At the writing of this book, it is unclear whether Enron's executives will be able successfully to hide behind a combination of exotic mathematical models and exotic accounting practices to avoid criminal as well as civil prosecution. Again and again, we hear the *dies irae* of the market: "If it sounds too good to be true, it probably is."

The collapse in 1998 of the Scholes-Merton backed Long Term Capital Management hedge fund should have given pause to those who take the efficient market based formulae for "fair prices" in options and derivatives as rigid laws. There should surely be some embarrassment to the neoclassical establishment in the obvious failure of a Black-Scholes based fund, one indeed conspicuously organized and consulted by Merton and Scholes, within six months of their having received the Nobel Prize for the discovery of the "law" used to manage the LTCM. And, indeed, some embarrassment should accrue to the neoclassical Chairman of the Federal Reserve Board, Alan Greenspan, who organized a $3.5 billion bailout of the fund as if those who followed efficient market orthodoxy should be rescued when their religion proved (temporarily, no doubt) false. The discipline of the "free" market is not, apparently, to be suffered by all.

There is an old joke illustrating the difference between communism and capitalism:

> Suppose you have two cows. Under communism, the government takes both from you, shoots one and milks the other, giving the milk to your neighbor who doesn't work. Under capitalism, you sell one cow and buy a bull. Your herd multiplies and the economy grows. You sell the herd and retire on the income.

A revised version of this joke has recently been circulating on the internet. It goes as follows:

Under the new capitalism of Enron, Scholes-Merton, and LTCM, the following applies: You have two cows. You sell three of them to your publicly listed company, using letters of credit opened by your brother-in-law at the bank, then execute a debt/equity swap with an associated general offer so that you get all four cows back with a tax deduction for five cows. The milk rights of the six cows are transferred via an intermediary to a Cayman Island company secretly owned by the Chief Financial Officer who sells the rights to all seven cows back to your listed company. The annual report says the company owns eight cows, with an option on one more. The securities analysts buy your bull.

Now, do you see why a company (Enron) with $62 billion in assets can be brought to the brink of bankruptcy? The fact that market analysts across the board tended to keep Enron as a "strong buy" almost until the time of its bankruptcy, explains the new market destabilizing skepticism of the investing public in accounting firms and stock analysts. If the highly respected Emperor Enron has no clothes, then what about all the other potentates of the market?

One might sympathize somewhat with Alan Greenspan rushing in to quell the loss of confidence which would have occurred had LTCM been allowed to go bust in 1998 by arranging a bail-out by a spectrum of brokerage houses and investment bankers. Had LTCM gone bust in 1998 the way Enron did in 2001, then the kind of investor paranoia which is devastating the market in 2002 would possibly have come earlier in 1998.

On the other hand, from the standpoint of the dollars involved, the crash of LTCM (a $3.5 billion bubble) was an order of magnitude less significant than that of the $62 billion Enron debacle in late 2001. Had nature been allowed to take its course with LTCM in 1998, it is likely that a general scrutiny of accounting practices might have precluded the devastating crash of Enron in 2001. Chairman Greenspan could have seen a natural dampening of "irrational exuberance" in 1998 had he simply let LTCM fail naturally. By bailing out LTCM on the one hand, and stifling investment capital on the other, it appears he acted with the wisdom one tends to associate with the economic planning of, say, Argentina.

Believers in free markets do not believe that all the players are honest. But they believe that allowing the market to work its will tends to weed out fictitious accounting practices and pyramid schemes sooner rather than later. Amidst all the Congressional inquisitions being carried out against the managers of Enron and its accounting consultants, nobody is likely to point the finger of accusation toward the Chairman of the Federal Reserve Bank. But by covering up the manageable LTCM brush fire, Alan Greenspan may well have facilitated the conflagration which has followed the collapse of Enron.

The world is a complex place, and economic theory does not admit, and probably never will admit, of the same precise models which characterize, say, mechanics or electromagnetic flux. Any market systematology which fails constantly to question and stress its models with "what if?" scenarios will almost surely bring about disappointment, sooner rather than later.

The general orthodoxy of derivative pricing is based on *risk neutral probabilities*. The dealer sells a derivative for such a price that risk is eliminated. Like a bookmaker in a horse race, he makes his money whichever horse wins, on trading commissions and the like. Persons who believe in such a world must stop their ears, lest they hear the continuing *memento mori*, "Enron! Enron! Enron!"

There is much to fault the current orthodoxy of neo-classical economists. For example, the assumption of security progression according to a geometric Brownian walk is hugely unrealistic. Such models do not realistically admit of a large (say 5%) 24-hour change in a stock price. Yet, in the real world changes of such magnitude happen to stock prices very frequently. If these random walk assumptions were really the whole story, then one could simply invest in a broadly based index fund and be confident that his/her capital would grow at the 10% long-term average which, indeed, the market demonstrates. An index fund would become an irresistible alternative to a savings account at the bank paying, typically, 5%.

Broadly based bear markets do happen, as the 2000−2001 collapse indicates. Such bear markets tend to make diversification an ineffective prophylactic. Hence, there is a price to be paid for putting money into the market. This price is *risk*. Whereas gain is easily defined as dividends plus stock price appreciation, risk is hard to grasp. The usual means of defining risk is based on *volatility*, which is some version of the time averaged standard deviation of the stock price. Harry Markowitz won a Nobel Prize for posing the portfolio selection problem as setting the volatility at the edge of the portfolio owner's comfort zone, and then picking stocks which would maximize expected gain subject to the volatility constraint. The solution of the Markowitz portfolio problem in this age of fast computing is child's play. However, as Peter Bernstein has made abundantly clear in his iconoclastic *Against the Gods: The Remarkable Story of Risk*,

$$\text{risk} \neq \text{volatility}.$$

It is insufficient (or, rather, it should be insufficient) for a fund manager to claim that he is not to blame for the disastrous performance of the fund.

He had computer code to prove that he had followed Markowitz to the letter. It is not his fault if something (e.g., terrorism, Greenspan, the SEC) untoward had happened to the market.

Events such as the attack of terrorists on the United States on September 11, 2001, are excellent excuses for failed investments. How can one be expected to deal with such hits on the market? And the Osama bin Laden strike was only one of the more obvious causes of "bear glitches" on the market. So far as the market is concerned, the decision of Alan Greenspan dramatically to increase interest rates was probably a major cause of the high-tech crash in 2000.[1] Then there was the decision of the Federal Trade Commission to oppose mergers. It appears that companies advancing new technologies, as they seek for appropriate markets and applications, are particularly needful of the ability to create combinations and spinoffs. Stifling this ability causes a significant hit on the market.

The fact is that it is unwise for investors to assume that America is safe from enemies, foreign or domestic. Moreover, it is also unwise for investors to assume that their government will carry out uniformly intelligent economic policies. There is ample historical evidence to indicate that both assumptions are wrong. We cannot readily forecast wars or the ability of bureaucrats to make big blunders. But we should rely on the inevitability of both and, to the extent possible, make investments in the light of such inevitabilities. There are many ways one can characterize the security markets starting in the 1920s. Our research shows that geometric Brownian motion is only part of the story. A really significant part, which is not readily explained by Brownian walks, is the fact that bear markets appear to impact themselves rather quickly on the market. It is true that the market has grown at roughly a 10% compounded rate over this time period. However, if one looks at the years of movement on the exchanges, it is almost as though the major indices behaved like an insured bank account in a country subject to random devaluations of the currency. That is, in years in which no bear hits on the index occur, the market grows very much. Investors over longer periods of time, have the advantage of the fact that in roughly 70% of the years, the index of large cap U.S. stocks rises rather than falls. And there is the further encouraging news that in over 40% of the years, the index has risen by over 20%. In 30% of the years, the market rises by over 25%. And in 25% of the years, the index has risen by over 30%. Over the roughly 75-year period such records have been kept, the United States has lived through the Great Depression, the Second World War, the Cold War, Korea, Vietnam, assorted massive sociological changes, shifts toward

---

[1]The most recent collapse was probably caused by a large increase in the monetary base (in 1999)) in anticipation of Y2K. This undoubtedly fueled the bubble, which then burst when the monetary base was restored to more normal levels.

and away from free markets, and assorted epidemics.

For general bear movements across the market, a portfolio of stocks will not spare the investor from the decline. The manner in which bear declines occur is, of course, a matter of disagreement. There are many rules of thumb which might be followed. We will largely use the following one: 10% declines in the market occur, on the average, once a year. Larger declines (say 20%) occur, on the average, once every five years. We do not know precisely when these across the board declines will disturb our expected upward progress. But we know that they are very likely to happen with a fairly predictable frequency and make our plans accordingly. The prudent investor has two weapons for the diminution of risk. The first is diversification across a broad sector of the stock market. This will protect against Enrons that go bad. The second is time. That will protect against bear markets caused by terrorists, bureaucrats, and things that go bump in the night.

Laurence Siegel, treasurer of the Ford Foundation, defines risk rather forcefully, if imprecisely:

> ... risk is the possibility that, in the long run, stock returns will be terrible.

The fact is that volatility is not a good summary of risk. Its use as such enabled Markowitz to reduce portfolio design to a straightforward problem in constrained optimization. But, as the late John Tukey, Donner Professor of Statistics at Princeton and Associate Director of Bell Labs, reminded us a decade before Markowitz's portfolio optimization:

> Far better an approximate answer to the right question, which is often vague, than an exact answer to the wrong question, which can always be made precise.

For many years, analysts have tried to forecast the value of a particular stock at a time in the future. Generally speaking, this task has proved intractable. In this book, we are not forecasting the value of the stock at a future time. Rather we are forecasting a probability distribution of values of the stock at that future time. It might seem that such a task is even more hopeless than forecasting the stock value. Our experience indicates that, with a fast computer, the task is not hopeless at all.

We have also discovered that using a nonparametric resampling (not quite the same as the bootstrap) technique for forecasting the distribution of an index fund (based on the 75 years of the Ibbotson index) gives essentially the same results that one obtains using a geometric Brownian model (with parameters estimated from the same data base) for time periods beyond one or two years. The concordance between the two dramatically different approaches would seem to be mutually supportive of the efficacy of both.

One of the contributions of this book is an intuitive definition of *risk*. We create a mechanism whereby the investor can put in assumptions that apply to a security's progression in time, and then view, say, one year out the probabilities of one's position being worth at least 150% of its current value or, say, less than 30% of its current value. We refer to this mechanism as *risk profile simulation*. The progression of a stock or portfolio is a *stochastic process*, that is, for any value of time, there is an associated probability distribution of portfolio values. We want to give the user the opportunity to view the entire gamut of what, given the model assumptions, can happen from the very good to the very bad with associated probabilities throughout. These time sliced displays of the probability distribution function associated with the model assumptions we term *simugrams*. In other words, we will not be proposing to show what can happen on the average, but the entire risk profile.

For longer term forecasting of the probability distribution function of a stock (say more than three years) the plain geometric model unpatched by the Poissonian add-ons may well be sufficient. However for short-term forecasts it may be less volatile than justified by realism. Rather than starting from scratch, we take the tact of "patching" the less than realistic albeit mathematically tractable models beloved by the believers in Efficient Market Theory, in the light of things that can go wrong with them, and thus bring about better (though never true) models. The mathematics of adding in factors such as Poissonian bear market jumps downward is complex, and we bypass the difficulties of running down closed-form solutions to, say, likelihood functions, by demonstrating how simulation can be used to answer pertinent "what if?" questions.

We now summarize briefly the contents of this book. In Chapter 1, we outline some of the basic institutional factors associated with how stock markets operate and raise questions about whether these markets are efficient. We suggest that, in an efficient market, there would be very little that an investor or securities analyst could do to provide superior (above average) returns for his or her own accounts or for those of clients. We imply that, if markets are perfectly efficient, it makes no sense to study a particular security or, for that matter, read the typical 800-page "Investments" textbook required in college undergraduate or graduate classes. Further, it makes little sense for professionals to acquire such designations as the CFA (Chartered Financial Analyst). All securities should be correctly priced at any time, and it should make no difference when or what stocks are purchased.

In Chapter 2, the reader is introduced to the basic concepts of utility theory. The essence of utility analysis rests on the basic economic premise that people prefer more good things to less, and that a continued flow of good things will have increasingly reduced marginal value as larger quantities are enjoyed. The law of diminishing marginal utility underlies this conceptualization, and it gives shape to utility of income (wealth) functions that have positive first and negative second derivatives. Utility theory forms the cornerstone of portfolio theory, and is the underlying explanation for why most individual investors are presumed to be risk averse.

Chapter 3 illustrates why diversification is an important fundamental requirement of rational investment choice. The chapter begins by examining the traditional Markowitz constrained optimization model, which depends for its solution on notions of mean returns from individual stocks, variances of those returns, and covariance among the returns. We ultimately conclude that what is essentially a sound procedure (diversification) applied to a concept (risk) that is, at best, difficult to quantify, all put together with historical data, may have no relationship at all to the underlying facts in the future! Chapter 3 provides a better way of looking at risk through simugrams of future stock prices.

The materials in Chapters 4 and 5 provide a summary of the current theory as generally accepted in financial economics along with an extensive critique of that theory. Chapter 4 contains an explicit discussion of the capital asset pricing model (CAPM) and its more recently hatched cousin the arbitrage pricing model (APM). The chapter outlines various methods to make the CAPM and the APM operational, but leaves for Chapter 5 to provide a specific critique of both. It turns out that neither model is much accepted any place but in textbooks but there is nothing left to occupy the vacuum that remains if either (or both) models is (are) rejected. Chapter 5 is really a *coup de grace*, a stinging critique of modern financial economic theory. After one reads this chapter, it becomes clear that what is done in many investments textbooks really has no meaning, and it provides the rationale for *Models for Investors in Real World Markets*. We leave the chapter convinced that securities markets are not efficient and understand the need for a new approach for investing in the real world.

In Chapter 6, we examine the basic valuation equations derived by J.B. Williams seventy years ago. In this very simple model, the "value" of a security is simply the present value of the future income stream expected to be generated by the security in question. Since aspects of this model will appear at almost every turn in the work to follow, we reproduce it here.

$$P(0) \ = \ \frac{D(0)}{(1+r)^0} + \frac{D(1)}{(1+r)^1} + \frac{D(2)}{(1+r)^2} + \cdots \tag{1}$$

$$= \sum_{t=0}^{\infty} \frac{D(t)}{(1+r)^t},$$

where

$$
\begin{aligned}
P(0) &= \quad \text{Price (Value) today} \\
D(t) &= \quad \text{Dividend } t \text{ years in the future} \\
r &= \quad \text{Market's (investor's) required rate of return.}
\end{aligned}
$$

Next, we amplify these equations by assuming various probability distributions for the underlying variables (sales, expenses, earnings, dividends, etc.), and compute means and variances for them. We suggest that this sort of methodology could be used to generate price or "value" distributions where a given market (or individual) rate of return, r, is required. In this chapter, we do not specifically point out to the reader that this is essentially the methodology used by securities analysts from the 1930s at least until the early 1960s to determine "undervalued" stocks (those where the present value of the future income stream exceeds the existing market price) and "overvalued" stocks (those where the present value of the future income stream is less than the existing market price). Variations on this methodology are still used today by many analysts where the market rate of return, r, is taken from the capital asset pricing model (see discussion in Chapters 4 and 5).

In Chapters 7 and 8, we provide a detailed procedure for analyzing securities. Borrowing from concepts developed by Benjamin Graham and others from the 1930s to the 1960s, we take concepts that were often only verbally constructed and build a more quantitative framework for those who have the patience (and inclination) to study financial statements carefully. Some accounting background is useful for a thorough understanding of the materials in Chapters 7 and 8.

In Chapter 9, we examine the importance of compound interest and observe that growth is merely noisy compound interest. Albert Einstein once said that his theory of relativity was of no great particular significance in the overall scheme of the universe; but of compound interest he felt "now there is a truly important revelation." Chapter 10 continues to develop our conceptualization of risk profiling and applies the idea to bundles of securities (called portfolios). We argue that looking at the expected (present) values of an investment in and of itself is seldom a good idea.

We introduce and analyze the options market in Chapter 11 as a way of considering the notion of risk. It is thus our view that a simple application of equation (1) will not allow an investor to make a rational choice. This is an extremely important point since many modern-day analysts and investors try to make decisions using variations of the equation by discounting essentially an uncertain (in the sense used in Chapter 5) income stream by means of taking a discount rate from the CAPM. Since the notion of "risk" is specifically incorporated into the evaluation by assuming that risk can be determined by means, variances, and covariances of historical stock returns (see discussion of the CAPM in Chapter 4), we see why the procedure is inappropriate. In Chapter 11, we also reproduce the Black-Scholes-Merton option pricing equation and suggest some serious problems associated with its "search" for risk neutrality. Confining oneself to looking only at mean values from a geometric Brownian walk model is reckless in the extreme. One must develop more realistic models and examine the entire realm of possible outcomes given input assumptions. This will occupy our thoughts throughout the book. Finally, in Chapter 12, we offer a short summary of the book and draw our conclusions.

This brief summary may also suggest the philosophy which underlies this book. We believe that the writers during the 1930s−1960s period (a diverse group running from Ben Graham to J.M. Keynes) were trying to confront real problems and had good ideas. They lacked modern analytical tools, large databases, and the cheap and vast computational power currently available. As the latter became available, the investments literature took a strange turn. Instead of solving these problems, they were increasingly viewed as having been eliminated by assumptions such as efficiency. equilibrium, and no arbitrage conditions. This has not caused the problems to go away, it has simply allowed them to be ignored. Although we make no pretense of having solved all the problems in this book, our goal is at least to address them using modern tools.

We acknowledge the support of the Army Research Office (Durham) under (ARO-DAAD19-99-1-0150). We acknowledge the permission of Ibbotson Associates to use information from their large cap data base in Chapter 9. The support of Dr. Scott Baggett in the writing of programs for optimizing portfolios using the simugram time-slice approach is gratefully acknowledged. We wish to thank Dr. Robert Launer, Director of Applied Mathematics of the Army Research Office and Professor Katherine Ensor, Director of the Rice Center for the Study of Computational Finance and Economic Systems, for their interest and support of this project.We also wish to thank Leticia Gonzales, Sarah Gonzales, Diane Brown, Fernando Gonzalez, Jason Deines, Rick Ott, Gretchen Fix, Ginger Davis, Matthias Mathaes, Ita Cirovic, Chad Bhatti, Olivier Pollak, Rick Russell, Melissa Yanuzzi, Heather Haselkorn, and Jamie Story. We give our special thanks to John Dobelman for his penetrating comments on the manuscript.

James R. Thompson
Edward E. Williams
M. Chapman Findlay, III

Houston and Los Angeles
September 23, 2002

Chapter 1

Introduction and the Institutional Environment

1.1 Introduction

Every individual who has more money than required for current consumption is potentially an investor. Whether a person places his or her surplus funds in the bank at a guaranteed rate of interest or speculates by purchasing raw land near a growing metropolis, he or she has made an investment decision. The intelligent investor will seek a rational, consistent approach to personal money management. The best method for some is simply to turn their funds over to someone else for management. A significant number of investors do indeed follow this policy, and it is quite likely the correct decision for many. Others, however, manage their own money or even become professionals who manage other people's funds. The latter groups are the audience for whom this book was written.

In the discussion that follows, we shall not unveil any mysteries about getting rich quickly. Indeed, if such secrets existed, it is doubtful that the authors would be willing to reveal them. Nevertheless, there are systematic procedures for making investment decisions that can enable the rational investor to maximize his or her economic position given whatever constraints he or she wishes to impose. Economic position is tacitly assumed to be the primary goal of the investor, although there may well be those who have other central goals. The purchaser of an art collection, for example, may be more interested in the aesthetic aspects of his investment than the financial appreciation that might be realized. There is nothing irrational about this, and it is not difficult to construct optimal decision models for such an individual. Similarly, another person may be strongly concerned about pollution, or human rights. Such a person may refuse to buy shares in companies that pollute or that do business in countries where ethnic cleansing is practiced. Again, this can be perfectly reasonable behavior, but these investors should at least be aware of the opportunity costs of their decisions. That is, they should realize that such an investment policy may have economic costs. The polluter may be a very profitable company whose stock could produce exceptional returns for its holders.

Maximizing economic position cannot usually be taken as the only objective of the investor. There is some correlation between the returns one expects from an investment and the amount of risk that must be borne. Thus, decisions must be made that reflect the ability and desire of the individual to assume risk. In this volume, we shall be very specific in both theoretical and practical terms about risk bearing and the optimal portfolio for the investor.

Although intelligence is about the only important personal attribute requisite for any kind of decision making, there are other traits that may be helpful to the investor. In particular, a certain amount of scientific curiosity may be very important to successful investors. By scientific curiosity we do not mean knowledge or even interest in disciplines generally considered "science," such as biology or chemistry, although the scientifically trained analyst may have an advantage in scrutinizing the stocks of high-technology companies. Rather, scientific curiosity refers to the systematic pursuit of understanding. An investor should be willing to take the time and spend the energy to know himself or herself and the investing environment.

It is unfortunately true that many otherwise successful people make poor investors simply because they do not have a logical investment policy. They have only vague objectives about what they want (such as "capital appreciation" or "safety of principal"), and they often substitute general impressions for solid fact gathering. How many highly competent doctors, for example, go beyond the recommendations of their brokers (friends, parents, relatives, or drug company sales people) when selecting a security? How many businesspersons take the time to familiarize themselves with the income statements and balance sheets of the firms in which they hold stock? How many professional portfolio managers make purchases based on a well-researched, documented effort to uncover those securities that others have passed over? Even in the case of portfolio managers, the number may be surprising. Of course, it could be reasoned that the doctor may not have the time or knowledge to make a thorough investigation of his or her investments and that the businessperson is too occupied with his or her own company to do a systematic search for information. If this is the case, then both doctor and businessperson should seek expert counsel.

Although knowledge of what other managers are doing is important and an experienced person's market "feel" may be superior to any professor's theoretical model, too often even the professional tends to substitute rumor and hunch for sound analysis and thorough investigation. This point will become more obvious in the pages to follow.

In addition to intelligence and scientific curiosity, the modern investor needs to be reasonably versed in mathematics and statistics. In this book, we provide an Appendix for review by those who are somewhat hazy on these subjects.

There are any number of investment possibilities that the investor may consider. The simplest is the commercial bank savings account, or certificate of deposit insured by the United States government. Next, in terms of simplicity, are the U.S. Treasury bills issued by the federal government in maturities of one year or less. These media provide safety of principal, liquidity, and yields that are not unattractive by historical standards. Nevertheless, they require little analysis as an investment vehicle, and any discussion of them must perforce be brief. There is a place for such investments in the portfolio of most investors, however, and the role of liquidity in investment strategy will be a focal point in the portfolio chapters coming later in this book.

At the other end of the investments spectrum are such highly illiquid assets as real estate, oil well interests, paintings, coins, stamps, antiques, and even ownership of business enterprises. These investments require a very specialized analysis, and anyone who is contemplating the purchase of such assets should be even more careful than the investor in securities. The unique nature of the aforementioned investment possibilities precludes them from extensive discussion in this book, although many of the techniques that are described in the pages that follow could be equally well applied to these investments.

In between the savings account, certificate of deposit, or U.S. Treasury bill and the illiquid assets mentioned above, are a host of investments that can generally be described as securities. A *security* is an instrument signifying either ownership or indebtedness that is negotiable and that may or may not be marketable. Securities are by far the most popular form of semiliquid investment (i.e., investment that is marketable but that may not be salable near the price at which the asset was purchased), and they can be analyzed in a systematic, consistent fashion. For this reason, the focus of this book will be on securities. We shall mostly be concerned with *common stocks* which are securities signifying ownership in a corporation or similar entity.

It was mentioned before that the investor should be well aware of the investment environment before he or she makes a decision. The environment for securities includes such important variables as the general state of the economy, the expected risk and return levels from purchasing a specific security, and the economic position of the investor. It also includes the more specific aspects of securities regulations, financial information flows, the securities markets, and general measures of security price performance (such as the Dow Jones averages). There are entire books devoted to each of these topics, and we will not purport to examine any of them in much detail. Nevertheless, the more important elements of these subjects will be discussed later in this chapter.

1.2   The Stock Market Efficiency Question

Of all the forms of securities, common stocks (and derivatives of common stocks) are the most romantic. Although the bond markets are quite important to both issuing corporations and many investors (pension fund money, life insurance reserves, bank portfolio funds, and so on in the aggregate are more heavily invested in bonds than equities), it is the stock market that engenders the interest of most investors. This is undoubtedly true because the rewards (and penalties) of stock market investment well exceed those obtainable in the bond market. Furthermore, equity analysis is more complicated than bond appraisal, and greater skill is required in selecting common stocks than fixed income securities. This is not to say that bond analysis is simple or even uninteresting. Indeed, some of the more sophisticated minds in the investments business are engaged in the bond market. Nevertheless, few people spend their lunch hours talking about the bond market, and the future performance of a bond rarely comes up in bridge table or golf course discussions. It is common stocks that entice most investors, and some investors have been known to feel a greater empathy for their stocks than their wives (or husbands). Thus, common stocks (and derivatives of common stocks) will be our focal point.

There is a school of thought that maintains that only insiders and those privileged to have information not known to the rest of us can make large profits in the stock market. These people subscribe to a theory of stock prices called the *efficient market hypothesis* (EMH). EMH advocates argue that the current price of a stock contains all available information possessed by investors and only new information can change equity returns. Because new information becomes available randomly, there should be no reason to expect any systematic movements in stock returns. Advocates of the EMH feel that the stock market is perfectly efficient and the cost of research and investigation would not be justified by any "bargains" (i.e., undervalued stocks) found.

The efficient market hypothesis has been tested by a number of scholars. These researchers have considered various hypotheses about the behavior of the stock market, from notions that past stock prices can be used to forecast future prices (the belief held by stock market chartists or "technicians") to reasoned opinions that stocks exist that are undervalued by the market and that these stocks can be uncovered by a thorough investigation of such fundamental variables as reported earnings, sales, price to earnings (P/E) multiples, and other pieces of economic or accounting data. The latter view of the market has long been held by most investors, and the whole profession of security analysis depends upon it. From the early days of the first edition of Graham and Dodd [10] down to the present, analysts have been taught that there are overpriced stocks and underpriced stocks and it is the job of the analyst to determine which are which. The EMH advocates have found, however, that the presence of so many individuals trying to find bargains (and overpriced stocks to sell short) makes it impossible for any one of them to outperform the general market consistently. Thus, as the economy grows and earnings increase, it is possible to make money in the stock market, but it is impossible to expect more than "average" returns. This is true, they say, because there are many buyers and many sellers in the market who have a great deal of similar information about stocks. If any one stock were "worth" more than the price for which it was currently selling, sharp analysts would recommend buying until its price rose to the point at which it was no longer a bargain. Similarly, if a stock were selling for more than its intrinsic value, analysts would recommend selling. The price of the security would fall until it was no longer overpriced.

The efficient market hypothesis has gained great currency in many quarters, particularly among academic economists, beginning in the early 1970s (see [5]). Nevertheless, it has not convinced too many practitioners, and many financial economists today no longer accept the EMH unequivocally (see [6] and especially [11]). This may be for two reasons. In the first place, if the EMH were believed, it would be hard for professionals to justify the salaries that they are paid to find better-than-average performers. Second, many analysts have suggested that their very presence is required for the EMH to work. If they could not find undervalued stocks, they would not come to their desks each day; and if they did not appear, there would no longer be that vast army of competitors to make the stock market efficient and competitive! Moreover, many analysts point out that there are substantial differences of opinion over the same information. Thus, although every investor may have available similar information, some see favorable signs where others see unfavorable ones. Furthermore, various analysts can do different things with the same data. Some may be able to forecast future earnings, for example, far more accurately than others simply because they employ a better analytical and more systematic approach. It is these differences of opinion and analytical abilities that make a horse race, and most practitioners (and an increasing number of financial economists) believe that this is what the market is all about.

1.3 Some History

Long before the EMH began to gain advocates, many economists (and almost all practitioners) believed that the stock market was neither competitive nor efficient (see [20]). These individuals viewed the market historically as an expression of the whim and fancy of the select few, a large gambling casino for the rich, so to speak. It has been observed that securities speculation in the past has been far from scientific and that emotion rather than reason has often guided the path of stock prices. Inefficiency proponents believed that people are governed principally by their emotions and that bull and bear markets are merely reflections of the optimism or pessimism of the day. They argued that economics plays a slight role in the market and that investor psychology is more important. This view traces back over 130 years. Charles Mackay, in a famous book published in 1869, entitled *Memoirs of Extraordinary Popular Delusions and the Madness of Crowds* ([16], pp. vii–viii) argued:

> In reading the history of nations, we find that, like individuals, they have their whims and their peculiarities-, their seasons of excitement and recklessness, when they care not what they do. We find that whole communities suddenly fix their minds upon one object, and go mad in its pursuit; that millions of people become simultaneously impressed with one delusion, and run after it, till their attention is caught by some new folly more captivating than the first. We see one nation suddenly seized, from its highest to its lowest members, with a fierce desire of military glory; another as suddenly becoming crazed upon a religious scruple; and neither of them recovering its senses until it has shed rivers of blood and sowed a harvest of groans and tears, to be reaped by its posterity .... Money, again, has often been a cause of the delusion of multitudes. Sober nations have all at once become desperate gamblers, and risked almost their existence upon the turn of a piece of paper.

Mackay's fascinating story details some of the most unbelievable financial events in history: (1) John Law's Mississippi scheme, which sold shares, to the French public, in a company that was to have a monopoly of trade in the province of Louisiana. Mississippi shares were eagerly bought up by French investors who knew that this "growth stock" could not help but make them rich.   After all, it was common knowledge that Louisiana abounded in precious metals. (2) The South-Sea Bubble, which induced Englishmen to speculate on a trading monopoly in an area (the South Atlantic) owned by a foreign power (Spain) that had no intention of allowing the English into the area for free trading purposes. The fevers produced by the South-Sea spilled over into other "bubbles," one of which proposed to build cannons capable of discharging square and round cannon balls ("guaranteed to revolutionize the art of war") and another that sought share subscribers to "a company for carrying on an undertaking of great advantage, but nobody to know what it is"([16], p. 53). (3) The Tulipomania, which engulfed seventeenth-century Holland.   Fortunes were made (and later lost) on the belief that every rich man would wish to possess a fine tulip garden (and many did, for a while at least). Tulip bulb prices reached astronomical levels, as one speculator bought bulbs to sell at higher prices to a second speculator who purchased to sell at even higher prices to yet another speculator.

In fact, as Mackay was writing, Jay Gould and Jim Fisk were busily manipulating the value of the shares of the Erie Railroad in the United States (see [1]. It was common practice for directors in many companies to issue information causing the price of their firm's stock to rise. They then sold their shares at inflated prices to the unsuspecting public. Some months later, they would release discouraging information about the company's prospects, in the meanwhile selling short the shares of their company. When the new information drove the price of the shares down, the directors would cover their short positions, again reaping nice profits at the expense of the unaware.

These practices continued on into the 1920s, an era when everybody believed that the life style of a J. P. Morgan or a Harvey Firestone could be within his reach. As Frederick Lewis Allen has pointed out in his wonderfully nostalgic yet penetrating *Only Yesterday*, it was a time when "the abounding confidence engendered by Coolidge Prosperity ... persuaded the four-thousand-dollar-a-year salesman that in some magical way he too might tomorrow be able to buy a fine house and all the good things of earth"([2], p. 11). A speculative binge started in 1924 with the Florida land boom (where "investors" paid large sums of money for plots that turned out in many cases to be undeveloped swampland) and continued on throughout most of the rest of the decade. As historian David Kennedy has pointed out, "Theory has it that that the bond and equity markets reflect and even anticipate the underlying realities of making and marketing goods and services, but by 1928 the American stock markets had slipped the bonds of surly reality. They catapulted into a phantasmagorical realm where the laws of rational economic behavior went unpromulgated and prices had no discernible relation to values. While business activity steadily subsided, stock prices levitated giddily" ([13, p. 35). All this came to an end with the stock market crash (and Great Depression that followed) in October, 1929.

From the Gilded Age to the 1920s, stock values were based upon dividends and book value (i.e., net asset value per share). In other words, stocks were valued much like bonds, based upon collateral and yield. Since it was hard to "fake" a dividend, manipulators often resorted to "watering the balance sheet" (writing up asset values to unreasonable levels so as to raise the book value). This discussion seems quaint today with the SEC recently *requiring* an accounting change (eliminating pooling on acquisitions) which will result in "watering" balance sheets in much the same way as in days of yore.

Earnings and even earnings growth as a basis for value were touted during the 1920s, especially to justify the ever-higher prices. The book often cited for providing intellectual justification for the stock market excesses of the 1920s was Edgar Lawrence Smith's, *Common Stocks as Long Term Investments* [18]. This is really an unfair assessment. What Smith did show, going back a century over periods of severe inflation and deflation, was that a diversified portfolio of stocks held over long periods might expect to earn a current yield of the short (e.g., commercial paper) rate and appreciate (from the retention of earnings) at about 2.5%, which can probably be interpreted as a "real" (inflation adjusted) return. At current values, this would translate to a shareholder rational required return of about 10% for the average stock. Not only is that estimate quite reasonable but it also is consistent with studies of long-run equity returns since the 1920s [19]. Furthermore, it contrasts with the arguments of the Dow Jones 40,000 crowd who contended, at the beginning of the year 2000 at least, that the average stock should currently be selling at 100 times earnings (see [9]). The market correction which began in March, 2000 has pretty much laid this notion to rest.

With the 1930s came the first edition of Graham and Dodd's classic *Security Analysis* [10]. Graham and Dodd devoted much of their attention to adjustments to make financial statements more conservative. They would allow conservative price to earnings multiples (P/Es) on demonstrated, historical earning power, with considerable attention again paid to book value and dividends. Finally, a much-neglected book, J.B. Williams' *The Theory of Investment Value* also appeared [21]. In it he developed most of the valuation formulations of financial mathematics. Along the way, he demonstrated that share price could be expressed as the discounted present value of the future dividend stream (see further discussion in later chapters). Such currently trendy notions as free cash flow analysis and EVA flow from his work.

Thus, in normal markets, the valuation approaches basically argue that investors should pay for what they can see (net assets, dividends) or reasonably expect to continue (demonstrated earning power). As markets boom, more and more optimistic future scenarios need to be factored into rational valuation models to obtain existing prices. Beyond some point the assumptions become so extreme that explanations other than rational valuation suggest themselves.

Although the EMH was yet decades away, economists in the 1920s and 1930s did advance the notion that markets should conform to some rationality which, with its great rise and collapse, the stock market seemed not to obey. John Maynard Keynes was one of the more perceptive observers of this phenomenon; and in 1936, he wrote *The General Theory of Employment, Interest, and Money* [14]. Keynes believed that much of man's social activities (including the stock market) were better explained by disciplines other than economics. He felt that "animal spirits" exercised greater influence on many economic decisions than the "invisible hand." His reasoning was based partly on a very acute understanding of human nature. It also depended on Keynes' lack of faith in the credibility of many of the inputs that go into economic decision making. He argued, "... our existing knowledge does not provide a sufficient basis for a calculated mathematical expectation. In point of fact, all sorts of considerations enter into the market valuation which are in no way relevant to the prospective yield" ([14], p. 152). He argued further that "... the assumption of arithmetically equal probabilities based on a state of ignorance leads to absurdities."

To Keynes, the market was more a battle of wits like a game of Snap, Old Maid, or Musical Chairs than a serious means of resource allocation. One of the most often quoted metaphors in the *General Theory* tells us (p. 156):

> ...[P]rofessional investment may be likened to those newspaper competitions in which the competitors have to pick out the six prettiest faces from a hundred photographs, the prize being awarded to the competitor whose choice most nearly corresponds to the average preferences of the competitors as a whole; so that each competitor has to pick, not those faces which he himself finds prettiest, but those which he thinks likeliest to catch the fancy of the other competitors, all of whom are looking at the problem from the same point of view. It is not a case of choosing those which, to the best of one's judgment, are really the prettiest, nor even those which average opinion genuinely thinks the prettiest. We have reached the third degree where we devote our intelligences to anticipating what average opinion expects the average opinion to be.

In the stock exchange, pretty girls are replaced by equities that one speculator believes will appeal to other speculators. Thus, ([14], p. 154):

> A conventional valuation which is established as the outcome of the mass psychology of a large number of ignorant individuals is liable to change violently as the result of a sudden fluctuation of opinion due to factors which do not really make much difference to the prospective yield; since there will be no strong roots of conviction to hold it steady.

During this period, economists would discuss efficiency in two ways— neither of which is related to the efficiency in the EMH. Allocational efficiency related to transmitting saving into investment and also investment funds to their highest and best use. Such issues arose in the debate over central planning versus market economies. Transactional efficiency related to the costs of trading (e.g., commissions and taxes, bid-asked spread, round trip costs, etc.) Keynes favored high trading costs, both to keep the poor out of the casino and to force a longer-term commitment to investments. He was also not optimistic about the ability of the market to allocate capital well ("when the investment activity of a nation becomes the byproduct of a casino, the job is likely to be ill done.")

Keynes's view was generally adopted by economists at least until the 1950s. A leading post Keynesian, John Kenneth Galbraith, argued along

Keynesian lines in his book *The Great Crash* [8] that stock market instability (inefficiency) had been important in economic cycles in the United
States since the War Between the States and that the 1929 collapse was
a major factor leading to the Great Depression of the 1930s. Attitudes
about the nature of the stock market began to change in the 1960s, and
many financial economists began to interpret stock price movements as a
"random walk".[1]

In sum, the 1929 Crash and Great Depression were still very recent
memories as the United States emerged from World War II. The Securities Acts (see discussion in Section 1.7 below) passed during the 1930s
were predicated upon the view, rightly or wrongly, that fraud pervaded the
markets. Financial institutions were either prohibited or generally severely
restricted in their ability to hold common shares. A "prudent man's" portfolio would still be mostly (if not all) in bonds. The stock market was, at
worst, a disreputable place and, at best, a place where one not only cut the
cards but also distrusted the dealer.

Viewed in this context, the EMH becomes one of the most remarkable
examples of image rehabilitation in history. From a very humble start,
within a decade or two it had converted numerous academics. By the
end of its third decade, it had converted a plurality of the U.S. Supreme
Court! Whether these people (especially the latter) know it or not, they
have adopted the position that an investor, having undertaken no analysis,
can place a market buy order for any stock at any time and expect to pay
a price which equals the true value of the shares. In other words, they now
have so much trust in the dealer that they do not even bother to cut the
cards!

1.4   The Role of Financial Information in the Market Efficiency Question

A fundamental postulation of the efficient market hypothesis is that investors (except insiders) have similar information with which to work. Indeed, the entire foundation of the EMH is based upon the presumption that
all data have been digested by the market and that the current price of a

---

[1]Technically, a one-dimensional random walk (here, the dimension is stock price) is a
real-valued quantity which is time indexed. Thus, suppose $S(t)$ is the price of the stock
on day t. Standing at time t, we know the value of the stock. But as we look from the
present time t to time $t + \Delta t$, the price is uncertain. A random walk representation of
$S(t + \Delta t)$ might be written as $S(t)[1 + \Delta t \eta(t)]$. Here, $\eta(t)$ is a random dilation factor
whose units are \$/time. This random dilation factor may itself be time changing. In
order to get some handle on the random walk, we will have to come up with some
compact representation of $\eta(t)$. Later in this book, we will indicate some ways this can
be done. Finally, as we suggested earlier in this chapter, random walk notions led to
the EMH by the 1970s. We shall return to the EMH in a more rigorous fashion in later
chapters.

security reflects all available information. Opponents of the EMH believe that information is not perfectly disseminated among investors and that investors may tend to interpret information differently. Because the ability to make above-average returns in the market depends upon differences in the flow and understanding of information, it is very important that the purchaser of securities appreciate all the possible sources of financial data.

A large quantity of information is generated by agencies and services that put together reports for the investing public (institutional and personal). These reports vary from general economic prognostications to very concrete analyses of industry and corporate prospects. Both Moody's and Standard & Poor's supply numerous bulletins and reports on a daily, weekly, and monthly basis. The Value Line Investment Survey publishes reports on hundreds of companies and ranks stocks in terms of quality, potential short-term and long-term price performance, and yield. Brokerage and investment banking houses also publish numerous reports that analyze and evaluate individual companies and securities. The bigger firms maintain substantial staffs of analysts, and it is not unusual for major entities to have analysts who cover the securities in only one industry.

In addition to the services listed above, there are a number of private investment letters that are distributed to a clientele of paid subscribers. These letters cost varying amounts, from a few dollars per year to thousands of dollars per year, depending on the nature of the information provided and the previous track record of the publishers of the letter. Some investment letters have been prepared for years and are widely respected.

An unfortunate feature of the aforementioned reports, however, is that the information that they contain is available to a large audience. Thus, if one of the leading brokerage houses recommends a particular stock, that knowledge is immediately transmitted to all participants and incorporated into stock prices. If only these data existed, there would be a good reason to accept the efficient market hypothesis on *a priori* grounds. There are other pieces of information, however, that may not be easily transmitted or understood. Important data are not always available to the general public, and even when widely disseminated some information is not easily interpreted. Data in the latter category often appear in the financial press and require specialized knowledge for proper understanding. Although many of the articles that are found in such publications as *The Wall Street Journal*, *Barron's*, *Forbes*, *Fortune*, and *Business Week* are very specific and provide obvious input in the appraisal of a firm, frequently it is not easy to make sense of an isolated fact that may be reported about a particular company. For example, suppose a firm reports its third quarter earnings and the statement is immediately carried by *The Wall Street Journal*. Suppose further that reported earnings are up significantly from the previous quarter and from the same period (quarter) in the previous year. What does this mean? The average reader might assume that the report is bullish and that the firm has done well. However, he looks at the price of the firm's stock and finds that it went *down* with the publication of the information. How can this be explained? One possibility is that professional analysts who had been scrutinizing the company very carefully for years had expected the firm to do *even better* than the reported figures and were disappointed by the result. Another possibility is that the market had discounted the good news (i.e., the improvement was expected and the price of the stock had been previously bid up accordingly). When the news was not quite as good as expected, the market realized that it had over-anticipated the result, and the price thus had to fall.

Information of this sort can often bewilder the "small" investor, and even the seasoned analyst is sometimes surprised by the way the market reacts to certain new inputs. Nevertheless, it is situations such as these that make possible above-average stock market performance. The investor who went against the crowd with the belief that the firm was not going to do as well as others expected and sold his shares (or established a "short" position) would have profited. Here, of course, superior forecasting or a better understanding of the situation would have enabled the shrewd investor to realize better-than-average returns. Hence, it is clear that the appropriate evaluation of financial information is the key to long-run investment success, and the trained analyst will have a decided advantage. Because the published investigations of others will very likely be generally available, and hence already included in the current price of a security, it should be obvious that the portfolio manager or private investor who really desires above-normal profits will have to make independent evaluations. The inputs that go into independent appraisals may include publicly available information, such as a corporation's annual report and Form 10-K, but these data will be uniquely interpreted. Moreover, the professional analyst may have access to certain inputs that are not generally known to the public. Competent securities analysts spend many hours each week interviewing corporate officers and employees of the firms that they follow. The capable analyst learns how to take clues from what managers say and use these clues to good advantage. This is not quite as good as inside information, but the really top analysts can sometimes deduce facts from evidence garnered that equals that possessed by management. These people are rare, however, and they are very highly paid. Moreover, this source is becoming increasingly limited by the new S.E.C. Rule FD, which requires companies to specify certain officers (employees) who can communicate with the investing public and requires immediate public dissemination in the event of inadvertent disclosure of material information.

Computers are used to perform tasks in a matter of seconds that previously required thousands of man-hours, and a whole new industry (financial information technology) has come into existence that has offered its own stock market high fliers in recent years. Significant amounts of data are now "on-line," including "First Call" earnings estimates of analysts, "ProVestor Plus" company reports, Standard & Poor's Stock Reports,

Vickers "Insider Trading Chronology," and virtually all press releases made by every publicly-held company. SEC filings are immediately available from EDGAR (Electronic Data Gathering and Research) On Line (edgar-online.com) which also provides a "People Search" with information on directors and officers, including salary and stock option data. Other similar information is provide by YAHOO! (finance.yahoo.com). Other useful websites include the following: smartmoney.com; premierinvestor.com; moneycentral.msn.com; fool.com; CBS.marketwatch.com; bloomberg.com; Kiplinger.com; bigcharts.com; and Morningstar.com. Also, most publicly held companies maintain web sites which post all major information about the company. Of course, one should not overlook the various "chat" room sources (which are mostly "gripe" sessions from disgruntled employees—some stockholders as well), the information content of which have only recently become the subject of academic research.

1.5   The Role of Organized Markets in the Market Efficiency Question

There is another prerequisite even more important than widely available information to the efficient market hypothesis: the existence of large, well-behaved securities markets. From economics, it will be recalled that a perfectly competitive market is one in which: (1) there are many buyers and sellers, no one of which can influence price; (2) there exists perfect information that is costless and equally available to all participants in the market; (3) there is a homogeneous commodity traded in the market; (4) there is free entry of sellers and buyers into the market (no barriers to entry); and (5) there is an absence of taxes and other transaction costs (e.g., brokerage commissions). The price observed in such a perfect market would be not only efficient but also in equilibrium. Clearly, however, no market meets this sufficient condition for the EMH. Advocates of the EMH would contend that the price could behave "as though" it were efficient if *none* of the above conditions were even approximated. This issue is addressed in Chapter 5.

The discussion in Section 1.4 above (and Section 1.7 below) deals with the descriptive validity of the information availability assumption. This section and Section 1.6 deal with the question of which "price" is assumed to be efficient. For example, if a stock is quoted $5 bid, $10 asked, what is the price which supposedly unbiasedly estimates the fully informed value? Likewise, if uninformed trading pushes the price up or down from the last trade, which is the right price? Finally, Section 1.8 begins the discussion (continued in later chapters) of trying to identify the market return which one cannot expect to beat.

For a market to be competitive, it usually must be sufficiently large so that any buyer (or seller) can purchase (sell) whatever quantity he or she wishes at the "going" price (i.e., the price set through the negotiations of all buyers and sellers together). The securities markets generally satisfy the size requirement in that literally billions of dollars worth of stocks and bonds are traded daily just in the United States. This does not mean that there is a good market for every single stock or bond in the hands of the public, however. If there is sufficient trading depth in a particular security, it will be possible to trade at a price very near the most recent past transaction price. As many investors can testify, however, there are numerous stocks (and many more bonds) that trade in very *thin* markets. That is, the number of participants in the market is so small that one may have to bid well above the last price to buy or ask well under that price to sell. Such a market is clearly not even nearly perfect.

To a large extent, whether or not a particular stock or bond is traded in a broad market depends on the "floating supply" of the issue outstanding. A stock with only a million shares in the hands of the public that "turns over" only 10 percent of the supply annually (i.e., 100,000 shares annually, or less than 500 shares on average each business day) will probably not trade often enough or in sufficient quantity to have market depth. Such a security may show rather substantial price volatility from one transaction to the next since the time span between transactions may be several hours or even days. Thus, no one buyer could accumulate more than a few shares at any one time without driving the price higher. Similarly, no seller could liquidate much of a position without pushing the price down.

One way to be sure that a stock does have a reasonably large floating supply and regular transaction patterns is to make certain that it is traded in an organized market, and organized securities markets have existed for centuries. Records show that securities were trading as early as 1602 in Antwerp and that an organized exchange existed in Amsterdam by 1611. Today, most of the leading capitalist countries have at least one major exchange, and the United States boasts several. In North America, the New York Stock Exchange (NYSE) is still the most important market, but the NASDAQ (a computer-based automatic quotation service) list competes with many newer, "high-tech" stocks. Securities listed on the American Stock Exchange and the regional exchanges do not generally have the market depth (or the earnings and assets) of stocks listed on the NYSE. Options on common stocks are traded on the Chicago Board Options Exchange (CBOE) and on the American, Pacific, and Philadelphia stock exchanges.

Many securities are traded on more than one exchange. An advantage of *dual* listing is that extra trading hours may be secured for a firm's stock. Thus, a company's shares listed on the NYSE, the London and the Tokyo Stock Exchanges could be traded almost around the clock. In fact, there is almost continuous trading in most major stocks held world-wide even when an organized exchange is not open. As discussed in Chapter 5, a disadvantage of this extended trading may be increased volatility.

On an exchange, there is a record of each transaction, and an investor can observe the last price at which the security was traded. The investor may call his or her broker (or check an on-line trading account) and find out at what price a particular stock *opened* (its first price for the day), find the high and low for the day, and obtain a current quotation with market size. A *bid* price is the amount that is offered for the purchase of a security and the *ask* price is the amount demanded by a seller. There will customarily be a "spread" between the bid and ask prices that serves to compensate those who make a market in the security (a specialist on most exchanges and various dealers on the NASDAQ).

The *primary* market is the first sale or new issue market. When a firm (or government) sells its securities to the public, it is a primary transaction. The first public sale of stock by a firm is the initial public offering (IPO); a subsequent sale by the firm is called a seasoned equity offering (SEO). After a bond or share is in the hands of the public, any trading in the security is said to be in the *secondary* market. Purchases of securities in the primary markets are usually made through *investment bankers* who originate, underwrite, and sell new issues to the public. In the usual case, a firm will arrange to "float" an issue through its bankers. The firm will be given a price for its securities that reflects market conditions, yields on equivalent securities (either the company's or those of similar concerns), and the costs involved to the investment bankers to distribute the stocks or bonds. Title to the securities is customarily taken by the underwriting syndicate (several banking houses), although small corporations usually have to settle for a *best-efforts* distribution in which the bankers merely act as selling agents for the company.

The primary (new issue) market has been quite popular in recent years for speculative investors. The reason for this popularity is the fantastic price movements experienced by many stocks after initial sale (IPOs). In the late 1990s, it was not unusual for newly public stocks to double or even quadruple in the first day of trading. Some "high tech" stocks went up by a factor of ten or more within days or weeks of their IPO. It was no wonder that just the mention of a new issue was often enough to get investors clamoring for "a piece of the action." It is interesting to note that this is not a new phenomena, and nearly all bull (rising price) markets for decades (actually centuries) have featured startling performers that rose to unbelievable levels even though these were brand new (or, in any case, not very seasoned) companies. For a while in 1999 and early 2000, just about any company with "dot com" or "e" or "i" in its name seemed to be able to go public and have the stock price sky-rocket within hours or, at most, days. Companies that never made money (and some that had never made a sale!) were accorded market capitalizations (number of shares outstanding times price per share) that often exceed those of old-line companies that have been in business for decades.

One "tech" stock that played the game with a vengeance was Aether

Systems which provides "wireless data services and software enabling people to use handheld devices for mobile data communications and real-time
transactions" (Aether Systems, Inc. Form 10-K for 1999, p. 2). Aether
went public at $16 per share in October 1999. On March 9, 2000, the stock
closed at $315! During the "tech crash" in April, 2000, the stock fell to
$65. It rebounded in only a few weeks to well over $100, but subsequently
fell to below $5. History suggests that these "high-fliers" would eventually
collapse in price (with many going bankrupt) when more sober market conditions reappear (as they always must, and, post 2000, did). Many have
offered these examples as *prima facie* evidence of market inefficiency.[2]

1.6 The Role of Trading in the Market Efficiency Question

In economic theory, traders play an important role in effecting a market
equilibrating process. That is, if the price of A is "low" relative to its "real
value" and the price of B is "high" relative to its real value, traders will
buy A and sell B until "equilibrium" values are established. Of course,
there must be some common agreement on just what "real value" means
and how such is determined, and the economics literature has searched for
this answer for over 200 years. Suppose Sam can buy a pound of chocolate
on Fifth St. for $5 and sell it on Sixth St. for $6. He would be advised
to buy all the chocolate he could on Fifth, run over to Sixth and sell all
he could. Now the very process of Sam engaging in this activity causes the
price on Fifth St. to rise (excess demand) and the price on Sixth St. to
fall (excess supply). Indeed, in theory at least, others besides Sam should
enter this market, and the price of chocolate should eventually settle (say
at $5.50) unless there were transactions costs involved in moving chocolate
from Fifth St. to Sixth St. (say, Fifth St. is in New York and Sixth
St. is in Houston). Economists have debated at length on how long this
process should take (in theory, quite rapidly), and under what conditions
others would join Sam in this endeavor. Suppose Sam is the only one who
knows chocolate can be bought on Fifth for $5 and sold on Sixth for $6
with virtually no transactions costs. The existence of *imperfect knowledge*
may provide Sam with quite an opportunity. Of course, information then
takes on its own value. Issues such as "Why does Sam know about this
opportunity?" and "Why don't others?" come into play. Also, suppose it
takes equipment to move chocolate from Fifth to Sixth. The requirement
of having capital investment may create a *barrier to entry* (and impute
an "opportunity cost" for the alternative use of the equipment) which may
prevent others from joining the market. Economists are generally suspicious

---

[2]Peter Bernstein provides numerous other historical examples of people paying ridiculous prices for "growth" in his delightful book *Against the Gods: the Remarkable Story
of Risk*. ([3], pp. 108−109).

of "free lunches" and usually search for reasons why the world looks the way it does. Suppose the chocolate on Fifth St. is actually inferior to that for sale on Sixth St. This may well explain the price difference, and it may mean that Sam does not have such a good opportunity after all. Thus, the existence of fairly *homogeneous* products may be required for this arbitrage opportunity to really exist.

Now let us return from chocolate to stock. In order to evaluate whether opportunities may exist, one should know something about just how trading takes place. Just as Sam had to know how to find Fifth and Sixth Sts. and buy and sell chocolate and judge the quality of chocolate, so must the intelligent investor know about how the stock market functions and who the players are. We have already established some of this above, but it would be wise to identify a few more important elements. First, a securities *dealer* maintains an inventory of securities in which he or she makes a market by purchasing and selling from his or her own account as a principal. A broker acts as an agent for his or her customers in purchasing or selling securities. On the floor of an exchange this is done through a *specialist* (a trader charged by the exchange with maintaining a market in the stock). A broker may also act by buying from and selling to dealers. As an agent, the broker is expected to obtain the best available price for the customer and is paid a commission for this service.[3]

The simplest order to a broker to buy or sell a security at the best available price is called a *market order*. In the dealer market (sometimes called the *over-the-counter, or OTC,* market), the broker would check dealer quotes and execute the order at the best available price. On the floor of the NYSE, the floor broker for the customer's firm would walk (or transmit electronically)to the post where the security is traded. From the assembled group of other floor brokers, floor traders (who trade for their own account), and the specialist, the customer's broker would determine the best available price and execute the order. There are, however, other types of orders. A *limit* order includes a maximum (minimum) price that the customer is willing to pay (receive) to buy (sell) the stock. A stop order contains a price above (below) the current market price of the stock that, if reached by the market, the customer desires to trigger a buy (sell) market order. Since it is quite likely that neither stop nor limit orders could be executed immediately, the floor broker would instruct the specialist to enter them in his "book" for execution when their terms were met. A *stop limit order* performs like a stop order with one major exception. Once the order is activated (by the stock trading at or "through" the stop price), it does not become a market order. Instead, it becomes a limit order with a limit price equal to the former stop price. For example, Smith places a stop limit

---

[3]Options and futures exchanges operate somewhat differently. Futures exchanges have daily price limits (upwards and downwards) that may not be exceeded (and therefore have no need for a specialist). The CBOE has a book order official responsible for the limit order book, and one or more dealers charged with maintaining an orderly market.

order to sell stock with a stop price of $50 a share. As with the stop order, once the stock trades at $50, the order is triggered. However, *the broker cannot sell it below $50 a share no matter what happens.* The advantage of this order is that the buyer sets a minimum price at which the order can be filled. The disadvantage is that the buyer's order may not be filled in certain fast market conditions. In this case, if the stock keeps moving down, Smith will keep losing money.

After a stock split or a dividend payout, the price on all buy limit orders and the stop price on sell stop and sell stop limit orders is adjusted. For example, if Jones places an order to buy 100 shares of XYZ Corp. at $100 a share and the stock splits 2 for 1, the order will automatically be adjusted to show that he wants to buy 200 shares of XYZ at $50, reflecting the split. Other restricted orders include the following:

- *Good-until-canceled (GTC)* orders remain in effect until they are filled, canceled, or until the last day of the month following their placement. For example, a GTC order placed on March 12th, left unfilled and uncancelled, would be canceled automatically on April 30th.

- A *day order* is a limit order that will expire if it is not filled by the end of the trading day. If one wants the same order the next day, it must be placed again.

- *All-or-none* is an optional instruction that indicates that one does not wish to complete just a portion of a trade if all of the shares are not available.

- *Do-not-reduce* means that the order price should not be adjusted in the case of a stock split or a dividend payout.

- *Fill or kill* is an instruction to either fill the entire order at the limit price given or better or cancel it.

The major functions of the specialist (mentioned above) are to execute the orders in his book and to buy and sell for his or her own account in order to maintain an orderly market. To limit possible abuses, the specialist is required to give the orders in the book priority over trades for his or her own account and to engage in the latter in a stabilizing manner. The larger blocks of stock being traded on the exchanges in recent years have caused the capital requirements for specialists to be increased, and rule violations (such as destabilizing trading) are investigated. Even the EMH advocates, however, agree that the specialist's book constitutes "inside information" and this group can earn above-normal profits.

In the past, NYSE designated commissions were charged by member firms on a 100 share (*round* lot) basis. No discounts were given. Since May 1, 1975, (known as "May Day" to many retail brokers), discounting has been allowed in a deregulated environment. For large transactions, as much as 75 percent of a commission might be discounted by the larger retail brokers (such as Merrill Lynch). Even larger discounts are now provided by firms that call themselves discount brokers (such as Charles Schwab), and some deep discount brokers are charging as little as $5 *per transaction* for market order trades done over the internet. Some are even free if a large enough balance is kept in the brokerage account; and for larger individual accounts (e.g., $100,000), many brokers are now allowing unlimited free trading for an annual fee of 1% to 1.5% of the portfolio value.

Reduced revenues resulting from negotiated commissions coupled with the higher costs of doing business altered the structure of the brokerage industry. A number of less efficient houses collapsed, were merged with stronger concerns, or undertook voluntary liquidation. During the 1970s, several large houses failed and millions of dollars in customer accounts were jeopardized. In order to prevent loss of investor confidence, the Securities Investor Protection Corporation (SIPC) was established to protect the customers of SIPC member firms. The SIPC provides $500,000 ($100,000 cash) protection per account in the event of failure. This arrangement does not protect against trading losses but rather in the event of failure of brokers to satisfy their agency responsibilities. Suppose Mr. X is a customer of ABC & Co. which is a member of the SIPC. Suppose further that X keeps his stocks in custody with ABC (i.e., "street name") and the firm goes bankrupt. X would be able to look to the SIPC to make good on the value of his investments up to $500,000. Many brokerage firms purchase insurance to provide protection above $500,000 and the capital requirement for firms has been increased. Thus, a more concentrated, hopefully stronger, and more efficient brokerage community has emerged over the past three decades. This has made it possible for the transactions costs to be reduced tremendously; and this, in turn, should have made markets relatively more efficient than they were, say, 30 years ago.

A caveat should be noted here: Substantially lower commissions and computer internet trading have inevitably led to the phenomenon of the under-capitalized "day trader." These are individuals who may have as little as $5,000 who speculate on small price movements in a particular stock within a single day's trading. This phenomenon was not possible when commissions were large, but with $5 trades almost anyone with a computer and internet access can play the game. The evidence suggests that most of these traders get wiped out, or suffer large percentage losses to their portfolios, within months of initiation of trading. Thus, even in the biggest bull market in history (ending in the year 2000), there were traders who lost most of their money by treating the stock market like a computerized Las Vegas. Interestingly, there are economists who contend that this added "liquidity" has actually made the markets more efficient! See [17] for an advocate of such reasoning.

A final note on the mechanics of securities trading: From the beginning of the New York Stock Exchange in the late 18th Century, stocks (originally

U.S. government securities) were traded in eighths, quarters and halves. Some stock even traded in fraction of eighths, but the basic trading unit was the 1/8, which was 1/8 of a dollar or $.125. This peculiarity was a result of having the old Spanish dollar (which was divided into eighths and thus called "pieces of eight") being a major currency during the U.S. colonial era. After the decimal U.S. currency system was effected, securities continued to trade in eighths, first as a matter of convenience and later because it increased bid/ask spreads where dealers make most of their money (buying at $11 7/8 and selling at $12 is much more profitable than buying at $11.99 and selling at $12). In late 1999 and early 2000, there was a movement initiated by the Securities and Exchange Commission and adopted by the stock exchanges (and the NASDAQ) to change trading to decimal units. Thus, we no longer buy (and sell) stocks at prices such as $16 3/8 or $30 1/8; rather, we may buy or sell at the more sensible $16.37 (or $16.38) or $30.12 (or $30.13). This may not seem like a big change, but has made the arithmetic of trading much simpler. (Quick Mental Check: XYZ goes from $10 5/64 to $10 13/32. What is your profit? How much easier is it to calculate an advance from $10.08 to $10.41!) Also, the greater competition has reduced spreads such that dealer margins have been reduced in favor of investors.

1.7   The Role of Securities Market Regulation in the Market Efficiency Question

Many people feel that a major element contributing to the efficiency of the U.S. securities markets is the regulation of those markets by the federal government. Before 1933, there were no laws governing stock-exchange or investment-house activities, and widespread manipulation and questionable practices abounded. Corporations were not required to provide information to investors, and fraudulent statements (or no statements at all) were issued by any number of companies.[4] As securities speculator Joseph P. Kennedy (father of future President John F. Kennedy) once remarked to one of his partners, "It's easy to make money in this market .... We'd better get in before they pass a law against it" ([13], p. 367). The excesses of the 1920s were attributed in part to the lack of comprehensive legislation regulating the securities industry, and with the coming of the New Deal a number of laws were indeed passed to prevent a recurrence of the events that led to the 1929 crash.

The Securities Act of 1933 (the '33 Act) requires full and complete disclosure of all important information about a firm that plans to sell securities in interstate commerce. Issues of securities exceeding certain dollar limits, and all issues sold in more than one state, must be registered. A *prospectus*

---

[4]State laws against fraud existed at this time, of course, but enforcement of these laws was difficult. Corrupt judges did not help matters either (see [1]).

must be prepared by the issuing company and distributed to anyone who is solicited to buy the securities in question. The prospectus must include all pertinent facts about the company, such as recent financial reports, current position, a statement about what will be done with the funds raised, and a history of the company. Details about the officers and directors of the company are also required.

The Securities Exchange Act of 1934 (the '34 Act) established the Securities and Exchange Commission (the "S.E.C." or the "Commission"). It also regulates the securities markets and institutional participants in the market, such as brokers and dealers. All exchanges are required to register with the Commission, although much of the supervision of individual exchanges is left up to the governing bodies of each exchange. Amendments to the act (e.g., the Maloney Act of 1938) now also include the OTC markets, although broker-dealer associations are accorded the same self-regulatory authority as the exchanges enjoy. (See discussion under "Self-regulation" below.) The '34 Act also calls for the continual reporting of financial information by firms that have "gone public" and a major part of the financial (and other) information flow from reporting companies to the public is done pursuant to this act and amendments to it. Interestingly, President Franklin Roosevelt appointed Joseph P. Kennedy (the speculator mentioned above) to be the first Chairman of the S.E.C. ("a choice often compared to putting the fox in the henhouse or setting a thief to catch a thief" ([13], p 367)!

The Investment Company Act of 1940 regulates the management and disclosure policies of companies that invest in the securities of other firms. Under the act, investment companies may be organized as *unit trusts, face-amount certificate companies,* or *management investment companies.* Only the last classification is currently significant, and it is further subdivided into open-end and closed-end management investment companies. *Closed-end companies* sell a fixed number of shares, and these shares then trade (often at a discount to net asset value) just like other shares. Many closed-end funds are listed on the NYSE. Some even issue preferred stock (or income shares) and borrow money. *Open-end* companies are better known as *mutual funds* and are required to redeem their shares at net asset value upon demand; because of this requirement, they may not issue long-term debt or preferred stock. If a fund registers with the S.E.C., agrees to abide by the above rules, invests no more than 5 percent of its assets in the securities of any one issuer, holds no more than 10 percent of the securities of any issuer, pays at least 90 percent of its income out to fund shareholders, and meets other rules, it may pay taxes only on earnings retained. Capital gains and dividends (interest) earned are paid out to the holders of the fund's shares who pay taxes at their respective individual or institutional rates.

Other acts that are important include the *Public Utility Holding Company Act of 1935*, which regulates the operations and financial structure of gas and electric holding companies; the *Trust Indenture Act of 1939*, which requires that bonds (and similar forms of indebtedness) be issued under an *indenture* that specifies the obligations of the issuer to the lender and that names an independent trustee to look after the interests of lenders; and the *Investment Advisors Act of 1940*, which requires the registration of investment counselors and others who propose to advise the public on securities investment.

One of the major excesses of the pre-1933 era was the practice of buying stocks on low *margin*. Margin purchases are those for which the investor does not advance the full value of the securities that he or she buys. Thus, if an investor bought one hundred shares of General Motors at 60 on 50 percent margin, he would only put up $3,000. The remaining $3,000 would be borrowed either from his broker or a bank. Margin purchases may increase the rate of return earned by an investor. If General Motors went up 10 percent to 66, the investor in the above example would have made $600/$3,000 = 20 percent. They also increase the degree of risk exposure. If GM went down 10 percent, the loss would be 20 percent. During the late 1920s, investors were buying stocks on less than 10 percent margin. Large rates of return were earned by everyone so long as stock prices were advancing. When prices began to skid in late 1929, however, many people were wiped out in a matter of days. Investors tended to build their margin positions as prices rose by buying more shares with the profits earned. Thus, a man might have put $1,000 into stock worth $10,000 in January 1929. As prices advanced by 10 percent, say, in February, he might have used his profit to buy another $9,000 worth of stock. His commitment was still $1,000, but he controlled $20,000 in stock. As prices rose further, he might have increased his position to $25,000. But suppose prices fell just a little, say 4 percent. This decline would be enough to wipe out his investment completely! Such a decline began during October 1929, and many investors were sold out of their stocks as prices fell. The process of liquidating shares as prices went below margin levels caused further price deterioration that, in turn, forced more liquidations. The snowballing effects of this phenomena produced the major crash of October 29, 1929 and contributed to the subsequent collapse of both the stock market and the American economy.

Because of the problems directly traceable to margin purchases, the Securities Exchange Act of 1934 gave the Board of Governors of the Federal Reserve System the power to set margin requirements for all stocks and bonds. Since 1934, margins have been allowed as low as 40 percent but have also been as high as 100 percent (no borrowing permitted). To some extent, the sobering experiences of 1929 caused a natural reaction against margin purchases in subsequent years. Nevertheless, most participants in the market today have only read about 1929 and would, if given the chance, follow their forefathers down the same speculative path. To protect them and society from such excesses, extremely low margins are no longer permitted.

Another practice that caused problems prior to 1933 was the *short sale*.

When one sells a security he or she does not own but borrows from someone else to make delivery, he or she is said to sell that security short. The device has been used for years and can be defended on economic grounds even though it does sound a bit immoral to be selling something one does not own. In practice, the short sale is consummated by specialists (who have responsibility for maintaining an orderly market on the NYSE) and dealers far more frequently than by the investing public. The average investor might consider selling short a security if she believed its price were going to decline. She would simply call her broker and ask to sell so many shares of such and such company short at a given price. If the broker could find the shares for the customer to borrow (usually from securities held by the broker in his own account or securities margined with the broker), the transaction could be effected. Because short selling can tend to exacerbate downward movements in stock prices, it is easy to see how speculative excesses could occur through unregulated use of the device. Thus, the Securities Exchange Act of 1934 allows the S.E.C. to set rules for short selling. There are several regulations in effect now governing the practice, the most important being the "up-tick" requirement. This rule prevents a short sale while a stock is falling in price. Thus, if a stock sells at $40, then $39.50, then $39, no short sale could be effected until there is an advance above $39.

Since the average securities firm functions as investment banker (representing the issuing firm), broker (representing the customer), and dealer (representing itself) simultaneously, the potential for conflict of interest is great. Many of the laws previously discussed were passed to protect the general public when such conflicts arise. These laws, in turn, provide for substantial *self-regulation.* This is manifested by exchange regulations for member firms and NASD (National Association of Securities Dealers) rules for most others, who subject themselves to such rules in order to obtain securities from other NASD firms at less than the price to the general public. NYSE members must restrict all transactions in listed stocks to the floor of the exchange, even though the larger firms could merely match buy and sell orders in their own offices. NASD firms may only trade with their customers if their price is the best obtainable and must reveal if they acted as principal on the other side of the transaction. Any research recommendations by broker-dealers must indicate if the firm holds a position in the stock. Other regulations call for ethical behavior on the part of members by prohibiting such practices as: (1) the spreading of rumors; (2) the recommending of securities clearly inappropriate for a given customer; and (3) the encouraging of excessive numbers of transactions (called "churning") in a given account. Although many of these rules have protected the public, others are clearly designed to protect the economic position of the broker-dealer community itself.

The Securities Exchange Act of 1934 defines officers, directors, and holders of more than five percent of the shares of a firm as *insiders.* Such persons are required to file a statement of their holdings of the firm's stock and any changes in such holdings (within a month) with the S.E.C. Profits made by insiders on shares held less than six months must also be reported and may be legally recovered by the firm (through a shareholders' derivative suit if necessary); in addition, malfeasance suits could be filed for other injuries to shareholder interests. Over the years, holdings of related persons have come to be included in the determination of whether the 5 percent rule were met and persons related to insiders were also considered to be insiders for the purpose of the law. In general, the principle was established that insiders and their relatives should not gain a special benefit over other shareholders by virtue of the information about the firm they possess. The Texas Gulf Sulphur case of the mid-1960s, in which corporate insiders withheld information about a minerals discovery until they could obtain stock, clearly reestablished this point through both civil and criminal action.

Several other cases expanded the concept of insider information. In the cases of Douglas Aircraft and Penn-Central in the 1970s, brokerage houses obtained insider information (of bad earnings and impending bankruptcy, respectively) and informed selected institutional investors before the general public. Subsequent suits and exchange disciplinary actions against the brokerage houses involved, and suits against the institutions, indicate that second- and third-hand possessors of inside information may also be classed as insiders. A securities analyst was charged in the Equity Funding case some years ago for providing information to selected investors that did not even originate from the company itself but rather from former employees. We clearly have moved in the direction of classifying insiders more on the basis of the information they possess than the position they hold in regard to the firm. Although such a situation would tend to validate the EMH by default, its long-run implications for investigative research analysis and individualistic portfolio management are not encouraging.

1.8 The Role of Stock Market Indicators in the Market Efficiency Question

When the EMH postulates that only "normal" returns can be earned in the stock market, an implicit assumption is made that there is some sort of average that summarizes stock market performance in general. In fact, it is extremely difficult to calculate measures of this sort. Perhaps the most widely used average is the Dow Jones Industrial Average (DJIA) that appears in *The Wall Street Journal* each day. The DJIA is computed by taking the price of each of thirty selected blue-chip stocks, adding them, and dividing by a divisor. The divisor initially was the number of stocks in the average (originally twelve), but because of the obvious biases of stock splits (a two-for-one split would tend to cause the price of a share to fall by one half), the divisor was adjusted downward for each split. The divisor now is well below one, which, in reality, makes it a multiplier. In addition to the DJIA, there is a Dow Jones Transportation (formerly rail) average of twenty stocks, a Dow Jones Utility average of fifteen stocks, and a composite average of all sixty-five stocks.

Dow Jones also calculates market indices for a number of foreign markets, an Asia/Pacific index and two World Indices (one with U.S. stocks and another without). Each is computed in the same manner as the DJIA. For many investors, the Dow Jones averages are the market. When an investor calls his broker to ask what the market is doing, he is very likely to get a response such as "down 56.75." The broker means, of course, that the DJIA is down 56.75 points. This information may have very little to do with what the investor really wants to know (that is, how are *my* stocks doing?). The DJIA is not an overall indicator of market performance, although many use it as if it were. In fact, only blue-chip stocks are included in the average. The thousands of other stocks that are not blue chips are not represented. Moreover, the DJIA has been criticized by many even as a measure of blue-chip performance. Because the DJIA merely adds the prices of all included stocks before applying the divisor, a stock that sells for a higher price receives a larger weight in the measurement.

The difficulties associated with the Dow Jones averages have led to the development of other stock price averages and indexes. Standard & Poor's computes an industrial index, a transportation index, a utility index, and a composite index (500 stocks) that include both the price per share of each security *and* the number of shares outstanding. These figures thus reflect the total market value of all the stocks in each index. The aggregate number is expressed as a percentage of the average value existing during 1941–1943, and the percentage is divided by ten. The S&P indexes are better overall measures of stock market performance than the Dow Jones averages because more securities are included. Furthermore, the statistical computation of the S&P indexes is superior to the Dow Jones method.

There are a number of other indexes that are also prepared. Both the New York and American Stock Exchanges compute measures that include all their respective stocks. The NYSE Common Stock Index multiplies the market value of each NYSE common stock by the number of shares listed in that issue. The summation of this computation is indexed, given the summation value as of December 31, 1965. The American Stock Exchange average simply adds all positive and negative changes of all shares and divides by the number of shares listed. The result is added to or subtracted from the previous close. The National Association of Securities Dealers (NASD), with its NASDAQ automated quotation service, computes a composite index based on the market value of over 5,000 over-the-counter stocks plus indices for six categories representing industrials, banks, insurance, other finance, transportation and utilities. The broadest index is calculated by Wilshire Associates. Their Wilshire 5000 Equity Index is based on all stocks listed on the New York and American Stock Exchanges plus the most actively traded OTC stocks. Although there is no single perfect indicator of average performance, many analysts are tending to favor the Wilshire Index as the most broadly indicative. Nevertheless, most observers do not ignore the Dow Jones averages because so many investors are influenced by them. Fortunately (at least for measurement purposes), there is a high positive correlation between the price movements of all stocks. Thus, if most stocks are going up (or down), almost any stock price measure will indicate this.

1.9   Summary

It is important to understand the institutional aspects of the securities markets in order to be a successful participant in them. The various investments media and the environment in which they trade are important elements in this regard. Of all the forms of securities, common stock and derivatives of common stocks are the most romantic but are also the most difficult to analyze. For this reason, they are the focal point of this book.

There is a school of thought that maintains that the current price of a stock contains all available information about that stock and only new information can change equity returns. This theory of stock market behavior is called the efficient market hypothesis (EMH). The EMH has been tested over the years by a number of scholars. It was generally endorsed by financial economists in the 1970s, but most practitioners never accepted the theory. Today, even many financial economists no longer accept the EMH, at least without qualification.

A crucial assumption of the efficient market hypothesis is that stock prices reflect all public information. A major form of such information is that supplied by research agencies and services. Reports periodically prepared by brokers and investment advisory companies are designed to aid the investing public in making decisions. An important feature of the established service and agency reports is that they are available to large audiences. Thus, the data that are contained in them could be expected to be incorporated in stock prices just as soon as the information is published. Other information that is not so easily transmitted or interpreted appears in the financial press. Much of this information requires special expertise or training for proper understanding, and one of the goals of this book is to provide the tools and understanding required for the investor or analyst who must appraise these data.

The efficient market hypothesis postulates the existence of large, well-behaved securities markets. For a market to generate a unique (no less efficient) price, it must be sufficiently large so that any buyer (or seller) can purchase (sell) whatever quantity he or she wishes without affecting the market price. The securities markets satisfy this requirement for some securities but not for others. In general, stocks traded on an organized exchange will have a broader market than those that are traded over-the-counter, because exchanges have listing requirements designed to guarantee market depth. Securities trading in the over-the-counter market may be either primary or secondary in nature. The primary market exists for the distribution of new securities. The secondary markets include both listed and OTC securities that are in the hands of the public. New issue securities (primary market) are sold by investment bankers to investors. The *initial public offering* (IPO) market has been very popular among speculators in recent years (at least until the market collapse in 2000).

A major element often cited as contributing to the efficiency of the American securities markets is the regulation provided by the United States government. After the 1929 stock market crash, a number of laws were passed that were designed to correct some of the abuses that existed in the past. Disclosure requirements were established, and the Securities and Exchange Commission was created to supervise the investments business. One of the excesses of the pre-1933 era was the practice of buying stocks on margin. Use of this device is now regulated by the Board of Governors of the Federal Reserve System. Another practice that caused problems prior to 1933 was the short sale. This process is still permitted, although the S.E.C. may make rules governing its use.

In order to determine whether one has earned above (or below) market returns, one must have a good measure of the average performance of stocks in general. No perfect indicator exists. The most widely used average is the Dow Jones Industrial Average. The DJIA is primarily a blue-chip measure, although many investors use it for overall market activity. Standard & Poor's computes indices that have a larger base than the DJIA. The S&P measures are also statistically superior to the Dow calculation. There are a number of other indexes and averages that are also computed. Fortunately (at least for measurement purposes), there is a high positive correlation between the price movements of all stocks. Thus, if most stocks are going up (or down), almost any stock price measure will indicate this.

References

[1] Adams, C. F., and Adams, H. (1956). *Chapters of Erie.* Ithaca, NY: Great Seal Books.

[2] Allen, F. L. (1931). *Only Yesterday.* New York: Harper and Brothers.

[3] Bernstein, P.L. (1996). *Against the Gods: The Remarkable Story of Risk.* New York: John Wiley and Sons.

[4]_____(1992). *Capital Ideas: The Improbable Origins of Modern Wall Street.* New York: Free Press.

[5]Fama, E. F. (1970). "Efficient capital markets: A review of theory and empirical work," *Journal of Finance* (May), 383–423.

[6]_____(1996). "Multifactor portfolio efficiency and multifactor asset pricing," *Journal of Financial and Quantitative Analysis*, Vol. 31, No. 4, December, 441−446

[7]_____ (1965). "Random walks in stock market prices," *Financial Analysts Journal* (Sept.−Oct.), 55−59.

[8] Galbraith, J. K. (1955). *The Great Crash*. Boston: Houghton-Mifflin Co.

[9] Glassman, J. K. and Hassett, K. A. (1999). "Stock prices are still far too low," *Wall Street Journal*, March 17 (op. ed.).

[10] Graham, B., and Dodd, D. (1934). *Security Analysis*, 1st ed. New York: McGraw-Hill.

[11] Haugen, R. A. (1999). *The New Finance: The Case Against Efficient Markets*, Second Edition, Englewood Cliffs, NJ: Prentice-Hall.

[12] Jensen, M. (1978). "Some anomalous evidence regarding market efficiency," *Journal of Financial Economics*.

[13] Kennedy, D. M. (1999). *Freedom from Fear*. Oxford: Oxford University Press.

[14] Keynes, J. M. (1936). *The General Theory of Employment, Interest, and Money*. New York: Harcourt, Brace, and World.

[15] Lo, A. W. and MacKinlay, A. C. (1999), *A Non-Random Walk Down Wall Street*. Princeton, NJ: Princeton University Press.

[16] Mackay, C. (1869). *Memoirs of Extraordinary Popular Delusions and the Madness of Crowds*. London: George Routledge and Sons.

[17] Malkiel, B. G.(1999). *A Random Walk Down Wall Street: The Best Investment Advice for the New Century*. New York: W.W. Norton & Company.

[18] Smith, E. L. (1924). *Common Stocks as Long Term Investments*. New York: Macmillan.

[19] *Stocks, Bonds, Bills, and Inflation* (2000). Chicago: Ibbotson Associates.

[20] Williams, E. E. and Findlay, M. C. (1974). *Investment Analysis*. Englewood Cliffs, NJ: Prentice-Hall.

[21] Williams, J. B. (1938). *The Theory of Investment Value*. Amsterdam: North Holland.

Chapter 2

Some Conventional Building Blocks (With Various Reservations)

2.1 Introduction

If our analysis consisted simply in trying to see what the dollar value of a stock might be six months from today, that would be difficult enough. But analysis of the market is complicated by the fact that it is not simply the dollar value that is of interest. A thousand dollars is not necessarily of the same subjective value to investor A and investor B. It is actually this difference in personal utilities which helps create markets.

Narratives concerning the differing values of the same property to individuals of differing means go back into antiquity. Around 1035 B.C., the Prophet Nathan told about a rich man with many flocks who slaughtered the sole lamb of a poor man rather than kill one of his own. King David was enraged to hear of the deed and promised harsh justice to the evil rich man. One lamb means more to a poor man than to a man with many lambs. (Of course, then Nathan dropped the punch line which had to do with David having had Uriah killed, so that David could gain access to Uriah's wife Bethsheba. The parable was about King David himself.) In the New Testament Jesus tells of a poor widow whose gift to the Temple of two mites, a small multiple of lowest currency in the realm, had moral value more than the magnificent gifts of the very wealthy, since the widow had given away everything she had.

All this more or less resonates with us as a matter of common sense. We understand that the gain or loss of a small amount of property means much more to a poor person than it does to a wealthy one. Although the effect of present wealth on the utility of the gain of a certain amount of money had been clear for millenia, it seems that no attempt had been made until 1738 to come up with a quantitative measure of the relationship between present wealth and marginal gain. In that year, the Swiss protostatistician Daniel Bernoulli published "Exposition of a New Theory on the Measurement of Risk" [3]. In this paper Bernoulli laid the basis of utility theory. Before starting our discussion of Bernoulli's paper, we should note that utility theory today in market modeling does not have quite the relevance one might have supposed in the golden age of utility theory, which ended some time before 1970. To many, a dollar is a dollar, regardless of one's wealth. That is unfortunate, for as we shall emphasize repeatedly in this text, markets are made by diverse individuals who view the same commodity or security quite differently from one another. Prices are most efficiently determined by a jostling of buyers and sellers who come together with different notions of utility and arrive at a trading price by agreement.

When governments or agencies attempt to enforce fair pricing, disaster is generally the result, even when the government is really high-minded and has worthy goals. As an extreme example, during the general uprising of the population of Warsaw against the Nazis in August of 1944, things were going rather badly for the Poles. Even clean water was all but impossible to obtain. Some industrious peasants from the suburbs who had access to water carrier wagons loaded up from the family wells and, risking life and limb, delivered water to the freedom fighters for a price well above the peace time price of fresh water. What was a fair price for water delivered to patriots fighting to the death for the freedom of the nation? The freedom fighter high command decided it was zero and had the water vendors shot. Of course, this marked the end of fresh water, but at least a politically correct fair price for water for freedom fighters had been enforced. Fresh water was both free and unavailable. "Nothin' ain't worth nothin', but it's free."

2.2 The St. Petersburg Paradox

Many are familiar with the following apparently paradoxical question:

How much should you be willing to pay to play a game in which a coin is repeatedly tossed. If the first heads appears on the first toss, you will receive $2^1 = 2$ dollars. If the first heads appears on the second toss, you will receive $2^2 = 4$ dollars. If the first heads appears on the kth toss, you receive 2^k dollars. The game terminates on the round where the first heads is obtained.

Now, on the average the expectation of the pay-off in this game is

$$V = \sum_{k=1}^{k=\infty} 2^k \left(\frac{1}{2}\right)^k = 1 + 1 + \ldots = \infty. \tag{2.1}$$

The expected pay-off in the game is infinity. Would anybody pay a million dollars to play this game? Very likely, the answer is negative. Possibly somebody who was already incredibly rich, for the chances of tossing 19 straight tails is small (one in $2^{19} = 524,288$). A poor person might be unwilling to pay more than two dollars, the minimum possible pay-off of the game. There is, it would appear, a relationship between one's wealth and the amount one would pay to play this paradoxical game. But things are more complicated than that. It really is the case that simply looking at the expected value of the game does not tell the whole story unless the game is replayed a very large number of times.

As we shall discuss later on, it is customary to talk about the expected value of an investment and also about its volatility (standard deviation). But the fact is that these two numbers generally will not give an investor all the information to determine whether an investment is attractive or not. Now, for the St. Petersburg game, we know that the probability that the player will realize at least $(\$2)^{k+1}$ is given (for $k = 1, 2, \ldots$) by[1]

$$1 - \sum_{j=1}^{k} \left(\frac{1}{2}\right)^{j} = \left(\frac{1}{2}\right)^{k}. \tag{2.2}$$

The player cannot make less than \$2. But what is the probability that he will make at least, say, \$1024? From (2.2), the answer is $1/2^9 = 1/512 = .001953$. This would be roughly two chances in a thousand. Probably, the player would dismiss such an event as being very unlikely. In Figure 2.1, we give a profile showing the probabilities that the player will make at least various amounts of dollars.

The probability profile gives a reasonable insight as to why the St. Petersburg game is not worth the expectation of payoff, in this case ∞.[2] We believe that, for most players, most of the time, it is advisable to look at the entire probability profile when deciding whether an investment suits the investor. Most investors will not be as impressed with the fact that the expected value of the game is infinity as they would with the fact that in only two chances out of a thousand will the game produce winnings in excess of \$1000. Looking at the overall picture, half the time, a player will make at least \$4. He will never make less than \$2. One-fourth of the time, he will make at least \$8. One-eighth of the time he will make at least \$16, etc. In deciding how much he will wager to play the game, the player should have the entire probability profile at his disposal, not simply the mean and standard deviation. At the end of the day, the player must decide how much he is willing to pay to play. In other words, he must combine the entire probability profile into his decision, "Yea or nay." It is tempting to

---

[1] Here we are using the fact that a series of the form $1 + r + r^2 + r^3 + \ldots + r^n = (1 - r^n)/(1 - r)$ if r is greater than 0 and less than 1.

[2] It is interesting to note, of course, that no gambling house exists which could pay ∞!

create a "deus ex machina" which will automatically decide how much one should pay to play the game. The expected value is such a rule, and we have seen that it does not work.

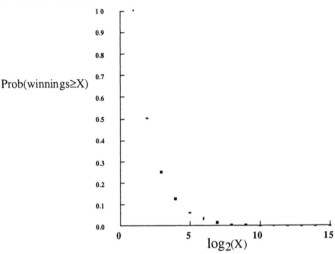

Figure 2.1. Probability Profile of Log$_2$(Winnings).

The situation where the *utility function* U is the dollar value of game payoff is a special case of the class of reasonable utilities. Here, the utility is *linear* in dollar pay-off, that is, the relation between utility U and payoff X is given by

$$U(X) = a + bX. \qquad (2.3)$$

We note that if we used any function which grows more slowly than X, then, for the coin flipping example under consideration,

$$E(U(X)) = V = \sum_{k=1}^{k=\infty} U(2^k)(\frac{1}{2})^k \qquad (2.4)$$

is finite. For example, suppose we consider $U(X) = \sqrt{X}$, then we have

$$E(U(X)) = \sum_{k=1}^{k=\infty} \sqrt{2^k} \left(\frac{1}{2}\right)^k \qquad (2.5)$$

$$= \sum_{k=1}^{k=\infty} 2^{-k/2} \qquad (2.6)$$

$$= \frac{1}{\sqrt{2}} \left[1 + \frac{1}{\sqrt{2}} + \frac{1}{2} + \frac{1}{2^{3/2}} + \ldots\right] \qquad (2.7)$$

$$= \frac{1}{\sqrt{2}} \frac{1}{1 - 1/\sqrt{2}}$$

$$= \frac{1}{\sqrt{2} - 1}. \qquad (2.8)$$

Another popular utility function is the logarithm $U(X) = \log(X)$.[3]
Here, for the coin-flipping problem [4]

$$E(U(X)) = \sum_{k=1}^{k=\infty} \log((2^k)) \left(\frac{1}{2}\right)^k \qquad (2.9)$$

$$= \log(2) \sum_{k=1}^{k=\infty} k \left(\frac{1}{2}\right)^k$$

$$= 4\log(2).$$

Again, it should be emphasized that for most investors, looking at the expected utility will be at best a poor substitute for looking at the entire probability profile of the utilities. We consider such a profile in Figure 2.2 where the probability of making at least a value of utiles as those given on the abscissa is plotted for the case where log(winnings) is the utility. We note that we have the same probability masses as in Figure 2.1. Only the abscissa axis has changed, since we are looking at $\log_e$(winnings) as opposed to $\log_2$(winnings) (where $e = 2.7183$). Most investors could work just as well with Figure 2.1 as with Figure 2.2, even if their utility function were $\log_e$. We really can get into trouble if we do not look at the entire probability profile (which could appropriately also be referred to as a *risk profile*). If we decide to make our decision based on one summary number, such as the expected value of the utility, we can quite easily lose our grasp of what the numbers are telling us.

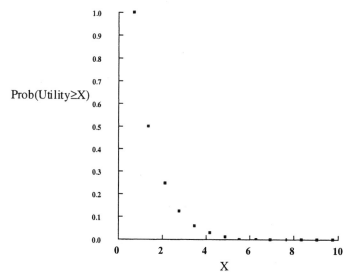

Figure 2.2. Probability Profile of Log(Winnings).

---

[3] The unit of measure of utility is termed the *utile*.
[4] Here we use the fact that $1/(1-y)^2 = \sum_{k=1}^{k=\infty} ky^k$.

Generally speaking, reasonable candidates for utility should be nondecreasing in capital. That is, by increasing one's capital, one's utility does not decrease. There are, of course, exceptions to this rather reasonable assumption. As with most exceptions to rationality, these are generally state imposed. For example, during the Socialist government of Olaf Palme in Sweden, graduated tax rates actually got over 100%. In Lenin's war to break the power of free agriculture in Ukraine, "kulaks" (*kulak* means "fist") were defined to be peasants with over a certain modest amount of holdings. If one had a bit less than the boundary value, one was (for the time) left alone. But over the value, one was shot. Similarly, in the Red controlled areas of Spain during the Civil War (1936–1939), there was a critical boundary between who was a peasant and who was a "landlord."

The marginal utility of a new increment or decrement of wealth is clearly a personal matter related to one's wealth. The loss of ten thousand dollars is trivial to a wealthy person. To someone in the lower income brackets, it may be ruinous. "A billion here, a billion there," can be easily be doled out by an American politician at the national level. An incremental dollar has utility dependent, somehow, on the assets of the person(s) considered. This rather obvious fact can be used for many purposes, including graduated taxation. Here we simply wish to consider the matter from the standpoint of its practical implications.

One can argue, as did Bernoulli, that the marginal increase in a person's wealth by a profit should be measured as a ratio of the new profit to the assets already in hand before the profit was realized. Expressing this in symbols, where U is the utility, k is a constant of proportionality, X is the base amount of wealth, and ΔX is the change in that wealth, one can write[5]

$$\Delta U = k \frac{\Delta X}{X}. \tag{2.10}$$

This, then, says that an increase in wealth of one dollar changes the utility of a person with beginning wealth of $100 the same as an increase in wealth of $1,000 has for a person with beginning wealth of $100,000. This "law" of Bernoulli's is, of course, not really a law but rather an assumption with shortcomings. For example, if both of these hypothetical persons have a child being held for a ransom of $101,000, then the wealthier individual can buy his child's freedom, whereas the poorer one is as far away from achieving the goal with $101 as with $100. On the other hand, if the ransom is $101, then the poorer person's utility goes up much more with the addition of one dollar than that of the richer one with the addition of $1,000. These are both examples of "step function" utilities, and Bernoulli wanted to look at a smooth utility. Objections can be raised that utility functions which have critical jumps up or down are not realistic. The thousands of bankruptcies experienced in the United States yearly would seem to be an example of step function realities.

---

[5]If the initial capital is X, then (2.9) is satisfied by $U(X, \Delta X) = \log[(X + \Delta X)/X]$.

Having noted that Bernoulli discovered an insight rather than a law, we must concede that his insight was valuable. Generally speaking, when starting to understand a new concept, it is good to try and reason from an example, even a hypothetical one. Bernoulli gave a hypothetical example based on Caius, a fictitious merchant of St. Petersburg, Russia, who was contemplating whether he should take insurance on a shipment from Amsterdam to St. Petersburg. The shipment, upon delivery, provides Caius with 10,000 rubles. But storms are such that, on the average, 5% will be lost at sea. The Amsterdam underwriters want a full covered policy payment of 800 rubles, or 8% of the profit if no mishap occurs. Should Caius buy the policy? His expected value if he does not is 9,500 rubles. The underwriters are clearly demanding a premium of 300 rubles above their expected payout. Is it worth it for Caius to purchase the policy or to "self-insure"? If we go by naive ruble values, then he should self-insure. But if Caius follows Bernoulli's advice, he should buy the policy if his expected utility for insuring is greater than that for not insuring. Let us suppose Caius's capital is X rubles. We shall, without loss of generality, take $k = 1$, since any k will change both the insured and self-insured options in the same way. Then Caius's expected utility of not insuring minus that of insuring is given by

$$f(X) = .95 \log \left[\frac{X + 10,000}{X} \right] - \log \left[\frac{X + 9,200}{X} \right]. \qquad (2.11)$$

Setting $f(X) = 0$, we can easily solve the resulting equation using Newton's Method (see Appendix A at the end of this book).

$$X_{n+1} = X_n - \frac{f(X_n)}{f'(X_n)}. \qquad (2.12)$$

Starting with 5,000 rubles as our first guess, we arrive at the indifference value of 5,042 rubles. If Caius has less than this amount, he should (according to Bernoulli), buy the 800 ruble policy. If he has more, he should self-insure (i.e., not buy the insurance).

Next, let us ask the question as to how much the underwriter (insurer) should have in hand in order to sell the 10,000 ruble policy for the amount of 800 rubles. Let Y be the assets of the underwriter. Then he should sell the policy if his assets exceed the Y value in

$$g(Y) = .95 \log \left[\frac{Y + 800}{Y} \right] + .05 \log \left[\frac{Y - 9,200}{Y} \right]. \qquad (2.13)$$

Again, using Newton's Method, we find that if the underwriter has a stake of 14,242 rubles or more, Bernoulli's rule tells us that selling the policy will improve the underwriter's position. Going further, let us ask the question as to what is the minimum price the underwriter might reasonably sell the

10,000 ruble policy if the underwriter has capital of one million rubles. To achieve this, we simply solve

$$h(W) = .95 \log \left[\frac{10^6 + W}{10^6} \right] + .05 \log \left[\frac{10^6 + W - 10^4}{10^6} \right]. \qquad (2.14)$$

The solution here tells us that the underwriter with capital of 1,000,000 rubles might reasonably sell the 10,000 ruble policy for 502.4 rubles. But Caius, if he has capital of 5,042 rubles, say, would find it reasonable to pay up to 800 rubles. Thus, the underwriter is in a position to get more than his own indifference value for the trade (502.4). Naturally, the underwriter is largely looking at things from his own standpoint, rather than that of potential clients. He is only interested in the amount of the policy, the risk to the underwriter, and charging whatever rate the market will bear. If the underwriter has no competition, he must remember that there is always competition from the merchant himself who can decide to self-insure. This is simply a manifestation of substitutability of one service by another. Still, there is the likelihood that the presence of a second underwriter in the Amsterdam to St. Petersburg run will drive prices downward. There is a spread of several hundred rubles where the purchase of the policy is a good deal, from a utility standpoint, for both underwriter and the insured merchant.

Note that in the above example, if the underwriter sells Caius a policy for 650 rubles, then Caius has a good deal. He has the policy for less than his utility function would make him willing to pay. And the underwriter has a good deal also, for he is well above the minimum rate his utility function dictates. Here is an example of the reason transactions take place, for the deal is good for both parties from the standpoints of their respective utilities. In a true free trade situation where there are a number of merchants and several underwriters, there will be a jostling back and forth of rates. The change in riskiness of the transit due to weather, war, pirates, etc., will be a major driver in the change of rates. They will never stop their fluctuation, though at any given point in time, the cost of a 10,000 ruble policy will be similar from all of the underwriters. It is the difference in utility functions (driven in part by wealth status) as well as the difference in personal views as to the value of a commodity or service that cause markets to exist at all. If Caius were willing to buy the policy for no more than 800 rubles and the underwriter were only willing to sell it for 850 rubles, then no transaction would take place. By jostling and haggling, any given selling price will be in the "comfort intervals" of both buyer and seller. The buyer would always have been willing to have paid a bit more than he did, the seller always to have taken a bit less.

Human nature being what it is, there will be an attempt by the underwriters to combine together to set the rates at an unnaturally high level. They may even decide "for the good of the public" to have the Czar set the rates. These will tend to be rates determined by the St. Petersburg

Association of Underwriters, and thus on the high side. On the other hand, the Patriotic Association of Merchants will try and get the rates lowered, particularly for favored clients. But the creation of fixed rates will indeed "stabilize" the market by fixing the rates. Statist intervention, then, is really the only means of arriving at a "stable" market. Otherwise, the rates will fluctuate, and there will always be a market opportunity for an insurance agent continually to negotiate with all the underwriters to obtain better deals for the merchants the agent represents.

The presence of variation in market prices over time is woefully misunderstood. So far from being evidence of chaos in the market, variation represents a stabilizing continuum of adjustments to the ever changing realities of the market, including the goals, expectations and situations of the persons and institutions taking part in the market. One example of the destabilizing effects of statist efforts to "stabilize" markets is rationing. Another is the fixing of prices at arbitrary levels. In "People's" Poland, before the Communists lost power on June 4, 1989, the retail price of milk was set by the government at one generally below the cost of production. There was little collectivization in Poland, so the state was forced to buy milk from independent farmers. It could, naturally, have paid the farmers something greater than the cost of production by using state funds. It chose, rather, to force each farmer to deliver a certain quantity of milk to the state at a price which was generally below the cost of production. Naturally, the milk sold to the state had a water content considerably in excess of that of milk which comes straight from the cow. The price stabilized milk that was made available to the populace had some similarities with milk, but was something rather different, and *it varied greatly in quality from day to day, from location to location.* On the other hand, the state turned more or less a blind eye to farmers selling a portion of their milk on the black (aka free) market. Occasionally (once every several years) a "speculator" was shot, but this was pro forma. Even under Russian control, some realism was ever present in the European satellites. Black market milk was expensive, and, since it was produced and vended under irregular conditions, the quality was variable. Those who could afford to do so generally would strike a deal with one particular farmer for deliveries of milk on a regular basis. Most city dwellers, however, were stuck with price stabilized "milk" for the fifty years of Soviet occupation.

There is much to fault with Bernoulli's treatment of utility. First of all, we can observe that the conditions of trade he posed were somewhat strange. Generally speaking, Caius would have to buy the goods for shipment. A more accurate way to pose the problem, perhaps, would be one in which Caius has acquired goods in Amsterdam for which he paid, say, 6,000 rubles. He can sell the goods in St. Petersburg for 10,000 rubles. When he buys his insurance policy, he may well have to settle for insuring at the amount of purchase. Caius may own his own ship, in which case, insuring the value of the ship is another consideration. And so on. But

Bernoulli has revealed a most important point: the marginal value of a dollar varies depending on the financial status of the person involved. Such differences in utility from person to person should not pass unnoticed. It is one (though not the only one) reason that markets exist at all. Another problem with Bernoulli's treatment is that everything is based on the expected value of the utility. In order for an investment to have low risk, we need to know that the probability of a large loss is small. The risk profile cannot be captured by the expected utility or any other single number.[6]

2.3 von Neumann-Morgenstern Utility

It is generally not a very good idea to assume a utility function for a particular individual concerning a set of possible transactions contemplated by an individual. A financial advisor who simply assumed, say, a logarithmic or square root relation between various returns and the utility of a client is making an unnecessary assumption. Utility being a matter of personal assessment, it is frequently possible to come up with a series of questions which would enable us to extract the implied utility of the various choices.

von Neumann and Morgenstern [7] have given us a set of axioms which should be satisfied by the preferences of a rational person. These are:

1. **Transitivity.** If the subject is indifferent between outcomes A and B, and also between B and C, he must be indifferent between A and C. Symbolically, $A \underset{\prec}{\succeq} B$ and $B \underset{\prec}{\succeq} C \Rightarrow A \underset{\prec}{\succeq} C$.

2. **Continuity of preferences.** If A is preferred to B and B is preferred to no change, then there is a probability α $(0 < \alpha < 1)$, such that the subject is indifferent between αA and B.

3. **Independence.** If $\mathbf{A}$ is preferred to $\mathbf{B}$, then for any probability α $(0 < \alpha < 1)$ $\alpha A + (1 - \alpha)B$ is preferred to B. (An equivalent axiom says that if $A \underset{\prec}{\succeq} B$, then $\alpha A \underset{\prec}{\succeq} \alpha B$.)

4. **Desire for high probability of success.** If A is preferred to no change, and if $\alpha_1 > \alpha_2$, then $\alpha_1 A$ is preferred to $\alpha_2 A$.

5. **Compound probabilities.** If one is indifferent between αA and B, and if $\alpha = \alpha_1 \alpha_2$, then one is indifferent between $\alpha_1 \alpha_2 A$ and B. In other words, if the outcomes of one risky event are other risky events, the subject should act only on the basis of final outcomes and their associated probabilities.

Let us now go through the largely psychometric exercise for determining a subject's utility function and his or her willingness to accept risk. First

---

[6]For an excellent and quite readable discussion of the world of "risk" considered from both economic and philosophical perspectives, see [4].

of all, notice that all of this discussion abandons the world where utility is linear in dollars. We shall talk of a new currency, called *utiles*. This is, by the way, an interpolation rule. We will not feel very comfortable about extrapolating outside the interval where our client can answer the questions we shall pose. We will start, then, with a range of dollar assets, say $0 to $1,000,000. We need to define a utility nondecreasing in dollars. It turns out that our hypothetical client has a utility function equal to the square root of the amount of dollars. (He does not necessarily realize this, but the answers to our questions will reveal this to be the case.) We need to define the utility at the two endpoints. So, we decide (ourselves, without consulting yet the client) that $U(\$0) = 0$ utiles and $U(\$1,000,000) = 1,000$ utiles.

Q. How much would you be willing to pay for a lottery ticket offering a 50-50 chance of $1,000,000 or $ 0?
A. The client responds "$250,000." (Of course, we recognize the expectation in utile scale as being .5× 0 utiles + .5× 1,000 utiles = 500 utiles.)
Q. How much would you pay for a 50-50 chance of $250,000 or nothing?
A. The client responds "$62,500" (which we recognize to be the expectation on the utile scale, .5× 0 utiles + .5 × 500 utiles = 250 utiles).
Q. How much would you pay for a 50-50 chance of $62,500 or nothing?
A. The client responds "$15,625" (which we recognize to be the expectation on the utile scale, .5× 0 utiles + .5×250 utiles = 125 utiles).

The above analysis is consistent with the

Utility Maxim of von Neumann and Morgenstern. The utility of a game (risky event) is not the utility of the expected value of the game but rather the expected value of the utilities associated with the outcomes of the game.

In Figure 2.3, we give a spline smoothed plot using the Q&A. We note that if all we had was the three questions and their answers, we would see a plot virtually indistinguishable from what we know the functional relationship is between utiles and dollars, namely,

$$U(X) = \sqrt{X}.$$

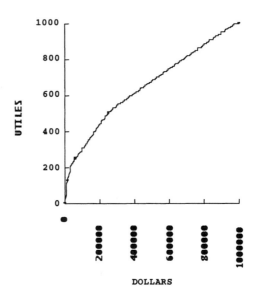

Figure 2.3. Empirically Determined Utility Function.

Now, we recall that our client may not know that his utility is the square root of dollars. But his answers to our three questions give us a graph which is essentially equivalent to the square root.[7]

We can now ask the following question: How much is it worth to the client to receive the payoffs $10,000 with probability .3, $90,000 with probability .5, and $490,000 with probability .2? Now, the naive answer would be that associated with the assumption that the utility of the client is simply the dollar amount:

$$E(X) = .3 \times 10,000 + .5 \times 90,000 + .2 \times 490,000 = 146,000.$$

But we have already determined that the utility function of the client is not dollars but the square root of dollars. This gives us:

$$E(U(X)) = .3 \times \sqrt{10,000} + .5 \times \sqrt{90,000} + .2 \times \sqrt{490,000} = 320 \text{ utiles.}$$

Going to Figure 2.3 (or recalling the $X = U^2$), we have that our client should consider a sure payment of $320^2 = \$102,400$ to be the value of the game to himself or herself.

Let us consider several scenarios each having expected utility value 500 utiles.

- A. A cash gift of $250,000 = \sqrt{250,000}= 500$ utiles. Here $E(X)$ = \$250,000. The standard deviation of the dollar payout is $\sigma_A = \sqrt{(250,000 - 250,000)^2} = \$0$.

---

[7]Economists derive the notion of "diminishing marginal utility" from functions of this sort.

- B. A game which pays \$90,000 with probability .5 and \$490,000 with probability .5. Here $E(U) = .5 \times \sqrt{90,000} + .5 \times \sqrt{490,000} = .5 \times 300 + .5 \times 700 = 500$ utiles. $E(X) = .5 \times \$90,000 + .5 \times \$490,000 = \$290,000$. $\sigma_B = \sqrt{.5(90,000 - 290,000)^2 + .5(490,000 - 290,000)^2} = \$200,000$.

- C. A game which pays \$90,000 with probability .5 and \$490,000 with probability .5. Here, $E(U) = .5 \times \sqrt{90,000} + .5 \times \sqrt{890,000} = 500$ utiles. $E(X) = .5 \times \$90,000 + .5 \times \$810,000 = \$400,000$. The standard deviation of the dollar output is given by $\sigma_C = \sqrt{.5(90,000 - 400,000)^2 + .5(810,000 - 400,000)^2} = \$400,000$.

- D. A game which pays \$1,000,000 with probability .5 and \$0 with probability .5. Here $E(U) = .5 \times 0 + .5 \times \sqrt{1,000,000} = 500$ utiles. $E(X) = .5 \times \$1,000,000 = \$500,000$.
$\sigma_D = \sqrt{.5(1,000,000 - 500,000)^2 + .5(0 - 500,000)^2} = \$500,000$.

In Figure 2.3, we show the *indifference curve* for various games each having a value of 500 utiles.

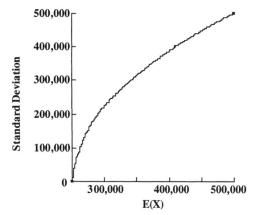

Figure 2.4. σ versus Expected Payoff with $E(U) = 500$ utiles.

We note the typical increase in expected payoff as the games become more risky. We note that the von Neumann-Morgenstern utility paradigm enables us to assess a rational choice system for a client who has no real notion of mathematical modeling, provided he or she can answer a few questions concerning indifference to choice between a sure thing and a set of particular games.

One major reason for the functioning of a market is that individuals and corporations will have different utilities for the same thing. Suppose that everybody valued a stock at \$100. What would be the incentive for the owner of such a stock to sell it at less than \$100? He might need money for some other purpose. Perhaps he knows of another stock that is selling for \$99 which is worth to him \$110. In that case, he might be willing to sell his share of stock for $99 because he could then use the money to make a net profit. But if all stocks, services and other properties were evaluated by all people the same, that would stifle the market.

Let us consider a bar of soap selling in a shop for one dollar. To the owner of the shop, the bar of soap is worth less than a dollar, or the owner would not sell it for a dollar. To the customer who buys the bar of soap, it is probably worth more than a dollar (to see that this is so, we need only ask whether the customer would buy the soap bar for $1.01.) The price of sale, from the standpoints of the buyer and the seller, is in the interval between what the vendor values the soap and that which the buyer values the soap (assuming the former is less than the latter, for otherwise no voluntary transaction will occur). A transaction at any price within the interval will be *Pareto preferred* to no transaction, as one or both parties will be better off without any party being worse off. However, there is no "scientific" basis for selecting a price within the interval, because each price is *Pareto optimal* (a movement from any price to any other price would make one of the parties worse off). A price outside of the interval will only generate transactions through coercion, because one of the parties will be worse off than not transacting at all. For every value within that interval, we have achieved *Pareto efficiency*, that is, the vendor is getting more than he values the bar of soap and the purchaser is buying it for less than he values it. The task of the merchant is to raise the price as high as possible without exceeding the valuation of a significant fraction of his customers. Any notion of *fair market price*, a one price for the same item, is achievable only in a state controlled situation. And such statist controlled prices generally produce absurdities.

In Soviet-controlled Poland, the state suppressed the price of bread to well below the cost of production. It made this work, in part, by forcing the farmers (Poland's farms were privately owned and operated, for the most part, even during the Russian occupation) to sell a certain fraction of their wheat below the cost of production. The price of pork, however, was not so artificially depressed. So some clever peasants became rich by buying stale bread and feeding it to their pigs. It was difficult for the authorities to overcome this strategy, since there was really no other good set of buyers for stale bread. The best that could be done was to raise the price of stale bread to nearly that of fresh bread. Ultimately, the only good solution (to escape embarrassment of the officials) was to raise the price of fresh bread as well. Actually, the statist strategies became much more complicated than the relatively simple scenario stated. A market with set prices is very much like a cardiac patient who takes one medication for his heart plus three other medications to counteract bad side effects of the heart medication plus five additional medications to counteract the bad side effects of the three medications taken to counteract the effects of the primary heart medication.

2.4 Creating a "St. Petersburg Trust"

In the Enlightenment world of Daniel Bernoulli, neat and concise answers to virtually any problem were deemed possible if one only had the right insight. In our modern world, we understand that there is likely to be some arbitrariness in most simple solutions. All the notions of utility, at which we have looked, have some underlying assumption that if one only looks at the proper function of wealth, then decisions become rather clear, utilizing a simple *stochastic* (probabilistic) model, relying on a few summary numbers, such as expected value. By constructing a utility function and then looking at its expected value in the light of a particular policy, Bernoulli thought he could capture the entire profile of risk and gain. Simply looking at a utility linear in money came up, in the case of the St. Petersburg game, with an absurd result, infinite gain. So, Bernoulli constructed other utility functions which had other than infinite expectation for the St. Petersburg game. He really did not, however, achieve his goal: reducing the entire prospectus of the gain to one scalar number. Moreover, market decisions these days are still based on dollars—not logarithms of dollars or square roots of dollars. This is surely evidence that utility theory has not lived up to the hope some have held for it.

The real essence of constructing a risk profile has to do with looking at the stochastic process which underlies the investment being considered. (For detailed information concerning stochastic processes, the reader is referred to Appendix A at the end of this book.) We need to look at the probabilities of various results which might be obtained and decide, in the aggregate, whether the deal appears good. Of course, at the end of the day, we must make a decision whether to take the deal or turn it down. However, we believe that this is better done from the risk profile (time slices of the cumulative distribution function of the payoff) than from a one dimensional summary number (which, of course, expectation is).

In our postmodern world, in which few things are really simple, we should take a different view. It could be argued that Bernoulli was enchanted with the notion that a benevolent Providence had constructed the universe in structures selected to make them easy for human beings to understand. Or, he might have taken the nominalist (essentially, postmodern) view of William of Ockham that "truth" was a matter simply of fashion, so one might as well pick the simplest model that seemed, more or less, to work. After all, Bernoulli had no notion of fast computing, a primitive slide rule being the hottest computer available.

The view we shall take is that one should try to use any means available to get close to the truth (and we do not put quotes around the word), realizing we will generally use a model at variance with reality, but hopefully not too far from it. Bernoulli's consideration of a game with infinite expected payoff was unfortunate. In the first place, one should ask who the croupier would be for such a game. Beyond that, games with infinite expectation do not correspond to any real-world economic situation (again, we notice that no casino in the world can offer a payout of ∞. Everyone places some maximum betting limit—perhaps high for "high rollers" but not ∞.) Finally, as we show below, a game with an infinite expected value can be diddled in such a way than a clever player can come up with a very high gain with high probability.

Below we will perform some computer simulations to try and understand better the St. Petersburg Paradox. First of all, we will follow the common policy of not using a utility function formally, but only implicitly. If we play the game for $2 payoffs, we get a false sense of value. So let us make the payoffs to be in the $2 million range. Suppose we take a reasonably well-to-do individual and ask how much he would pay to play the game using the profile in Figure 2.1. We note that the investor cannot make less than $2 million dollars. So that is an obvious floor. We recall that the expected payoff of the game is infinite, but the chance of making more than $2 million is only 50%. The chance of making more than $4 million is only 25%. The chance of making more than, say, $5 million is also 25%. If the person's total assets are, say, $5 million, it is hard to believe he will be willing to pay the full $5 million to play the game, even knowing that there is a $2 million floor below which he cannot fall. Rather clearly, he would be willing to pay, say, $2.1 million to play. For figures between, say $2.4 million and $5 million, we would see a variety of decisions made by the variety of possible players, depending upon their incomes, ages, psychology, etc. Will the decision be made on the basis of expected winnings? Of course not. It will be made utilizing the entire information given in Figure 2.1.

Next, let us see whether the investor can structure the game somewhat differently. Suppose he notes the possibility of arranging things so that another investor and he form a trust and strike an agreement with the house that each will play the game for half the payoffs and the winnings will be pooled and divided by two. In this case, we note that both players may get heads first toss. In that case, the pooled winnings will be $2,000,000. This occurs with probability $.50 \times .50 = .25$. Suppose the first player gets heads first time, but the second gets tails first time immediately followed by heads. That would result in net winnings of $3,000,000. Or it could be the first player gets tails first, then heads. Again, net winnings of $3,000,000. The probability of one or the other of these is $2 \times .5 \times .25 = .25$. Next, they could both toss TH for a total winning of $4,000,000. The probability of this is $.25 \times .25 = .125$. So, then, the probability that the trust wins more than $2,000,000 is $1 - .25 = .75$. The probability the trust wins more than $4,000,000 is $1 - .25 - .25 - .125 = .375$. According to the original game in which only one player plays, these probabilities were .50 and .25, respectively. Clearly, the idea of playing according to each player playing for half the pay-offs of the original game rules is a good one. We can carry out computer simulations for games with varying numbers of players: one, ten, one hundred, one thousand. We show the probability profile in Figure

2.5 where sums are in millions of dollars.

Here we see the advantage which can be achieved by pooling in the rather stylized St. Petersburg Paradox situation. With 1,000 investors participating, the probability of the trust winning a total in excess of nine million dollars is 90%. For the same game played by one investor, the probability of winning in excess of nine million dollars is less than 10%.

It would be a fine thing if there were some sort of way we might develop a portfolio or trust fund which mimicked the strategy above for dealing with Bernoulli's St. Petersburg scenario. Alas, the authors are unable to point out such a strategy. Essentially, our "trust strategy" uses the fact that sample means converge, with increasing sample size, to the mean (expected value) of the population. In the St. Petersburg case, this value is infinite. Nature and the market do not tend to provide such opportunities.

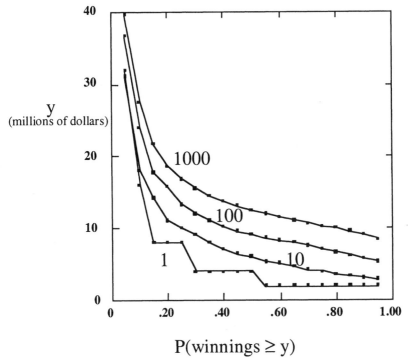

$$P(\text{winnings} \geq y)$$

Figure 2.5. Probability Profile for St. Petersburg Trust.

There are other flaws with the St. Petersburg scenario of Bernoulli. For example, the coin throws are taken to occur with constant probability. But storms in the North Atlantic and Baltic are not of constant probability of occurrence. A heavy storm period will increase the risks to all shipping during that period. This happens also in markets. There will be periods of bull market growth across the market. And there will be periods of bear market declines. There may be safe harbors during the bear market times.

But very frequently, as in the 2000–2001 case, a bear market will adversely affect most securities.

Perhaps the bottom line in our analysis of the St. Petersburg Paradox is that by examining the full probability profile instead of simply looking at the expected value of utility, we can gain something of an understanding of what is going on. We replace arbitrary formalism and a simple answer with a more complicated, albeit more insightful analysis. Moreover, the real world may not be one of risk but rather varying degrees of hazy uncertainty. Finally, almost all people violate the axioms anyway. That is why they purchase insurance on one day and go to Las Vegas on the next!

2.5 Some Problems with Aggregate Choice Behavior

So far, we have been concerned with the utility based choices made by individuals. We now look at the pooling of individual choices to obtain pooled choices of groups of individuals. A key part of our treatment will be to show Kenneth Arrow's proof [1] that rational rules for preferences among individuals do not translate into the same rules for the group. We essentially follow the treatment given by Thompson [6] in *Empirical Model Building*.

To begin, we have a collection of n individuals $\{1, 2, \ldots, n\} = G$. The task confronting the individuals is to rank their preferences amongst at least three decisions $D = \{a, b, c, \ldots\}$. By $a \succ_1 b$, we mean that the first individual in the group prefers a to b. By $a \succ_G b$, we mean that the group as a whole prefers a to b, i.e., whatever the underlying mechanism used to obtain group consensus, the group picks a over b.

Suppose we have the following four axioms for pooling individual preferences into group decision making.

1. For a particular set of individual preferences, suppose the group prefers a to b ($a \succ_G b$). Then, suppose that some of the individuals change their preferences in such a way that preferences for a over b are unchanged or increased in a's favor, and that each individual's preference between a and any alternative other than b are unchanged. Then, the group preference for a over b is maintained.

2. (Axiom of the Irrelevant Alternative.) Suppose that the group prefers a to b. Then, some of the individual preferences between alternatives other than a and b are changed, but the preferences between a and b are unchanged. Then, the group preference for a over b is maintained.

3. For any pair of alternatives a and b, there is some collection of individual preferences for which $a \succ_G b$.

4. (Axiom of Disallowing Dictators). No individual in the group has such influence that if he or she prefers a to b, and every other member of the group ranks b over a, then $a \succ_G b$.

Arrow's Impossibility Theorem. If Axioms 1–3 hold, then Axiom 4 cannot hold, assuming there are two or more decision makers and three or more possible decisions.[8]

Definition: Suppose we have $k \geq 3$ mutually exclusive decisions for which each of $n \geq 2$ voters have ordered preferences. Let $\mathcal{P} = \{P_1, P_2, \ldots, P_n\}$ represent the ordered preferences (*profile*) of each of the individual voters. Let J be a subset of the set of individual voters G and a and b be included among the set of decisions D. If, for all the individuals in J, $a P_i b$ (i.e., $a \succ_i b$), we then say that $\mathcal{P}$ is $J - favored$ for (a, b). If for all the individuals not in J, $b P_i a$, then we say that J is *strictly $J- favored$*.

Definition: Suppose that the fact that all members in J favor a over b implies that a F($\mathcal{P}$) b, where F is the group decision rule (a. k. a. "social utility function") which integrates the preferences of the individual voters into a group decision for $\mathcal{P}$ (i.e., we suppose that all the members in subset J deciding for a over b will cause the consensus of the entire group $\mathcal{P}$ to prefer a over b.) Then we say that J is *decisive* for (a, b).

Definition: A *minimal decisive set* J is a subset of G which is decisive for some (a, b) and which has the property that no subset of J is decisive for

---

[8]Before proving the theorem, we note that all the axioms appear quite reasonable at first glance, somewhat less so upon closer inspection. For example, even in North Korea, the Supreme Leader could not make policy opposed by everyone else on the Politburo. But then, it is not hard to consider examples where dictators are possible. For example, a person owning 51% of the stock of a company can make policy disagreed to by all the other shareholders. Then, again, looking at the Axiom of the Irrelevant Alternative, suppose there are two Republican candidates, one strongly pro-life, the other strongly pro-choice. 51% of the party strongly supports the pro-life position, the other 49% strongly supports the pro-choice position. If the vote were held between these two, the strongly pro-life candidate would win the primary. Of course, this would be done in the light of the general election, where the Democrats would then be able to capture the votes of many pro-choice Republicans. But then, a third feasible candidate appears, one who is mildly pro-life. In all probability, this third candidate will get the nomination. Clearly, in bringing forward this example, we have used the fact that decisions are seldom stand alone. It is the knowledge of the general election which will cause some of the strongly pro-life Republicans to consider voting for the mildly pro-life candidate. But let us suppose the four axioms above are all satisfied. Then, Arrow shows that they cannot be all satisfied. It is simply an impossibility. Fundamentally, the reason for this fact has to do with the fact that although in one-dimensional space, it is easy to make orderings (e.g., clearly 2>1), it is not nonarbitrarily possible in higher dimensional spaces (e.g., we cannot say that $(2,1)>(1,3)$ nor that $(1,3)>(2,1)$, unless we arbitrarily impose a one-dimensional structure onto the two-dimensional space; but if we use as the criterion the sum of squares of the components, then we can say that $(1,3)>(2,1)$, since $1^2 + 3^2 > 2^2 + 1^2$. Having said this, we shall go through *Arrow's Impossibility Theorem* in part to show how much more important are insight and conjecture than theorem proving ability in the acquisition of Nobel Prizes. The proof is rather easy. The conjecture (which is what a theorem is before it has been proved) is profound, and its consequences important in the consideration of efficient market theory.

any other pair of decisions.

Lemma 2.1. Assume Axioms 1–4. Then J is decisive for (a, b) if and only if there is a set of preferences which are strictly $J - favored$ for (a, b) and for which $aF(\mathcal{P})$ b.

Proof. Suppose J is decisive for (a, b). Then any strictly $J - favored$ for (a, b) profile has $aF(\mathcal{P})b$.

Next, suppose there is a profile $\mathcal{P}$ that is strictly $J - favored$ for (a, b) and for which $aF(\mathcal{P})b$. Then, every voter in J prefers a to b. But since $\mathcal{P}$ is strictly $J - favored$ for (a, b), we know that voters in $G - J$ all prefer b to a. Let $\mathcal{P}'$ be some other $J - favored$ for (a, b) profile. For all voters in J, a is preferred to b for this second profile. Thus, insofar as voters in J are concerned, they all prefer a to b for both $\mathcal{P}$ and $\mathcal{P}'$. But for $G - J$, all the voters, following profile $\mathcal{P}$ prefer b to a. Nevertheless, $aF(\mathcal{P})b$. However, for profile $\mathcal{P}'$ it is possible some voters in $G - J$ prefer a to b. By Axiom 1, then, it must be true that $aF(\mathcal{P}')b$. Hence, J is decisive for (a, b).

Lemma 2.2. Assuming all the four axioms to be true, the entire group G is decisive for every (a, b).

Proof. Suppose every voter in G prefers a to b, but that $bF(\mathcal{P})a$. By Axiom 3, there must be some profile $\mathcal{P}$ such that $aF(\mathcal{P})b$. Now, if every voter in G prefers a to b, but the group decision is for b over a, then changing some of the voters' preferences to b over a can only (by Axiom 1) strengthen the group resolve to prefer b over a. That would contradict Axiom 3, for then there would be no profile $\mathcal{P}$ such that $aF(\mathcal{P})b$. Hence, G is decisive for every (a, b).

Next, let J be a minimal decisive set. We know there is a decisive set, since the preceding lemma proved that G is decisive for every (a, b). So, we can remove individuals from G until one more removal would no longer give a decisive set. Pick one of the voters j from the minimal decisive set J. We shall prove that j must be a dictator, contrary to Axiom 4.

Suppose J is decisive for (a, b). Pick another decision c which is neither a nor b. Consider the profile shown in Table 2.1.

Table 2.1.		
P_i for i in $J - j$	P_i for i not in J	P_j
c	b	a
a	c	b
b	a	c
$D - \{a, b, c\}$	$D - \{a, b, c\}$	$D - \{a, b, c\}$

By construction, J is decisive for (a, b). Thus $aF(\mathcal{P})b$. We note that $\mathcal{P}$ is strictly $J - j$ favored for (c, b). Thus, if $cF(\mathcal{P})b$, then by Lemma 2.1, $J - j$ would be decisive for (c, b) contrary to the assumption that J is a minimal decisive set. Thus, c is not favored over b by the group, and a is favored over b by the group.

Consequently, we have two possible scenarios for the preference of the group: either a is preferred to b is preferred to c, or a is preferred to b and c and the group ties b and c. Then, in both cases, we have $aF(\mathcal{P})b$. But j is the only voter who prefers a to c. Thus, by Lemma 2.1, j is decisive for (a, c). Thus j cannot be a proper subset of j, that is, $j = J$. So far we have shown that j is decisive for (a, c) for any $c \neq a$.

Next, we shall establish that j is decisive for (d, c) for any d, c not equal to a. Consider the profile in Table 2.2.

Table 2.2.	
P_j	P_i for $i \neq j$
d	c
a	d
c	a
$D - \{a, c, d\}$	$D - \{a, c, d\}$

Note that the entire set G is decisive for any pair of decisions as we have proved in Lemma 2.2. Hence $dF(\mathcal{P})a$. Thus the group as a whole ranks d over a and ranks a over c. Thus $dF(\mathcal{P})c$. Therefore, by Lemma 2.1, j is decisive for (d, c).

Finally, we shall demonstrate that j is decisive for (d, a) whenever $d \neq a$. Consider the profile in Table 2.3.

Table 2.3.	
P_j	P_i for $i \neq j$
d	c
c	a
a	d
$D - \{a, c, d\}$	$D - \{a, c, d\}$

Since j is decisive for (d, c), we have $dF(\mathcal{P})a$. But j is the only individual preferring d to a. Hence, j is a dictator, contrary to Axiom 4 and the theorem is proved!

2.6   Jeffersonian Realities

Arrow's Impossibility Theorem was perceived intuitively by earlier social scientists. Vilfredo Pareto, for example, whose father had labored his entire life to bring forth an enlightened Jeffersonian democratic system to Italy, and who himself believed in the feasibility of such a system until well into middle age, finally opined that all social systems would naturally be controlled not by an orderly pooling of individual preferences, but rather by a circle of elites. In his youth, Pareto assumed that the twin pillars of Jeffersonian government and Adam Smith policies toward free trade would bring economies and governments into a state of efficiency and optimality. Let us look at Arrow's result in the context of Jeffersonian performance as opposed to Jeffersonian ideals. In 1792, the Congress was confronted with the task of deciding how many members of the House of Representatives would be allocated to each state. Alexander Hamilton, the alleged opponent of states' rights, proposed the following rule:

Hamilton's Rule. Pick the size of the House $= n$. Divide the voting population N_j of the jth state by the total population N to give a ratio r_j. Multiply this ratio by n to give the quota q_j of seats for the jth state. If this quota is less than one, give the state one seat. Give each state the number of representatives equal to the integer part of its quota. Then rank the remainders of the quotas in descending order. Proceeding down the list, give one additional seat to each state until the size of the House n has been equaled.

The method of Hamilton has firmly embodied in it the notion of the state as the basic entity of indirect democracy. Once the number of Representatives had been arrived at by a comparison of the populations of the several states, the congressional districts could be apportioned by the state legislatures within the states. But the indivisible unit of comparison was that of state population. If one conducts a poll of educated Americans and asks how seats in the House of Representatives are apportioned amongst the several states, much the most popular rule given is that of Hamilton. It is a very intuitive rule. Furthermore, if we let a_j be the ultimate allocation of seats to each state, then Hamilton's Rule minimizes

$$\sum_{j=1}^{k} |a_j - q_j|, \qquad (2.15)$$

that is, Hamilton's Rule minimizes the sum of the discrepancies between the allocations obtainable without consideration of states and those with the notion of noncrossover of state boundaries to obtain districts.[9]

---

[9]Census figures are from Balinski and Young.[2].

Table 2.4. Method of Hamilton.				
State	Population	Quota	Hamiltonian Allocation	Voters per Seat
Connecticut	236,841	7.860	8	29,605
Delaware	55,540	1.843	2	27,770
Georgia	70, 835	2.351	2	35,417
Kentucky	68,705	2.280	2	34,353
Maryland	278,514	9.243	9	30,946
Massachusetts	475,327	15.774	16	29,708
New Hampshire	141,822	4.707	5	28,364
New Jersey	179,570	5.959	6	29,928
New York	331,589	11.004	11	30,144
North Carolina	353,523	11.732	12	29,460
Pennsylvania	432,879	14.366	14	30,919
Rhode Island	68,446	2.271	2	34,223
South Carolina	206,236	6.844	7	29,462
Vermont	85,533	2.839	3	28,511
Virginia	630,560	20.926	21	30,027
Total	3,615,920	120	120	

It is interesting to note that the first Congressional Senate and House passed a bill embodying the method of Hamilton and the suggested size of 120 seats. That the bill was vetoed by President George Washington and a subsequent method, that of Thomas Jefferson, was ultimately passed and signed into law is an interesting exercise in Realpolitik. The most advantaged state by the Hamiltonian rule was Delaware, which received a seat for every 27,770 of its citizens. The most disadvantaged was Georgia which received a seat for every 35,417 of its citizens. Jefferson's Virginia was treated about the same as Hamilton's New York with a representative for every 30,027 and 30,144 citizens, respectively. The discrepancy between the most favored state and the least favored was around 28%, a large number of which the supporters of the bill were well aware. The allocation proposed by Hamilton in 1792 did not particularly favor small states. In general, however, if we assume that a state's likelihood of being rounded up or down is independent of its size, the method of Hamilton will favor somewhat the smaller states if our consideration is the number of voters per seat. But Hamilton, who was from one of the larger states and who is generally regarded as favoring a strong centralized government which de-emphasized the power of the states, is here to be seen as the clear principled champion of states' rights and was apparently willing to give some advantage to the smaller states as being consistent with, and an extension of, the notion that each state was to have at least one Representative. Now let us consider the position of Thomas Jefferson, the legendary defender of states' rights. Jefferson was, of course, from the largest of the States, Virginia. He was loathe to see a system instituted until it had been properly manipulated to enhance, to the maximum degree possible, the influence of Virginia. Unfortunately for Jefferson, one of the best scientific and mathematical minds in the United States who undoubtedly recognized at least an imprecisely stated version of (2.14), the result in (2.14) guaranteed that there was no other way than Hamilton's to come up with a reasonable allocation rule fully consistent with the notion of states' rights. Given a choice between states' rights and an enhancement of the power of Virginia, Jefferson came up with a rule which would help Virginia, even at some cost to his own principles. Jefferson arrived at his method of allocation by departing from the states as the indivisible political units. We consider the method of Jefferson below.

Jefferson's Rule. Pick the size of the House $= n$. Find a divisor d so that the integer parts of the quotients of the states when divided by d sum to n. Then assign to each state the integer part of N_J/d.

We note that the notion of a divisor d is an entity which points toward House allocation which could occur if state boundaries did not stand in the way of a national assembly without the hindrance of state boundaries. Let us note the effect of Jefferson's method using the same census figures as in Table 2.4.

We note that the discrepancy in the number of voters per representative varies more with Jefferson's method than with Hamilton's—94% versus 28%. In the first exercise of the Presidential veto, Washington, persuaded by Jefferson, killed the bill embodying the method of Hamilton, paving the way for the use of Jefferson's method using a divisor of 33,000 and a total House size of 105. Let us examine the differences between the method of Hamilton and that of Jefferson.

The only practical difference between the two allocation systems is to take away one of Delaware's two seats and give it to Virginia. The difference between the maximum and minimum number of voters per seat is not diminished using the Jeffersonian method which turns out to give a relative inequity of 88%; for the Hamiltonian method the difference is a more modest 57%. The method of Jefferson favors the larger states pure and simple. Jefferson essentially presented George Washington and Congress with a black box and the message that to use Hamilton's Rule would be unsophisticated, whereas Jefferson's Rule was somehow very politically correct.It worked. Congress approved the method of Jefferson, and this method was in use until after the census of 1850 at which time the method of Hamilton was installed and kept in use until it was modified by a Democratic Congress in 1941 in favor of yet another scheme.

Table 2.5. Method of Jefferson (divisor of 27,500).				
State	Population	Quotient	Jeffersonian Allocation	Voters Seat
CN	236,841	8.310	8	29,605
DE	55,540	1.949	1	55,540
GA	70, 835	2.485	2	35,417
KY	68,705	2.411	2	34,353
MD	278,514	9.772	9	30,946
MA	475,327	16.678	16	29,708
NH	141,822	4.976	4	35,456
NJ	179,570	6.301	6	29,928
NY	331,589	11.635	11	30,144
NC	353,523	12.404	12	29,460
PA	432,879	15.189	15	28,859
RI	68,446	2.402	2	34,223
SC	206,236	7.236	7	29,462
VE	85,533	3.001	3	28,511
VA	630,560	22.125	22	28,662
Total	3,615,920	120	120	

We have in the 1792 controversy a clear example of the influence on consensus of one individual who passionately and cleverly advances a policy about which his colleagues have little concern and less understanding. Jefferson was the best mathematician involved in the Congressional discussions, and he sold his colleagues on a plan in the fairness of which one must doubt he truly believed.

Table 2.6. Allocations of Hamilton and Jefferson.					
State	Population	H	J	Voters/Seat Hamilton	Voters/Seat Jefferson
CN	236,841	7	7	33,834	33,834
DE	55,540	2	1	27,220	55,440
GA	70, 835	2	2	35,417	35,417
KY	68,705	2	2	34,353	34,353
MD	278,514	8	8	34,814	34,814
MA	475,327	14	14	33,952	33,952
NH	141,822	4	4	35,455	35,455
NJ	179,570	5	5	35,914	35,914
NY	331,589	10	10	33,159	33,159
NC	353,523	10	10	35,352	35,352
PA	432,879	13	13	33,298	33,298
RI	68,446	2	2	34,223	34,223
SC	206,236	7	7	29,462	29,462
VE	85,533	2	2	42,766	42,776
VA	630,560	18	19	35,031	33,187
Total	3,615,920	105	105		

Naturally, as the large state bias of the method of Jefferson began to be understood, it was inevitable that someone would suggest a plan that, somewhat symmetrically to Jefferson's method, would favor the small states. We have such a scheme proposed by John Quincy Adams.

> **John Quincy Adams' Rule.** Pick the size of the House $= n$. Find a divisor d so that the integer parts of the quotients (when divided by d) plus 1 for each of the states sum to n. Then assign to each state the integer part of $N_j/d + 1$.

The plan of Adams gives the same kind of advantage to the small states that that of Jefferson gives to the large states. Needless to say, it has never been used in this country or any other (though amazingly, Jefferson's has). It is interesting to note that Adams, instead of saying, "I see what's going on. Let's go to Hamilton's Rule," tried to do for the small states the same thing Jefferson had done for the large states.

It is interesting to note that Daniel Webster attempted to come up with a plan which was intermediate to that of Jefferson's and that of Adams. He noted that whereas Jefferson rounded the quotient down to the next smallest integer, Adams rounded up to the next largest integer. Webster, who was a man of incredible intuition, suggested that fractions above .5 be rounded upward, those below .5 be rounded downward. From the 1830's until 1850 there was very active discussion about the unfairness of the method of Jefferson and a search for alternatives. It was finally decided to pick Hamilton's method, but Webster's was almost selected and it was a contender as recently as 1941. As it turns out, there is very little practical difference between the method of Hamilton and that of Webster. Both methods would have given identical allocations from the beginning of the Republic until 1900. Since that time, the differences between the two methods usually involve one seat per census. The method of Hamilton was replaced in 1941 by one advocated by Edward Huntington, Professor of Mathematics at Harvard. Huntington, instead of having the division point of fractions to be rounded up and rounded down to one half, advocated that if the size of the quotient of a state were denoted by N_j/d then the dividing point below which rounding down would be indicated would be the geometric mean $\sqrt{[N_j/d]([N_j/d] + 1)}$ where [.] denotes "integer part of." One might say that such a method violates the notion that such methods should be kept simple. Furthermore, the rounding boundaries do increase slightly as the size of the state increases, giving an apparent advantage to the smaller states. At the last minute, the more popular method of Webster was rejected in favor of that of Huntington, since its application using the 1940 census would give a seat to Democratic Arkansas rather than to Republican Michigan. The Huntington method is in use to this day, though not one American in a thousand is aware of the fact. And indeed, it is not a very important issue whether we use the method of Hamilton or that of Webster or that of Huntington or even that of Jefferson or that of

Adams. Not one significant piece of legislation would have changed during the course of the Republic if any one of them were chosen. The subject of apportionment possibly receives more attention than practicality warrants.

But looking over the history of the apportionment rule should give pause to those among us who believe in the pristine purity of the Founding Fathers and the idea that there was a time when Platonic philosophers ruled the land with no thought except virtue and fair play. Jefferson, the noblest and wisest and purest of them all, was working for the advantage for his state under the guise of fairness and sophistication. And George Washington, the Father of his Country, was tricked into using the first veto in the history of the Republic to stop a good rule and put one of lesser quality in its place. All this, in the name of being a good Enlightenment chief of state.

Arrow proved mathematically what Machiavelli had observed and Pareto had described as being part of a general taxonomy of political behavior: in economics, politics and society generally: important public policy decisions are not made as the orderly aggregation of collective wisdom. Group decisions are made using mechanisms we do not clearly understand (and, as Pareto urged, we really should try without passion to learn these mechanisms). It appears that insiders have a great deal to do with these decisions. Sometimes, as when Robert Rubin, Secretary of the Treasury, rushed in to secure Mexican loans made by American financial institutions, including that of which he had been boss, Goldman-Sachs, one may raise an eyebrow. Of course, he could respond that he was just following an example set by arguably the smartest and most idealistic President in the history of the Republic. Then again, we have the Federal Reserve Board in recent years presenting the stock market with frequent "Gotcha!" type surprises. What sort of efficiency is possible in a market where interest rates are changed at the whim of Alan Greenspan and his fellows? Then there is the matter of anti-trust law. Should a CEO be careful lest his company be so successful that it is dismembered by the Attorney General?

Some years ago, one of the authors (Thompson) was consultant to a firm that had a well-designed plan to raise chickens in Yucatan according to the notions of modern poultry husbandry. In order to purchase buildings, poultry, land, and equipment, the firm had to convert millions of dollars of cash from dollars into pesos. It did so on the Friday before the deals were to be paid for on Monday. During the week-end, the President of Mexico devalued the peso hugely, wiping out the cash reserves of the firm. Naturally, on Monday, all the Mexican parties from whom land, structures, equipment, poultry, etc., were to be purchased, changed their peso prices hugely upward to reflect the devaluation. (It turns out that prior to the devaluation, the President of Mexico had leveraged his own cash assets to purchase a number of villas in Mexico.) Such surprises are hardly helpful to facilitating market efficiency. One may well ask how an investor can cope with such inefficiency producing spikes in the market. Pareto's answer is that one had better learn how to do precisely that, for such is the real world. There is indeed a tendency towards efficiency in most markets. But this is only part of the market mechanism. In our study, we model an efficient market tendency and then look at what happens when varieties of contaminating mechanisms are superimposed upon it.

2.7 Conclusions

In this chapter on utility, we have started with Enlightenment assurance as to the way rational people should make choices. The buttressing of Bernoulli by von Neumann and Morgenstern seemed very promising indeed. But then we ran into Arrow and looked back to Pareto and (shudder) Thomas Jefferson. And we became less confident about the orderliness of the way aggregate decisions are made. In *Mind and Society*, Pareto [5] tells us that in reality many decisions are made with results which are the equal in consequence of any Adam Smith might have made. The market may not be efficient; politicians may not be clones of Lucius Quintus Cincinnatus; but in some societies (and Pareto always had great hopes for the United States), the market moves toward efficiency, and political decisions are frequently Cincinnatus-like in their consequences. The West and its economies are not at all chaotic. There is a tendency toward efficiency. But it is ridiculous if we assume efficient markets as an iron law, and then fudge when our illusions are challenged by reality. In this book, we shall assume a more or less efficient driver for a given market situation with departures therefrom taken cognizance of empirically by perturbations to the model.

Problems

Problem 2.1. Sempronius owns goods at home worth a total of 4,000 ducats and in addition possesses 8,000 ducats of commodities in foreign countries from where they can be transported only by sea. However, our daily experience teaches us that of ten ships, one perishes.

(a) What is Sempronius' expectation of the commodities?

(b) By how much would his expectation improve if he trusted them equally to two ships?

(c) What is the limit of his expectation as he trusted them to increasing numbers of ships?

Problem 2.2. Let us use Bernoulli's logarithmic utility function

$$U(X) = \log \left(\frac{X + \Theta}{\Theta} \right),$$

where Θ is one's initial wealth. Would it be rational to play a game where there is a finite probability of losing all one's wealth? Why? How might this result be rationalized? Why, in modern first world society, might it be said that is impossible to lose all one's wealth?

Problem 2.3. Suppose a woman were offered either a certain $230 or a 50-50 chance of $400 or $100. Which option should be taken if:

(a) She possesses a square root utility function,

(b) A Bernoulli utility function with initial wealth of $1,000,

(c) The same as 2, with initial wealth of $100?

Problem 2.4. Consider the section on von Neumann–Morgenstern utility.

(a) Determine the certain sum for which the subject in the example would relinquish the following opportunity:

W	Probability
40,000	.4
160,000	.4
250,000	.2

(b) Explain how von Neumann–Morgenstern Axiom 5 treats the utility derived from gambling itself (i.e., deals with the existence of casinos).

(c) Compare the mean and standard deviation of outcomes in the example and in (a).

(i) What are the mean and standard deviation in each case of the certain dollar value for which the subject would be indifferent?

(ii) We now have two sets of (σ, μ) for each of the two opportunities. Which of the two games (example or (a)) would the subject prefer to play?

(iii) For what probability α would he be indifferent between $\alpha \times$ (game he prefers) and the game he does not prefer?

Problem 2.5. For this problem, assume $U = \sqrt{W}$ and $U(0) = 0$ and $U(1,000,000) = 1,000$ utiles.

(a) For what $0 \leq \alpha \leq 1$ would the subject be indifferent between $\alpha \times \$1,000,000$ and a certain \$250,000?

(b) For what α would the subject be indifferent between $\alpha \times \$500,000$ and $(1 - \alpha) \times \$200,000$?

Problem 2.6. Consider the *Last Shall Be First Rule* for establishing group preferences

- Step 1. Select as the group choice, the candidate who was ranked number one by a majority of the decision makers, if such a candidate exists.

- Step 2. If no candidate received a majority of number one rankings, then take all the preference rankings of voters who voted for the candidate with the smallest number of first choice preferences and treat them as if their first choice candidate is simply deleted from their preference lists and all other candidates on their lists moved up one rung on the preference ladder.

- Step 3. Go to Step 1.

Which of Arrow's Axioms does this rule fail to satisfy? Discuss its advantages and disadvantages in three settings:

(a) A mayoral election in an American city of over 10,000 voters.

(b) An election for a position on a university committee.

(c) A decision making process for picking one of four potential new products by a board of directors (with 20 members) of a company.

Problem 2.7. Consider the *Borda Count Rule*

- Step 1. Rank the preferences amongst k choices for voter i, one of n voters.

- Step 2. For each preference a, count the number of choices below a in the preferences of voter i. This gives us $B_i(a)$.

- Step 3. For each choice sum the Borda counts, e.g., $\mathbf{B}(a) = \sum_i^n B_i(a)$.

- Step 4. The group chooses the preference with the highest Borda count sum.

A group decision rule defined on the set of all profiles (preferences) on the set of decisions is said to be *Pareto optimal* if for every a and b in the set of decisions whenever a is ranked over b in every ranking of a profile, then a is ranked over b in the corresponding group ranking.

Prove or disprove the following:

The Borda count rule is Pareto optimal.

Problem 2.8. Consider the following *Plurality Rule*: Rank a over b for the group if and only if a receives more first place votes than b. Is this rule Pareto optimal?

References

[1] Arrow, K. J. (1950). "A difficulty in the concept of social welfare," *Journal of Political Economy*, Vol. 58, 328–346.

[2] Balinski, M. L. and Young, H. P. (1982). *Fair Representation: Meeting the Ideal of One Man, One Vote*. New Haven, CT: Yale University Press.

[3] Bernoulli, D., "Exposition of a new theory on the measurement of risk" (Louise Sommer, trans.), *Econometrica*, January 1954, 23–36 (first published 1738).

[4] Bernstein, P.L. (1996). *Against the Gods: The Remarkable Story of Risk*. New York: John Wiley and Sons.

[5] Pareto, V. (1935). *Mind and Society*. New York: Harcourt, Brace, and Company.

[6] Thompson, J. R.(1989). *Empirical Model Building*. New York: John Wiley & Sons, pp. 158–177.

[7] von Neumann, J., and Morgenstern, O. (1944) *Theory of Games and Economic Behavior*. Princeton, NJ: Princeton University Press.

[8] Williams, E.E. and Findlay, M.C. (1974). *Investment Analysis*. Englewood Cliffs, NJ: Prentice-Hall.

Chapter 3

Diversification and Portfolio Selection

3.1 Introduction

In bygone days, many widows and orphans came to a bad financial end by having their stock assets invested in only one or two stocks. For example, at one time, railroad stocks were considered to be sure things. A trust manager who put all his client's assets into the Missouri Pacific Railroad was behaving in a completely acceptable fashion. Then, problems occurred. There was a depression, so that freight traffic was reduced. Then, Franklin Roosevelt had no good feelings for the railroad tycoons, who, with the exception of the Harrimans, had almost uniformly been strong supporters of the Republican Party. He allowed small truck freight companies to use the federal highways for practically no use tax, while the railways had to pay for maintenance of their railways, and had to pay property taxes for them. The greater flexibility of the trucking companies gave them a competitive advantage vis-à-vis the railroads in a time when economic depression had made shipping lots smaller. Insofar as passenger traffic was concerned, the Greyhound Bus line took advantage of free usage of the public roads to grab much of the clientele which had been the province of the railroads. Again, lack of the ability to thrive on small volume hurt the railroad companies. Most railroad companies had stock which steadily declined. Trust managers *knew* that it was only a matter of time until the railroad stocks rebounded, economic, political and technological realities notwithstanding. The trust managers managed better than their clients, many of whom lost everything.

Of course, history does have a way of repeating itself. At the writing of this book, the collapse of Enron is unfolding at flank speed. Enron employees had their pension funds in company stock. Moreover, they were seriously restricted in their abilities to replace these shares by other securities. When the company collapsed, most of the employees lost not only their jobs but their pension funds as well. Top management, happily, suffered no such restrictions and managed to avoid the devastation to which their workers were subjected.

These days, unless an investor knows of a *sure thing* he or she will not be content to stick with one particular stock. There is seldom such a thing as a free lunch in business. Diversification, however, comes rather close to being a no-cost insurance policy. There are many ways for an investor to spread his or her stock market investments around among several securities to build a portfolio. Let us start with a relatively mathematically simple procedure.

3.2   Portfolio Design as Constrained Optimization

We will now consider strategies for reducing the risk to an investor as a well defined constrained optimization problem. The argument follows roughly that of Markowitz [8,9]. Let us suppose that the *proportional gain* of a security is given the symbol X_i. Here we are not necessarily assuming X_i is the growth rate of the stock. X_i may include as well, for example, the dividends accruing to the security. And we will treat X_i as a Gaussian (normal) random variable rather than as a constant. We will then assume that the average (expected value) of X_i is μ_i, and its *variance* is σ_i^2. Then we shall assume that the *covariance* of X_i and X_j is given by σ_{ij} (alternatively, that the *correlation* is given by ρ_{ij}). That is, we assume that:

$$E(X_i) = \mu_i$$
$$E(X_i - \mu_i)^2 = \sigma_i^2 = \sigma_{ii}$$
$$E[(X_i - \mu_i)(X_j - \mu_j)] = \sigma_{ij}$$
$$\rho_{ij} = \frac{\sigma_{ij}}{\sigma_i \sigma_j}.$$

Our portfolio will be formed by a linear combination of n stocks where the fraction α_i of our portfolio will consist of shares of stock i. Clearly, it would be a fine thing to have a high value of

$$\mu_{ave} = \sum_{i=1}^{n} \alpha_i \mu_i. \tag{3.1}$$

On the other hand, we would like for the portfolio to be as close to a sure thing as possible, that is, we would like to minimize the volatility of the portfolio

$$S = E\left(\sum_{i=1}^{n} \alpha_i X_i - \sum_{i=1}^{n} \alpha_i \mu_i\right)^2 \tag{3.2}$$

$$= \sum_{i=1}^{n} \sum_{j=1}^{n} \alpha_i \alpha_j \sigma_{ij}.$$

Clearly, there is a problem, for minimizing S would logically drive us to something like Treasury bills, which strategy is not historically very good for maximizing μ_{ave}. It might be postulated that what should be done is to ask an investor what μ_{ave}^* he requires and then design a portfolio obtaining a mix of the n stocks so as to minimize S for that target value. This is not a very natural posing of the problem from the standpoint of the investor (picking a μ_{ave}^* is not natural for most). However, from the standpoint of the mathematician, it is a formulation easily solved. To see that this is so, consider the Lagrange multiplier formulation which seeks to minimize:

$$Z = \sum_{i=1}^{n} \sum_{j=1}^{n} \alpha_i \alpha_j \sigma_{ij} + \lambda_1 \left(\sum_{i=1}^{n} \alpha_i \mu_i - \mu_{ave}^* \right) + \lambda_2 \left(\sum_{i=1}^{n} \alpha_i - 1 \right) \quad (3.3)$$

Differentiating partially with respect to each of the α_i, λ_1, and λ_2 and setting the derivatives equal to zero gives us the $n+2$ equations (linear in $\alpha_1, \alpha_2, \ldots, \alpha_n, \lambda_1, \lambda_2$):

$$\frac{\partial Z}{\partial \alpha_i} = 2 \sum_{i=1}^{n} \alpha_i \sigma_i^2 + 2 \sum_{j>i} \alpha_j \sigma_{ij} + \lambda_1 \mu_i + \lambda_2 = 0$$

$$\frac{\partial Z}{\partial \lambda_1} = \sum_{i}^{n} \alpha_i \mu_i - \mu_{ave}^* = 0$$

$$\frac{\partial Z}{\partial \lambda_2} = \sum_{i=1}^{n} \alpha_i - 1 = 0.$$

This is an easy formulation to solve in the current era of cheap high-speed computing. Naturally, as formulated here, it is possible that some of the α_i may go negative, though it is easy to impose the additional constraint that all α_i be nonnegative. Furthermore, it will happen that the α_i will yield fractional shares of stocks. Generally rounding will give us a satisfactory approximation to the solution, though we can easily impose the restriction that all shares be bought as integer lots. All these little details can be dealt with easily. However, the formulation is not particularly relevant in practice, for few individuals will find it natural to come up with a hard number for μ_{ave}^*.

Well then, we could also pose the problem where we maximize the gain of the portfolio subject to some acceptable level of the volatility. But this is also not (for most people) a natural measure of portfolio riskiness. An investor probably would like to know his or her probabilities of achieving varying levels of value as time progresses. This is not an easy task, but we address it subsequently.

3.3   A Graphical Depiction

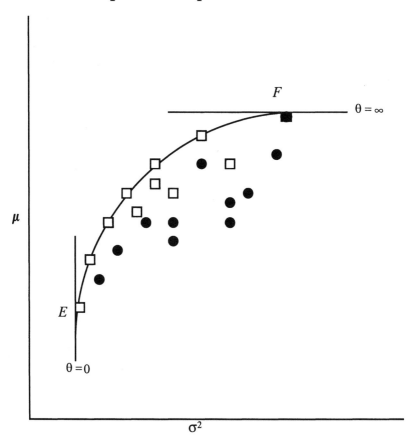

Figure 3.1. Efficient Frontier Generation.

The set of all portfolios with maximum expected gain at a given level of volatility (or minimum volatility at a given level of expected gain) was defined by Markowitz as the *efficient frontier*. His basic method, which is fairly similar conceptually to the other techniques discussed in this section, can perhaps be understood by reference to Figure 3.1. Here the dots represent security parameters and the boxes represent portfolio parameters. Markowitz set about to minimize a function of the type $\sigma^2 - \theta\mu$. By initializing the procedure at $\theta = \infty$, the highest return security (F) is obtained. Note that, because diversification cannot increase return, the highest return portfolio will be composed entirely of the highest return security. From this point, Markowitz employed a quadratic programming algorithm to trace the efficient frontier by allowing θ to decrease to 0 (at which point E, the minimum variance portfolio is obtained). In actuality, the iterative procedures only determine "corner" portfolios, which are those points at which a security enters or leaves the efficient portfolio. The efficient frontier between two corner portfolios is a linear combination of the corner portfolios. Aside from the objective function, these techniques also generally involve constraints, such as the requirement that the weights assigned to securities be nonnegative and/or sum to one.

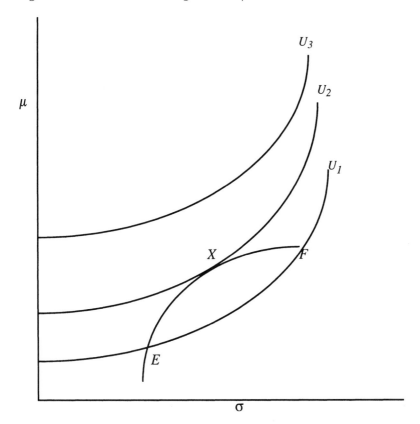

Figure 3.2. Portfolio Selection.

The selection of an optimal portfolio follows in a straightforward fashion. Indifference curves (derived in the last chapter) are applied, and the optimal portfolio is found at the point of tangency of EF with the highest attainable indifference curve (point X in Figure 3.2). This point is preferable to any other on the curve (e.g., E or F) because it provides a higher level of utility (i.e., $U_2 > U_1$). A point on a higher indifference curve (e.g., U_3) would be even better, but by the definition of an efficient frontier given above, points above EF do not exist.

Now, let us consider a two asset portfolio with one of the assets being Treasury bills (generally deemed to be riskless). Designating the riskless security with 2, we have

$$\mu = \alpha_1 \mu_1 + \alpha_2 \mu_2 \qquad (3.4)$$

and [1]

$$\sigma^2 = \alpha_1^2 \sigma_1^2. \tag{3.5}$$

Now the same conditions will hold if "1" were a portfolio, instead of a security, and a riskless security were added to it. Figure 3.3 designates all combinations of portfolio 1 and riskless security 2. The combination of portfolios so depicted is called, in the literature of financial economics, the *lending case,* because in essence a portion of the funds available is being lent at the riskless rate of interest r, while the remainder is invested in portfolio 1.

Using the simple construct developed in Figure 3.3, it is also possible to consider the case of borrowing against, or leveraging, the portfolio. Because positive α_i's are used to indicate securities bought and included in the portfolio, it does not seem unreasonable to use negative α_i's to indicate liabilities sold against the portfolio.

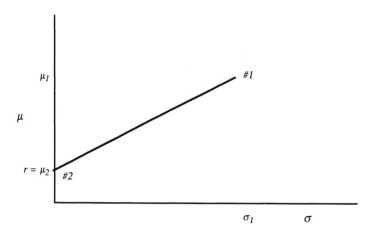

Figure 3.3. Lending Case.

In Figure 3.3, if α_2 becomes negative, the α_1 becomes > 1 and $\alpha_1 \sigma_1$ also increases linearly. Making the rather unrealistic assumption that the individual can also borrow at the riskless rate, our borrowing line becomes merely an extension of the lending line beyond 1, as shown in Figure 3.4:

---

[1]Recall that for a portfolio consisting of n stocks each with weight α_i and variance σ_i^2, the variance is given by $\sum_{i=1}^{n} \alpha_i^2 \sigma_i^2$.

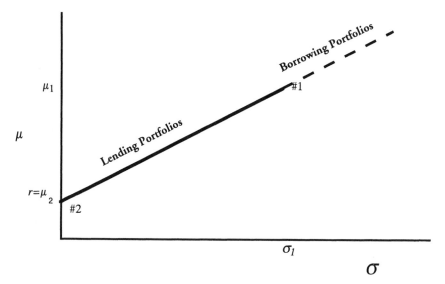

Figure 3.4. Borrowing and Lending Cases.

Therefore, beginning with all our funds in portfolio 1, we can either lend some of the funds at the riskless rate and move down the line in the direction of μ_2 or borrow at the riskless rate to invest more funds in portfolio 1 and move up the line. Where an individual should be on this line depends on his or her risk preference. A completely risk-averse individual would invest all of his or her funds in portfolio #2 (Treasury bills) and assume no risk. Another, more adventurous soul, might put all of his or her money in portfolio #1 (all stocks and no Treasury bills), while one not very risk adverse at all might actually go out and borrow (at r) to buy even more of the stocks in portfolio #1. The latter would be said to be buying stocks on *margin*.[2] Of course, in the real world, the investor "leveraging" his or her portfolio would most likely have to borrow at a rate higher than the Treasury bill (riskless) rate. This may be done by specifying a rate r^* for all *borrowing portfolios*.

The effects of lending part of the funds in the portfolio and leveraging (borrowing against) the portfolio may be illustrated with a simple example. Suppose a risky portfolio of securities has $\mu = 15\%$ and $\sigma = 10\%$, and the riskless rate of interest (r) is 6%. The parameters of a holding of (1) 50 percent in the portfolio and 50 percent lending, and (2) 150 percent in the portfolio and 50 percent borrowing would be determined as follows:

$$\mu = \alpha_1\mu_1 + \alpha_2\mu_2 \qquad (3.6)$$
$$\mu = .5(.15) + .5(.06) = .105 \text{ or } 10.5\%$$
$$\sigma = \alpha_1\sigma_1 = .5(.10) = .05 \text{ or } 5\%$$

---

[2]See earlier discussion in Chapter 1.

$$\mu \;=\; 1.5(.15)+(-.5)(.06) = .195 \text{ or } 19.5\% \qquad (3.7)$$
$$\sigma \;=\; 1.5(.10) = .15 \text{ or } 15\%.$$

We notice that, in the case of lending, both expected return and risk are reduced; conversely, in the case of borrowing each parameter is increased. If the borrowing rate (r_B) were higher than the lending rate (r_L) (say .08 rather than .06), (3.7) would become:

$$\mu = 1.5(.15)+(-.5)(.08) = .185 \text{ or } 18.5\%.$$

Given the effects of borrowing and lending at the riskless rate, it then becomes necessary to determine which of the available portfolios should be so employed. It would seem reasonable, as a first principle, to borrow or lend against that portfolio which provided the greatest increase in return for a given increase in risk. In other words, we should desire that borrowing-lending line with the greatest available slope. The cases are illustrated in Figure 3.5.

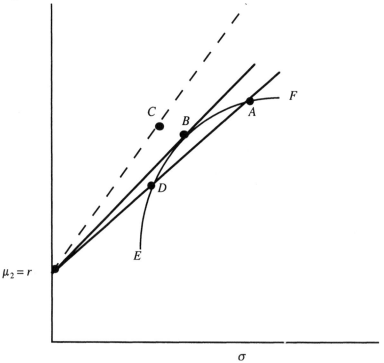

Figure 3.5. Determination of the Optimal Borrowing and Lending Portfolio $(r_L = r_B)$.

The efficient frontier is shown as EF. Borrowing and lending against either portfolio A or D (which gives the same line $= \mu_2 - D - A$) is inefficient because a better risk-return trade-off is given by B. Portfolio C would appear even better, except that by definition no security or portfolio exists above EF. Therefore, the optimal attainable borrowing-lending portfolio is found by extending a straight line from the pure rate of interest tangent to the efficient frontier.

Let us take another look at such an optimal line, shown in Figure 3.6 as $r - O - Z$, through portfolio O. It will be observed that for every portfolio on the efficient frontier between E and O, there exists some combination of portfolio O and the riskless security (given by the lending line $r - O$) that is more desirable (lower σ for the same μ, and so on). In like manner, every point on the efficient frontier between O and F is dominated by some borrowing against O (given by the line OZ). It would therefore appear that the revised efficient frontier is $r - O - Z$. If this were the case, then all efficient portfolios would be composed of portfolio O plus borrowing or lending. A very risk-averse person, such as Mr. A, might choose portfolio M (see Figure 3.6) composed of portfolio O plus riskless loans, while a less risk-averse person, such as Mr. B, might choose portfolio N, which involves leveraging portfolio O.[3]

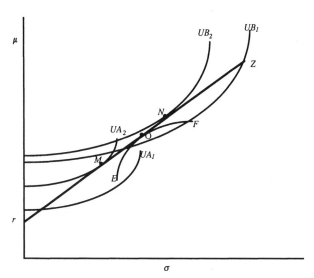

Figure 3.6. Determination of the Optimal Portfolio for Individuals with Varying Risk Preferences.

---

[3]This example illustrates a separation property. At least for risk averse investors, the optimal portfolio of risky assets (i.e., O in the example) can be selected without further knowledge of the investor's degree of risk aversion. The latter can be accommodated by the individual by means of borrowing or lending.

3.4   Other Approaches: Confidence Limits and Stochastic Dominance

Other portfolio selection (elimination) approaches have been proposed over the years. Baumol [1] has pointed out that portfolios with low μ and σ could not be eliminated by the concept of efficiency in comparison with portfolios having much higher μ and only slightly larger σ, even though the probability that the best outcome of the former would ever exceed the worst outcome of the latter was very small. He therefore suggested that a lower *confidence limit* be set at $\mu - K\sigma$ and that low return portfolios dominated by this rule be dropped from the efficient frontier. The parameter K would be a function of the investor's degree of risk aversion, and if return could be assumed to be normally distributed, could be determined by a one-tail test of the percentage of the time the investor would tolerate a shortfall of actual returns (e.g., 16 percent for $K = 1$, 2 percent for $K = 2$, etc.).

For example, suppose that an investor were prepared to ignore low returns that occurred no more than 2 percent of the time. This percentage would correspond to the portion remaining in the left tail of a normal (Gaussian) distribution at a point two standard deviations below the mean; thus $K = 2$. Next, consider the following points on an efficient frontier:

	A	B
μ	.04	.10
σ	.01	.03
$\mu - 2\sigma$	.02	.04

Neither point can be termed inefficient, as B has a higher return but at greater risk. It may be noted, however, that the investor's perceived worst outcome of B (the lower confidence limit of B) is greater than the same for A (i.e., .04>.02). Thus, portfolio A, and other portfolios in the lower range of the efficient frontier, could be eliminated by the confidence-limit criterion.

A more comprehensive procedure, called *stochastic dominance*, allows for the elimination from further consideration of securities or portfolios on the basis of known characteristics of entire classes of utility functions. *First-Degree Stochastic Dominance* (FSD) requires only the assumption that more wealth is preferable to less (i.e., $dU/dW > 0$). It may then be said that investment B exhibits FSD over investment A if, for each possible level of return, the cumulative probability of B falling below the level of return is $\leq$ the same for A (and the strict inequality holds in at least one case). Consider the following example:

Return r_i	Prob.A	Prob.B
.050	.2	.1
.100	.3	.2
.150	.4	.3
.200	.1	.4
	1.0	1.0
μ	.120	.150
σ	.046	.050

The cumulative probability distributions of returns for A and B are shown in Figure 3.7. The FSD decision rule is shown as follows:

r_i	Cumulative Probability $A = A(r_i)$	Cumulative Probability $B = B(r_i)$	$A(r_i) - B(r_i)$
.05	.2	.1	+.1
.10	.5	.3	+.2
.15	.9	.6	+.3
.20	1.0	1.0	0

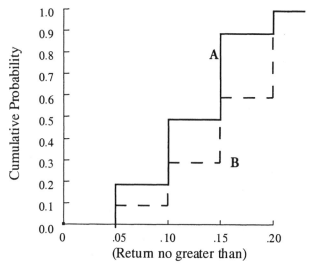

Figure 3.7. First-Degree Stochastic Dominance.

For investment B:

$$FSD(r_i) = A(r_i) - B(r_i) \quad \geq 0 \quad \text{for all } r_i, \text{ and}$$
$$> \quad 0 \text{ for at least one } r_i.$$

Thus, investment B exhibits FSD over investment A and would be preferred by anyone to whom more wealth was preferable to less (i.e., all rational persons, including risk seekers and neutrals, as well as averters).

Second-Degree Stochastic Dominance (SSD) requires the assumption of risk neutrality or aversion $(d^2U/dW^2 \leq 0)$ as well as rationality $(dU/dW > 0)$. Under these assumptions, it is possible to make selections between certain investments that do not exhibit FSD. Consider the following investment alternatives:

Return r_i	Prob.A	Prob.B
.05	.2	.1
.10	.3	.4
.15	.3	.4
.20	.2	.1
	1.0	1.0
μ	.125	.125
σ	.053	.040

The FSD analysis would provide:

r_i	Cumulative Probability $A = A(r_i)$	Cumulative Probability $B = B(r_i)$	$A(r_i) - B(r_i)$ $= FSD(r_i)$
.05	.2	.1	+.1
.10	.5	.5	0
.15	.8	.9	−.1
.20	1.0	1.0	0

Thus, neither investment would exhibit FSD over the other. SSD, in effect, considers the cumulative difference between the two cumulative probability distributions employed in FSD. Again considering investment B:

$$SSD(r_i) = SSD(r_{i-1}) + (r_i - r_{i-1})[A(r_{i-1} - B(r_{i-1}]$$
$$SSD(r_i) = SSD(r_{i-1}) + (r_i - r_{i-1})FSD(r_{i-1}).$$

(1) i	(2) r_i	(3) $r_i - r_{i-1}$	(4) $FSD(r_{i-1})$	(5) $(3)\times(4)$	(6) $SSD(r_{i-1})$	(7) $SSD(r_i)$
1	.05	—	—	—	—	0
2	.10	.05	+.1	+.005	0	+.005
3	.15	.05	0	0	+.005	+.005
4	.20	.05	−.1	−.005	+.005	0

Thus, investment B is preferred by SSD if $SSD(r_i) \geq 0$ for all i and the strict inequality holds in at least one case.

Finally, if neither FSD nor SSD will allow a choice to be made between two investments, *Third-Degree Stochastic Dominance* (TSD) may be employed [16, 17]. In addition to the first-order and second-order conditions for the utility of wealth function discussed above, TSD also requires that $d^3U/dW^3 \geq 0$. Following the parallelism previously established, TSD is essentially concerned with the cumulative functions employed in SSD. We present our final investment comparison with regard to showing the dominance of B:

Return r_i	Prob.A	Prob.B
.05	.1	0
.10	0	.3
.15	.9	.6
.20	0	.1
	1.0	1.0
μ	.14	.14
σ	.03	.03

r_i	$A = A(r_i)$	$B = B(r_i)$	$FSD(r_i)$
.05	.1	.0	+.1
.10	.1	.3	−.2
.15	1.0	.9	+.1
.20	1.0	1.0	0

(1)	(2)	(3)	(4)	(5)	(6)	(7)
i	r_i	$r_i - r_{i-1}$	$FSD(r_{i-1})$	(3)×(4)	$SSD(r_{i-1})$	$SSD(r_i)$
1	.05	—	—	—	—	0
2	.10	.05	+.1	+.005	0	+.005
3	.15	.05	−.2	−.010	+.005	−.005
4	.20	.05	+.1	+.005	−.005	0

Because neither FSD nor SSD can be shown, the TSD for B is given by:

$$TSD(r_i) = TSD(r_{i-1}) + (r_i - r_{i-1})[SSD(r_i) + SSD(r_{i-1})]/2.$$

For B to be preferred, then, $TSD(r_i) \geq 0$ for all i (with the strict inequality holding for at least one i) and, furthermore, $\mu_B \geq \mu_A$. The TSD of B over A is demonstrated below:

Worksheet for Third-Degree Stochastic Dominance					
(1)	(2)	(3)	(4)	(5)	(6)
i	r_i	$r_i - r_{i-1}$	$FSD(r_{i-1})$	(3)×(4)	$SSD(r_{i-1})$
1	.05	—	—	—	—
2	.10	.05	+.1	+.005	0
3	.15	.05	−.2	−.010	+.005
4	.20	.05	+.1	+.005	−.005

(7)	(8)	(9)	(10)	(11)
$SSD(r_i)$	$\frac{(6)+(7)}{2}$	(8)×(3)	$TSD(r_{i-1})$	$TSD(r_i) = (9) + (10)$
0	—	—	—	0
+.005	+.0025	+.000125	0	+.000125
−.005	0	0	+.000125	+.000125
0	−.0025	−.000125	+.000125	0

3.5   Non-Utility Techniques

A specific technique for portfolio selection, on the basis of minimizing the probability of the occurrence of some disastrous level of loss (D) was developed many years ago [11] employing Chebyshev's inequality, which does not require the assumption of a special form of probability distribution for returns. Expressed in our terminology, the relevant form of the inequality is:

$$P(X| \leq D) \leq \frac{\sigma^2}{(\mu - D)^2}, \tag{3.8}$$

which, when differentiated, set equal to zero, and solved provides a minimum probability at $\sigma/(\mu - D)$. This technique has been called "safety-first."

An approach similar in methodology has been suggested by Lintner [6, 7]. If we let r represent the rate to be earned on some riskless asset then Lintner defines the excess return (X) to be earned on a given risky asset as:

$$X_i = \mu_i - r. \tag{3.9}$$

It will be recalled that the parameters for the two-asset portfolio when security 2 is riskless are:

$$\mu = \alpha_1\mu_1 + \alpha_2\mu_2, \text{ and } \sigma = \alpha_1\sigma_1.$$

Thus:

$$\alpha_1 = \frac{\sigma}{\sigma_1},$$

but

$$\mu_2 = r, \text{ and } \alpha_2 = 1 - \alpha_1.$$

Therefore,

$$
\begin{aligned}
\mu &= \alpha_1\mu_1 + (1 - \alpha_1)r \\
\mu &= \alpha_1\mu_1 + r - \alpha_1 r \\
\mu &= r + \alpha_1(\mu_1 - r).
\end{aligned}
$$

Substituting the relationships defined above:

$$
\begin{aligned}
\mu &= r + \frac{\sigma}{\sigma_1}X_1 \\
\mu &= r + \frac{X_1}{\sigma_1}\sigma \\
\mu &= r + \theta\sigma, \text{ where } \theta = \frac{\mu_1 - r}{\sigma_1}. \tag{3.10}
\end{aligned}
$$

An examination of (3.10) indicates that a portfolio that maximizes θ will also maximize μ for a given σ and r. The verbal sense of this is that we are trying to maximize excess return per unit of risk (which is Lintner's criterion).

Comparing the approach in (3.8) with that of (3.9) and (3.10) suggests that the safety-first model, from its basis in Cheybyshev's inequality to its goal of minimization of probability of outcomes below one specific point, seems to come as close to decision making under true uncertainty as any technique presented in this book; very little is assumed to be known about either the risk preferences of the investor or the distribution of returns on investment. This lack of specification is an advantage in many real-world situations, and this is one reason the model has been included here. On the other hand, if more complete information is available, more precise models may be employed. Surprisingly enough, given their technical similarity, the Lintner model is on the other end of the spectrum in regard to equilibrium and perfection assumptions. It basically seeks portfolio B in Figure 3.5. To be universally applicable for risk averters, (i.e., portfolio O in Figure 3.6), the excess-return criterion requires the separation theorem to hold, which assumes, among other things, that the portfolio can borrow or lend at the same rate (specifically at r). The excess-return criterion essentially involves the application of the basic capital market theory model (discussed in the next chapter) to the selection of individual portfolios. Although this application does not require all the equilibrium assumptions of capital market theory, the likely inapplicability of the separation theorem is enough to raise doubts regarding its general adoption.

A final criterion for portfolio selection without specific reference to utility considerations that has been suggested is the maximization of the geometric mean. If it may be assumed that (1) the same investment opportunities are available in each investment period; (2) the outcomes of these opportunities are independent over time; and (3) all income is reinvested, then the terminal wealth at the end of n periods associated with investment in any given portfolio may be ascertained. Let

- $W_{i,n}$ = wealth in period n associated with investment in portfolio i;

- $\rho_{i,j}$ = return earned in period j on portfolio i;

- W_0 = intial wealth;

- n = number of periods.

Then

$$W_{i,n} = W_0 \prod_{j=1}^{n}(1 + \rho_{i,j})$$

$$\left[\frac{W_{i,n}}{W_0}\right]^{\frac{1}{n}} = \left[\prod_{j=1}^{n}(1 + \rho_{i,j})\right]^{\frac{1}{n}}$$

$$E\left[\left(\frac{W_{i,n}}{W_0}\right)^{\frac{1}{n}}-1\right] = E\left[\left(\prod_{j=1}^{n}(1+\rho_{i,j})\right)^{\frac{1}{n}}-1\right]. \quad (3.11)$$

The right side of (3.11) is the expected geometric mean return on portfolio i. The portfolio that maximizes this expectation thus will maximize the expectation of terminal wealth. It can also be shown by the Central Limit Theorem that the probability of the maximum geometric mean portfolio resulting in greater terminal wealth than any other available portfolio approaches unity as n approaches infinity.

All the above, however, is simply mathematical tautology. We note that if ρ_j is the return in period j then, the maximization of the geometric mean of the ρ_j's is equivalent to the maximization of the algebraic mean of the logarithms of the ρ_j's for each period, since

$$\frac{1}{n}\sum_{j=1}^{n}\log(\rho_j) = \frac{1}{n}\log\left[\prod_{j=1}^{n}\rho_j\right] = \log\left[\prod_{j=1}^{n}\rho_j\right]^{\frac{1}{n}}.$$

Thus, if the investor possesses a log utility function or, given the CLT, if the number of investment periods is infinite, the geometric mean criterion is sufficient. In some other cases, it has been found to be a reasonable approximation. However, it can be shown that as long as n is finite, there exist classes of utility functions for which the proposed criterion is not even a good approximation. Indeed, it can be demonstrated that there are classes of utility functions for which no uniform strategy is optimal. Moreover, there is good reason to question whether the assumptions $(1)-(3)$ are very realistic. The problems of multiperiod optimization are indeed great, however, and are considered at greater length subsequently. It will soon be apparent that the other techniques are sufficiently troublesome that geometric mean maximization becomes a very attractive alternative if the assumptions are anywhere close to being fulfilled.

3.6   Portfolio Rebalancing[4]

Over the course of time, expectations regarding the earning power and dividend-paying potential of the firm, the risk of the same, and the firm's stock's covariance with other securities will change. Furthermore, the market price of the stock and the level of interest rates will undoubtedly change. Finally, the investor's utility function itself may change or else changes in his level of wealth may cause his degree of risk aversion to change. Any of these factors will tend to cause the investor's portfolio to be suboptimal. It then becomes necessary to revise the portfolio.

---

[4]Much of this discussion is drawn from [3]. Also, valuable research in this area was pioneered by Fama [5] and Mossin [10].

Portfolio rebalancing may be viewed as the general case of asset acquisition and disposition, with portfolio selection representing the special case of revision of the all-cash portfolio. As such, revision involves all the problems of portfolio selection and some addtional difficulties as well. Any asset sold at a profit, for example may become subject to taxation that, in turn, reduces the amount of funds avilable for reinvestment; it might thus be desirable to maintain a position in a slightly inferior security if the capital gains taxes would wipe out any advantage. Also, the purchase or sale of any security will incur transactions costs, which also reduce the amount to be invested. This problem could be ignored in portfolio selection because it was assumed that the costs would be incurred no matter what was bought. A simple one-period revision model is presented below to illustrate these effects. Let

μ_i = expected return on ith asset

C_{ij} = covariance of return between ith and jth security

α_i = amount invested in ith security at beginning of period

β_i = amount invested in ith security at end of period before portfolio revision

γ_i = proportion in ith security after portfolio revision (decision variable)

$\delta_i(\omega_i)$ = increase (decrease) in amount of ith asset held from existing to revised portfolio

$b_i(s_i)$ = buying (selling) transfer cost of ith asset ($b_0 = s_0 = 0$ for cash) and assumed proportional for convenience

r_i = realized return on ith asset, composed of cash dividend return (r_i^c) and capital gain return (r_i^g). By definition,

$$r_i = r_i^c + r_i^g$$

t_c, t_g = ordinary income and capital gain tax rates for investor.

The cash account at the end of the year is equal to beginning cash plus all dividends after taxes:

$$\beta_0 = \alpha_0 + \sum_{i=1}^{n} r_i^c \alpha_i (1 - t_c). \tag{3.12}$$

The amount invested in each risk asset is

$$\beta_i = (1 + r_i^g)\alpha_i, \tag{3.13}$$

for $i = 1, 2, \ldots, n$.

From this point, short sales and borrowing are excluded:

$$\alpha_i, \beta_i, \gamma_i \geq 0, \text{ for all } i. \tag{3.14}$$

By definition, the net change in holdings of the ith asset is given by the difference between holdings in the end of period portfolio and the revised portfolio:

$$\delta_i - \omega_i = \gamma_i - \beta_i$$
$$\delta_i, \omega_i \geq 0$$
$$\delta_i \text{ or } \omega_i \text{ or both} = 0. \qquad (3.15)$$

The total value of the portfolio after revision must equal the value prior to revision less and transactions costs and taxes realized:

$$\sum_{i=0}^{n} \gamma_i = \sum_{i=0}^{n} \beta_i - \sum_{i=1}^{n} [\delta_i b_i + \omega_i (s_i + (1 - s_i) r_i^g t_g)]. \qquad (3.16)$$

Subject to these constraints, the objective function then becomes the maximization of Z, where

$$Z = \sum_{i=0}^{n} (1 + \mu_i) \gamma_i - \lambda \sum_{i=1}^{n} \sum_{j=1}^{n} \gamma_i \gamma_j C_{ij}. \qquad (3.17)$$

Multiperiod revision models generally involve dynamic programming, in which the last-period portfolio is optimized first and then the model works backward to the present. Various computational limitations have, in the past, prevented the practical applications of such models except for only very small portfolios over a limited number of periods. Today, these computational limitations are no longer a significant problem, and commercial software is available to carry out portfolio revision.

Two final points should be made. One of the reasons that dynamic programming is necessary stems from Mossin's contention [19] that one-period portfolio planning tends to be "myopic" over many periods unless the underlying utility function is a log or power function. Fama, however, has shown [5] that if investors may be assumed to be risk-averse and markets perfect, one-period investment planning is consistent with long-run optimization. Finally, models of the sort presented above include taxes and transactions costs, but not information costs. But for the latter, the model could be run (and revision take place) continuously.

Problems

Problem 3.1. Compute the parameters of the following holdings, given a risky portfolio (1) with $\mu = .20$ and $\sigma = .25$ and a riskless rate of .05.

	α_1	α_2
(a)	0.0	1.0
(b)	0.5	0.5
(c)	1.0	0.0
(d)	1.5	−0.5
(e)	2.0	−1.0

Problem 3.2. Assume an efficient frontier composed of only the following discrete points:

Portfolio	μ	σ
A	.10	.05
B	.15	.10
C	.20	.20
D	.25	.30
E	.30	.50

a. If $r_L = r_B = .06$, which of the above portfolios would be optimal for borrowing and lending?

b. If $r_L = .06$, but $r_B = .10$, how would your answer change?

Problem 3.3. What general relationship would you expect to find among r_L, r_B, and r? Why?

Problem 3.4. Determine which of the following portfolios would be an optimal choice by stochastic dominance for an investor whose utility function possesses the requisite first-order, second-order, and third-order characteristics:

Probability

Return	A	B	C	D
−.05	.2	.1	0	0
.00	.2	.3	.3	.4
.05	.0	.0	.2	.0
.10	.1	.1	.0	.0
.15	.0	.0	.1	.3
.20	.5	.5	.4	.3
μ	.100	.105	.105	.105

Problem 3.5. Reconsider the four portfolios described in Problem 3.4.

a. Which could be eliminated as inefficient?

b. Which would be selected by the geometric-mean criterion?

Problem 3.6. Assume that portfolios with the parameters given below lie along an efficient frontier. Which would be eliminated if a lower confidence limit (K) of 1.0 were applied? If limits of 0.5, 2.0, and 3.0 were applied?

Security	A	B	C	D	E	F	G	H	I
μ	.30	.25	.20	.15	.13	.11	.08	.06	.03
σ	.50	.31	.19	.10	.07	.05	.03	.02	.01

Problem 3.7. Graph the efficient frontier from the coordinates given in Problem 3.6.

a. If the riskless rate of interest were .02, which portfolio would be selected by the excess-return criterion?

b. Which would be the safety-first portfolio if the disaster level (D) were set at .10? At .05?

References

[1] Baumol, W. (1963). "An expected gain–confidence limit criterion for portfolio selection," *Management Science*, October, 174–182.

[2] Blume, M. (1971). "On the assessment of risk," *Journal of Finance*, March, 1–10.

[3] Chen, A., Jen, F. and Zionts, S. (1971). "The optimal portfolio revision policy." *Journal of Business*, January, 51–61.

[4] Fama, E. F. (1970). "Efficient capital markets: A review of theory and empirical work," *Journal of Finance*, May, 383–417.

[5] ___(1970). "Multi-period consumption–investment decisions," *American Economic Review*, March, 163–174.

[6] Lintner, J. (1965). "Security prices, risk, and maximal gains from diversification," *Journal of Finance*, December, 587–615.

[7] ___ (1965). "The valuation of risk assets and the selection of risky investments in stock portfolios and capital budgets," *Review of Economics and Statistics*, February, 13–37.

[8] Markowitz, H.(1952). "Portfolio selection," *Journal of Finance*, March, 77–91.

[9] _____(1959). *Portfolio Selection*, New York: John Wiley & Sons.

[10] Mossin, J. (1969). "Optimal multi-period portfolio policies," *Journal of Business*, April, 215–229.

[11] Roy, A. D. (1952). "Safety first and the holding of assets," *Econometrica*, July, 431–449.

[12] Sharpe, W. (1963). "A simplified model for portfolio analysis," *Management Science*, Jan., 277–293.

[13] ___(1966). "Mutual fund performance," *Journal of Business*, January, 119–138.

[14] ___ (1964). "Capital asset prices: A theory of market equilibrium under conditions of risk," *Journal of Finance*, September, 425-442.

[15] ___ (1972). "Risk, market sensitivity and diversification," *Financial Analysis Journal*, January-February 74–79.

[16] Whitmore, G. A. (1970). "Third-Degree stochastic dominance," *American Economic Review*, June, 457–459.

[17] Whitmore, G.A. and Findlay, M.C.(eds.)(1978). *Stochastic Dominance: An Approach to Decision-Making Under Risk.* (Lexington, MA: D.C. Heath.

[18] Williams, E.E. and Findlay, M.C. (1974). *Investment Analysis*, Englewood Cliffs, NJ: Prentice-Hall.

Chapter 4

Capital Market Equilibrium Theories

4.1   Introduction: The Capital Market Line

If we may assume that investors behave in a manner consistent with Chapters 2 and 3, then certain statements may be made about the nature of capital markets as a whole. Before a complete statement of capital market theory may be advanced, however, certain additional assumptions must be presented:

1. The μ and σ of a portfolio adequately describe it for the purpose of investor decision making $[U = f(\sigma, \mu)]$. [1]

2. Investors can borrow and lend as much as they want at the riskless rate of interest.

3. All investors have the same expectations regarding the future, the same portfolios available to them, and the same time horizon. [2]

4. Taxes, transactions costs, inflation, and changes in interest rates may be ignored.

Under the assumptions above, all investors will have identical opportunity sets, borrowing and lending rates ($r_L = r_B$) and thus identical optimal borrowing-lending portfolios, say X (see Figure 4.1). Because all investors

---

[1] Technically, this condition requires either that the distribution of security returns be Gaussian (normal) or that the investor's utility function be quadratic.

[2] Technically, homogeneity of expectations is not required if a weighted average of divergent expectations may be employed. In the latter case, however, it is necessary that the weights be stable as well as the proportional response of each divergent expectation to new information, and so on. This condition is often referred to as *idealized uncertainty* [17], a rather odd concept to which we shall return.

will be seeking to acquire the same portfolio (X), and will then borrow or lend to move along the *Capital Market Line* in Figure 4.1, it must follow for equilibrium to be achieved that all existing securities be contained in the portfolio (X). In other words, all securities must be owned by somebody, and any security not initially contained in X would drop in price until it did qualify. Therefore, the portfolio held by each individual would be identical to all others and a microcosm of the market, with each security holding bearing the same proportion to the total portfolio as that security's total market value would bear to the total market value of all securities. In no other way could equilibrium be achieved in the capital market under the assumptions stated above.

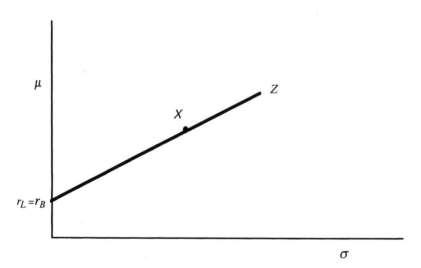

Figure 4.1. The Capital Market Line.

The borrowing-lending line for the market as whole is called the *Capital Market Line*. The securities portfolio (X) employed is the total universe of available securities (called the *market portfolio*) by the reasoning given above. The CML is linear by the logic of Chapter 3 because it represents the combination of a risky portfolio and a riskless security. One use made of the CML is that its slope provides the so-called *market price of risk*, or, that amount of increased return required by market conditions to justify the acceptance of an increment to risk, that is,

$$\text{slope} \; = \frac{\mu(X) - r}{\sigma(X)}.$$

The simple difference $\mu(X) - r$ is called the *equity premium*, or the expected return differential for investing in risky equities rather than riskless debt.

The intuition of the CML can be aided by noting that all we have done is to take Figure 3.6, applied it to everybody, and assumed the result to be in equilibrium. Not only must all securities be owned by somebody (which causes 0 in Figure 3.6 to become X in Figure 4.1 to become the market portfolio) but also the loan market must clear (i.e, borrowing = lending). The latter requires securities' prices (and thus μ, σ relations) to adjust until every investor's highest attainable indifference curve is tangent to the resulting CML (as illustrated for A and B in Figure 3.6). Only at this point is equilibrium attained. And, because every investor's curve is tangent to the CML at this point, every investor has a marginal risk-return tradeoff given by the slope of the line, which is the reason it is called the market price of risk.

4.2   The Security Market Line

One major question raised by the CML analysis of Section 4.1 involves the means by which individual securities would be priced if such a system were in equilibrium. Throughout this book, we generally assume that markets are *not* in equilibrium. Therefore, when we combine securities into a portfolio that has the average return of the component securities but less than the average risk, we simply ascribe this *gain from diversification* to our own shrewdness. In the type of market assumed in Section 4.1, however, everyone will be doing the same thing, and the prices of securities will adjust to eliminate the windfall gains from diversification.

Sharpe (see [24] and [27]) has suggested a logical way by which such security pricing might take place. If everyone were to adopt a portfolio theory approach to security analysis, then the risk of a given security might be viewed not as its risk in isolation but rather as the change in the total risk of the portfolio caused by adding this security. Furthermore, because capital market theory assumes everyone to hold a perfectly diversified (i.e., the market) portfolio, the addition to total portfolio risk caused by adding a particular security to the portfolio is that portion of the individual security's risk that cannot be eliminated through diversification with all other securities in the market.[3]

Because the concept of individual security pricing is rather elusive, let us restate it. Sharpe argued that the price (and thus return) of a given security should not be determined in relation to its total risk, because the security will be combined with other securities in a portfolio and some of the individual risk will be eliminated by diversification (unless all the securities

---

[3]If the standard deviation of the market as whole is σ_M and the standard deviation of security i is σ_i, and the correlation of security i with the market is ρ_{iM}, then the non-diversifiable portion of the individual security's risk is the covariance of returns between the security and the market as a whole, i.e., $C_{iM} = \sigma_i \sigma_M \rho_{iM}$.

have correlation equal to 1). Therefore, the return of the security should only contain a risk premium to the extent of the risk that will actually be borne (that is, that portion of the total risk which cannot be eliminated by diversification—which is variously called *nondiversifiable risk* or *systematic risk*).

If this logic is accepted, it is then possible to generate a Security Market Line as shown in Figure 4.2, where the return on individual securities is related to their covariance with the market. If capital markets are in equilibrium and all the other assumptions of this chapter hold, then the parameters of each security should lie on the SML. Furthermore, because the risk of a portfolio is the weighted sum of the nondiversifiable risk of its component securities, all portfolios should also fall on the SML in equilibrium. (It should be noted that although all portfolios will fall on the SML, as a general rule, no individual securities or portfolios, other than the market portfolio, will fall on the CML. They will all lie to the right of the CML.)

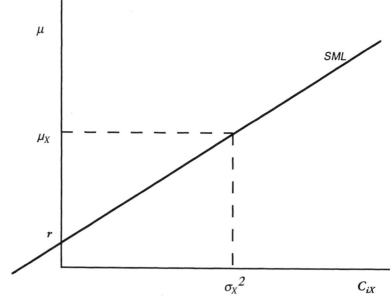

Figure 4.2. The Security Market Line.

4.3   The Sharpe Diagonal Model

As we discussed in Chapter 3, the Markowitz approach required that a variance-covariance matrix for all the securities under consideration for inclusion in the portfolio first be estimated and then be inverted. Neither of these exercises was feasible 50 years ago (and the first is no fun to this day). Sharpe proposed a simplified approach to allow implementation by assuming that all securities were only correlated with each other through a common index. If we assume the index is the market and add the other assumptions from Section 4.1 above, the result becomes (rather than the originally intended generator of efficient frontiers) the model used to estimate the position of securities and portfolios on the SML.

The Sharpe Diagonal Model assumes that the return on a security may be related to an index (such as the DJIA, S&P 500, Wilshire 5000, or whatever) as follows:

$$\text{Return}_i = a_i + b_i \, \text{Return}_I + c_i \qquad (4.1)$$

$$\mu_i = a_i + b_i \mu_I + c_i,$$

where:

a_i and b_i are constants
μ_I is the return (including dividends) on the index
c_i is an error term with $\mu_{c_i} = 0$ and $\sigma_{c_i} = $ a constant.

It is further assumed that c_i is not correlated with μ_I, with itself over time, nor with any other security's c (the last implying that securities are only correlated through their common relationship to the index). Therefore, μ_i can be estimated as $(a_i + b_i \mu_I)$. The parameters a_i and b_i can either be estimated, computed by regression analysis, or both. Furthermore, σ_{c_i} can be viewed as the variation in μ_i *not* caused by variation in μ_I.

The return of the portfolio becomes

$$
\begin{aligned}
\mu &= \sum_{i=1}^{n} \alpha_i (a_i + b_i \mu_I + c_i) \\
&= \sum_{i=1}^{n} \alpha_i (a_i + c_i) + \left(\sum_{i=1}^{n} \alpha_i b_i \right) \mu_I, \qquad (4.2)
\end{aligned}
$$

where the first term is viewed as an investment in the essential nature of the securities, and the second term is an investment in the index. The risk of the portfolio is

$$\sigma = \sqrt{ \sum_{i=1}^{n} (\alpha_i \sigma_{c_i})^2 + \left(\sum_{i=1}^{n} \alpha_i b_i \right)^2 \sigma_I^2 }, \qquad (4.3)$$

where, again, the first term under the radical may be viewed as the risk of the portfolio attributable to the particular characteristics of the individual securities, and the second term as the risk attributable to the index.

Thus, the Sharpe model simplifies the input problem by making it directly amenable to simple regression analysis. In addition, by assuming that securities are only related through the index, the nonzero elements in the covariance matrix are reduced to those on the diagonal, thus easing the computational burden.

To illustrate the Sharpe approach, assume that the index currently stands at 1000 and, with reinvestment of dividends, it is expected to be at 1100 at the end of the year. Given the following data, suppose we wished to determine portfolio μ and σ for $\alpha_1 = .2$, $\alpha_2 = .5$ and $\alpha_3 = .3$.

$$
\begin{aligned}
\sigma_I &= .10 \\
\mu_1 &= .06 + .1\mu_I; \sigma_{c_1} = .03 \\
\mu_2 &= -.03 + 2\mu_I; \sigma_{c_2} = .20 \\
\mu_3 &= .00 + \mu_I; \sigma_{c_3} = .10.
\end{aligned}
$$

Employing (4.2) we obtain

$$
\begin{aligned}
\mu &= (.2)(.06) + (.5)(-.03) + (.3)(.00) + [(.2)(.1) + (.5)(.2) + (.3)(1)](.10) \\
&= .012 - .015 + (.42)(.10) = .039 \text{ or } 3.9\%.
\end{aligned}
$$

Employing (4.3),

$$
\begin{aligned}
\sigma &= \sqrt{[(.2)(.03)]^2 + [(.5)(.2)]^2 + [(.3)(.1)]^2 + [(.2)(.1) + (.5)(.2) + (.3)(1)]^2(} \\
&= \sqrt{(.006)^2 + (.1)^2 + (.03)^2 + (.42)^2(.1)^2} \\
&= \sqrt{.000036 + .01 + .0009 + .001764} = \sqrt{.0127} = .1127 \text{ or } 11.27\% .
\end{aligned}
$$

It is also possible to discuss the SML in terms of Sharpe's index model

$$
\mu_i = a_i + b_i \mu_I + c_i. \tag{4.4}
$$

The b_i term (called Sharpe's *beta coefficient*), given μ_I, is equal to

$$
(b_i | \mu_I) = \frac{C_{iI}}{\sigma_I^2} = C_{iI} \frac{1}{\sigma_I^2}, \tag{4.5}
$$

which, if the index is a valid depiction of the market:

$$
(b_i | \mu_I) = \frac{C_{iX}}{\sigma_X^2} = C_{iX} \frac{1}{\sigma_X^2}. \tag{4.6}
$$

Under these assumptions, the abscissa of a point on the SML expressed in terms of b_i is merely $1/\sigma_X^2$ times that of the same point expressed in terms of C_{iX} and the two are directly comparable. Viewed another way, the risk premium an individual security would exhibit in equilibrium is

$$\mu_i - r = \frac{\mu_X - r}{\sigma_X^2} C_{iX} = (\mu_I - r) b_i. \tag{4.7}$$

A major advantage of transferring the discussion into beta terminology is that the regression coefficient can be used directly to estimate the systematic risk of the asset. Unfortunately, the beta concept also possesses serious pitfalls. In the first place, its very simplicity and popularity cause it to be used by many who fail to understand its limitations. Because the concept is subject to all the assumptions of both linear regression and equilibrium, including the efficient capital market hypothesis, statistical problems and economic imperfections may undermine its usefulness. Many investors are unaware of these limitations and have blithely assumed that one need only fill a portfolio with securities possessing large betas to get high returns. At best, the beta is a risk-measure surrogate and *not* an indicator of future returns. The idea that the assumption of large amounts of risk will generate large returns only approaches being correct over the long run in reasonably efficient markets in equilibrium. Even then it ignores utility considerations. A further difficulty with the beta concept follows from empirical findings that betas for small portfolios (and, of course, individual securities) over short periods can be highly unstable (see [8] and [16]), although there is evidence of reasonable stability for the betas of large portfolios over long holding periods (see [2,12,26] where, of course, beta approaches one by definition anyway). It would thus appear that one of the few valid applications of the beta concept would be as a risk-return measure for large portfolios. An example of how betas can be used in this regard is presented in the next section.

4.4 Portfolio Evaluation and the Capital Asset Pricing Model (CAPM)

Several measures directly related to capital market theory have been developed for the purpose of portfolio evaluation. The latter is essentially a retrospective view of how well a particular portfolio or portfolio manager did over a specified period in the past. Most of the published research in this area has dealt with mutual funds, seemingly because they are controversial, economically important in certain financial markets, and possessed of long life with readily available data. Much of the early work in this area (including the advertisements of the funds themselves) was of a simple time series nature, showing how well an investor could have done over a given period in the past if her or she had invested then or else comparing these results to what the investor could have earned in other funds, or the market as a whole. The more recent work considers both return and its variability, contending that mutual funds that invest in riskier securities should exhibit higher returns. One result of this work, considered subsequently, has been the finding that investors do as well or better on average by selecting securities at random as they could with the average mutual fund. Another implication, of more relevance here, is the growing feeling that the managers of any kind of portfolio should be rated not on the return they earn alone, but rather on the return they earn adjusted for the risk to which they subject the portfolio.

Before proceeding, however, a caveat is in order about the nature of *ex post* risk and return measures. As in any problem in measurement, one must delineate (1) why a measurement is being made, (2) what is to be measured, (3) which measurement technique is appropriate, and (4) the import of the results of the measurement. If one is not careful *ex post* return measurements can easily result in the "if only I had ...." syndrome, which is a waste of time and effort as far as making an investment in the present is concerned. For such measures to be of use, one must assume that the ability of a manager or fund to earn greater-than-average returns in the past is some indication of ability to do so in the future. As the empirical work cited below indicates, there is little evidence to support this contention. As far as risk in concerned, there is some doubt about what the concept of *ex post* risk means. Most of the writers in this area are careful to stress the term "return variability" instead of risk per se. Because the outcomes of all past events are currently known with certainty, the use of return variability as a measure of risk in this instance involves a different notion of risk than we have been using. Again, to make operational investment decisions, it would seem necessary to assume that past risk-return behavior of managers or portfolios either could or would be maintained in the future.

Deferring judgment for the moment on the above reservations, let us consider the proposed evaluation measures. Sharpe [24] has proposed the use of a reward-to-variability ratio related to the slope of the capital market line:

$$\text{Sharpe's Measure for } i\text{th portfolio} = \frac{\mu_i - r}{\sigma_i}. \qquad (4.8)$$

In effect, Sharpe is computing the slope of the borrowing-lending line going through the given portfolio and arguing that a greater slope is more desirable (which is similar to the Lintner criterion discussed in Chapter 3).

A second measure on the SML has been proposed by Treynor [31,32]:

$$\text{Treynor's Measure for the } i\text{th portfolio or security} = \frac{\mu_i - r}{b_i} \qquad (4.9)$$

and the line in (beta, return) space $= r + (\mu_i - r)/(b_i) = $ *characteristic line* of security or portfolio i. Treynor's methodology is fairly similar to that of Sharpe, except that by using the SML instead of the CML, the Treynor measure is capable of evaluating individual security holdings as well as portfolios. A disadvantage is that the accuracy the rankings depends in part upon the assumption (implicit in the use of the SML) that the fund evaluated would been held in an otherwise perfectly diversified portfolio.

A third measure, also based on the SML but different from Treynor's, has been proposed by Jensen [14]:

$$\text{Jensen's Measure} = (\mu_i - r) - b_i(\mu_X - r). \qquad (4.10)$$

This measure is expressed in units of return above or below the riskless rate of a line drawn through the parameters of the security or portfolio parallel to the SML. This measure does allow comparisons of a portfolio to the market and is also amenable to estimation by regression; because of its treatment of differential risk, however, direct comparisons between funds or portfolios generally cannot be made. Furthermore, it has been suggested that all three of the above measures are biased against high risk portfolios by failing to recognize the inequality of borrowing and lending rates and the resulting nonlinearity of the SML and CML.

Use of geometric means as an evaluation tool should not be overlooked as well. Over a given period of time, the geometric mean portfolio return could be compared to that of other portfolios or some market index. There are several advantages to such a measure. Assuming that the interval considered is "sufficiently" long (and if it is not, one may doubt the validity of any evaluation technique), then undue risk taking should manifest itself in numerous low period returns and, thus, a reduced geometric mean (or terminal wealth, which is an equivalent concept in this context). If such is not the case, then the equivalence of historical variability and risk becomes increasingly dubious. The geometric mean also facilitates the use of very short investment periods (because funds value their holdings several times a day, thousands of observations per year could be obtained) and provides a cumulative effect if desired (by simply including each new set of observations without discarding any of the old).

4.5   Arbitrage Pricing Theory (APT) and Fama-French (FF)

In its simplest form, the Capital Asset Pricing Model (CAPM) is the more common name for the SML. Over the years, however, efforts have been made to extend the CAPM to multiple periods, other investment media, foreign markets, and even human wealth. As discussed further in Chapter 5, however, it became increasingly apparent that very little of the cross section of securities' returns was reliably explained by beta.

The primary alternative to the CAPM advocated in the finance literature is the arbitrage pricing theory (APT) first proposed by Ross [21]. Similar to the CAPM in its intuitive appeal, the APT assumes that asset returns are generated by a linear factor model. There are no restrictions on the specific form of individual utility functions except general monotonicity and concavity. More importantly, there is no central role for a mean variance efficient market portfolio. The equilibrium condition is merely asset pricing such that no arbitrage profits exist.

The APT (defined in a frictionless and competitive market) assumes individuals possess homogeneous beliefs about asset returns described as a common k-factor generating model:

$$\tilde{r}_i = \chi_i + b_{i1}\tilde{\delta}_1 + \ldots + b_{ik}\tilde{\delta}_k + \tilde{\epsilon}_i \qquad (4.11)$$
$$i = 1, 2, \ldots, n,$$

where $\tilde{r}_i$ = random returns on asset i

$\tilde{\delta}_k$ = k (common) generating factors

b_{ik} = individual response coefficients for each of the k factors

χ_i = expected return on asset i

$\tilde{\epsilon}_i$ = random term idiosyncratic to asset i.

Whereas the factors $\tilde{\delta}_k$, common to all assets, represent the systematic risk components, the $\tilde{\epsilon}_i$ factors represents the unsystematic component particular to that asset. The values of the factors are assumed to follow:

$$E(\tilde{\delta}_k) = 0 \text{ for all } k$$
$$E(\tilde{\epsilon}_i|\tilde{\delta}_k) = 0 \text{ for all } i, k.$$

Further, the $\tilde{\epsilon}_k$'s are assumed to be mutually independent.

Since the underlying presumption of the APT is that no arbitrage profits exist, the equilibrium relationship is found by forming all possible arbitrage portfolios (i.e., portfolios that have a zero net investment and no risk). No arbitrage profits implies that the portfolios' expected returns are zero. The following conditions define the choice of arbitrage portfolios.

(i) $\displaystyle\sum_{i=1}^{n} \alpha_i = 0$ \hfill (4.12)

where α_i = proportion of wealth invested in asset i

(ii) $\displaystyle\sum_{i=1}^{n} \alpha_i b_{ik} = 0$ for all k \hfill (4.13)

(iii) Investment in each asset i is of the order $1/n$ where n is the number of assets in the economy.

Condition (i) is the formal statement that the portfolio of assets contains zero net wealth. Condition (ii) insures that the portfolio is formed so there is zero systematic risk; the latter is embodied in the common factors. The last condition suggests that by forming a portfolio of assets, where noise terms are mutually independent and the amount invested in each asset is of order $1/n$, unsystematic risk is diversified away (approximately). Thus the three conditions assure that an arbitrage portfolio has been formed. This portfolio return is

$$\sum_{i=1}^{n} \alpha_i \chi_i = 0. \tag{4.14}$$

Equations (4.12)–(4.14) imply that an asset's expected return χ_i can be written as

$$\chi_i = \omega_0 + \omega_1 b_{i1} + \ldots + \omega_k b_{ik}, \tag{4.15}$$

where $\omega_0, \omega_1, \ldots, \omega_k$ is a set of constant weights.
If a riskless asset with return χ_0 exists, then (4.15) can be written as

$$\chi_i - \chi_0 = \omega_1 b_{i1} + \ldots + \omega_k b_{ik}. \tag{4.16}$$

χ_0 is also the return on all zero beta assets.

Equation (4.16) is the crux of the APT. It is a linear multifactor model of asset returns which does not require use of a market portfolio as a factor. But as Roll and Ross [20] point out, the market portfolio is not *necessary* but may well be one of the factors predicting systematic return. Further, if a single-factor model using a market portfolio were a correct specification of asset return, (4.16) would reduce to

$$\chi_i - \chi_0 = \omega b_i. \tag{4.17}$$

Comparing (4.16) and (4.17), the CAPM can be seen as a special case (one factor model) of the APT.

The APT has never developed a popular following despite its popular treatment in textbooks (see [27]). If identifiable variables (e.g., the difference between long-term and short-term, or high-grade and low-grade, bond yields) are employed as factors, the results are generally not much better than the CAPM. If the factors are derived empirically, the results are generally devoid of intuitive interpretation and the coefficients tend to be unstable.

Over the last decade, Professors Fama and French (cf. [6] and [7]) have published a series of papers employing a three factor benchmark model. The first factor is beta from the CAPM. A problem with the CAPM was that firms with high book values relative to their price and those with small total market capitalizations tended to have higher returns than the model would imply. The other two FF factors simply correct for these excess returns. Whether the resulting benchmark describes a risk-based equilibrium model is a subject of heated debate (which we take up in Chapter 5).

4.6   Interaction of Equilibrium and Efficiency

One of the major assumptions underlying capital market theory presented above was that investors possessed either homogeneous (or a stable weighted average of) expectations regarding the future. One implication of this assumption is that a change in expectations will either occur unanimously or else in some stable average way that is an operational equivalent of unanimity. Thus, if markets are in equilibrium, a change in expectations will cause them to move in a rapid and unbiased fashion to a new equilibrium. Markets that behave in this manner will always "fully reflect" available information, which is the definition of an efficient market. Hence, in frictionless markets, efficiency is simply one of many attributes of equilibrium.

Even in a world of perfect certainty, prices of securities could be expected to change over time as long as the income stream was not instantaneous and constant. Indeed, price change would be necessary in such a case to cause the expected return for each holding period to be obtained, that is,

$$P_1 + D_1 = P_0(1 + r_1)$$

or

$$P_1 - P_0 = r_1(P_0) - D_1. \tag{4.18}$$

In a world of less-than-perfect certainty, changes in expected income or the required rate of return (caused, in turn, by changes in the riskless rate or the perceived risk of the issue) could also cause the prices of securities to change over time. It is argued, however, that only the availability of new information or better analysis of already available information should cause the expectations regarding income, risk, and interest rates to change and, thus, prices to change for these reasons. If it may be assumed (among other things discussed in Chapter 5) that (1) all relevant information is available to all market participants; (2) any new information is spread and assimilated immediately; and (3) vast numbers of market participants employ the most sophisticated analytical techniques, then the securities market may be viewed as a "fair game," where

$$E[(\tilde{P}_1 + \tilde{D}_1)|I_0] = P_0[1 + E(\tilde{r}_1|I_0)]. \tag{4.19}$$

The above merely states that the expected price of a security in period one given the information available in period zero (assuming all dividends or interest are reinvested) is equal to the price in period zero times one plus the expected rate of return, given available information. This hypothesis also implies that (1) the excess market value of a security or (2) the excess expected return to be earned by holding the security both have, given available information, an expected value of zero.

Several points should be noted. Equation (4.19) does not imply that prices need be stable. Indeed, to the extent that $E[\tilde{D}_1] \neq E[\tilde{r}_1](P_0)$, then $E[\tilde{P}_1] \neq P_0$. Thus, we have a probabilistic model with implications similar to the certainty model. A major difference, however, is indicated by the presence of the expectation operator and tildes for both the $(\tilde{P}_1 + \tilde{D}_1)$ and $\tilde{r}_1$ terms, implying that they are random variables at time zero. Thus, although the certainty model would indicate a constant rate of price change over time (i.e., linear on semilog graphs), the expectations model indicates that price could be expected to vary about such a line. Such a possibility is also reflected by the error term in the various capital-market estimation models.

Finally, it will be noted that $(\tilde{P}_1 + \tilde{D}_1)$ and $\tilde{r}_1$ are conditioned by available information, I_0. To the extent new information or better analysis becomes available $E[\tilde{P}_1 + \tilde{D}_1]$ and $E[\tilde{r}_1]$ can alter, and the expected rate of price change over time can shift. Adding the assumption that the timing of the arrival of new information in the market is a random variable, we see that shifts in the expected-price-change line, as well as movement about it, can be treated as random events.

4.7 Expectations, Convergence, and the Efficient Market Hypothesis[4]

In most areas of economics, dynamic theory is the most complicated form of analysis. This is true because expectations, rather than existing conditions, are the main determinants. Equilibrium is based on a given state of expectations and any changes in those expectations will force a movement to a new equilibrium. Unfortunately, "the adjustments needed to bring about equilibrium take time" ([11], p. 116). When markets are inefficient, time becomes the most important variable to the analysis. But when markets are efficient, the adjustments occur rather quickly and time is not terribly important. A perfectly efficient market will return to equilibrium immediately. Proponents of the EMH maintain that timing is not important in analyzing investments because the market digests new information instantaneously, redetermines expectations, and recalculates prices accordingly.

A condition of static equilibrium can cease to exist if new information concerning systematic risk, expected dollar returns, or the market price of systematic risk (i.e., the slope of the SML) is perceived. For example, let us assume a market in which all dividends are reinvested so that returns over the next period ($t = 1$) are given by

$$\tilde{r}_1 = \frac{\tilde{P}_1 - P_0}{P_0}, \qquad (4.20)$$

where

$\tilde{r}_1$ = the expected return one period hence
$\tilde{P}_1$ = the expected market price of the stock one period hence (with dividends reinvested)
P_0 = the current market price ($t = 0$).

---

[4]This section draws heavily from [29].

We further assume that time periods are very short (i.e., they are too short to act between $t = 0$ and $t = 1$). Finally, we assume that R is the appropriate return for a given level of systematic risk (taken from the SML). In equilibrium,

$$E(\tilde{r}_1|I_0) = R = r_0. \qquad (4.21)$$

If R changes either because of a shift in the SML or a change in the systematic risk of a stock, there exists a new $R^*(R^* \neq R)$ which will produce the potential for a windfall between $t = 0$ and $t = 1$ of ω.

$$
\begin{aligned}
E(\tilde{r}_1|I^*) &= R + \lambda\omega \\
&= R + \lambda\left(\frac{\tilde{P}_1^* - \tilde{P}_1}{P_0}\right) \qquad (4.22) \\
E(\tilde{r}_2|I^*, \lambda = 1) &= R^*, \qquad (4.23)
\end{aligned}
$$

where

I_0 = available information at period $t = 0$
I^* = new information becoming available between $t = 0$ and
$\quad t = 1$
R^* = the new equilibrium return, given new information about
$\quad$ either the level of risk evidenced by the shares or the
$\quad$ position of the SML
P_0 = the share price at $t = 0$
$\tilde{P}_1$ = the share price at $t = 1$ (corresponding to information I_0)
$\tilde{P}_1^*$ = the share price at $t = 1$ (corresponding to information I^*)
λ = a one-period price adjustment coefficient
ω = the potential windfall which would occur between
$\quad t = 0$ and $t = 1$ assuming complete adjustment (i.e.,
$\quad$ where $\lambda = 1$).

If markets are efficient, $\lambda = 1$ and adjustment takes place immediately. Because time periods are assumed to be short, only those holding the security when new information reaches the market will receive the windfall (i.e., ω will result too quickly for anyone to act on the information profitably). Beyond period $t = 1$, the only return expected would be the new equilibrium rate R^*. If markets are inefficient, $\lambda \neq 1$ and adjustment will not take place immediately. In the case $0 < \lambda < 1$, the entire windfall will not be allocated before profitable action can be taken. Indeed, if $\lambda < 0$ (which corresponds to a complete misinterpretation of the new information), then a potential windfall even greater than ω could be possible to the extent that the market eventually interprets the information correctly. If $\lambda > 1$, the market overreacts, and a windfall in the opposite direction is possible.

We have considered two of the conditions that will produce a disturbance in the static equilibrium of stock prices. The third condition (a change in expected dollar returns) will initially manifest itself as $\tilde{P}_1^* \neq \tilde{P}_1$ whereas, in the former conditions, the disturbance was introduced by $R^* \neq R$. The analysis proceeds along similar lines, with (4.22) remaining intact. The expected return for $t = 2$, however, becomes:

$$E(\tilde{r}_2 | I^*, \lambda = 1) = R \qquad (4.24)$$

rather than (4.23).

In all three instances, early empirical evidence suggested that λ was close to 1. In more recent years, a whole literature on market microstructure has developed over the issue of "overreaction" (see [29]), and many EMH advocates now view the equilibrating process as a "longer run" phenomenon. That is, the market is efficient and in equilibrium, but adjustments take time and often "overshoot" the correct level. Conversely, efforts to rationalize "momentum" investing assert an initial "undershoot" reaction. To the authors of this book, such arguments seem sophistic at best. Moreover, as we shall see in Chapter 5, all the EMH studies have examined I^* of a fairly obvious sort (earnings announcements, stock splits, and so on). They do not address themselves to the fact that analysts may obtain other kinds of I^* that cannot be neatly measured. The analyst who has a better forecasting method, for example, may be able to discern information that no simple sort of empirical investigation could detect, and his or her long-run returns could be expected to surpass the market return.

Of course, determining better I^* is what security analysis is all about. The analyst should be cautioned, however, that returns will be above the market's only if what one knows eventually becomes known by the market (hence, producing a new equilibrium). If analyst A envisages a disequilibrium that will produce a windfall gain (ω) and the market never "learns" ($\lambda = 0$ in all future periods), A is better off only to the extent that A's superior insight allows him or her to earn higher returns for a given level of risk than would be provided by the market's perception of equilibrium risk-return relationships. A does not earn the windfall adjustment return, however, that would follow if the market eventually came to agree with his or her superior insight. On the other hand, if the market received and correctly interpreted new information at the same time as the analyst ($\lambda = 1$), no above-normal returns would be expected. Thus, the analyst must be rather sure that his or her insight is indeed superior, and that the market will eventually agree in order to derive the most benefit from his or her prognostications.

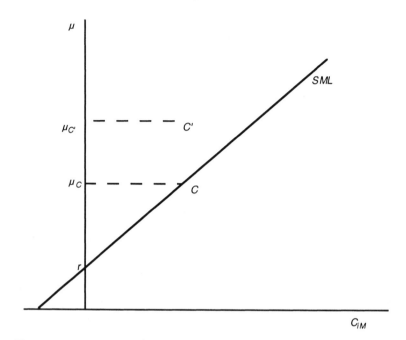

Figure 4.4. Differences in Expected Rates of Returns.

Expectations about future dollar returns (and estimated risk) may be even more important than expectations about future rates of return for making above-average profits. If analyst B foresees better prospects for a security than the market, B may get even greater returns if the market eventually comes around to his or her. Suppose an analyst expects security C (Figure 4.3) to produce a future stream that will yield 12 percent, given the current market price ($100), but the market only forecasts a 10 percent return, which is the equilibrium yield. The analyst expects a perpetual earnings stream (perhaps in the form of dividend payments) of $12, whereas the market only expects $10. If the analyst turns out to be correct, the price of the security will rise to ($12/0.1) = $120, assuming the stock is still in the 10 percent risk class. If the market moves toward the analyst's position within one year, his or her return would not only be better than the equilibrium 10 percent, but it would also exceed his expected 12 percent. His yield, in fact, would be

$$\frac{120 + 12 - 100}{100} = .32 \text{ or } 32\% \ .$$

This result can be found by using equation (4.22). Assuming the reinvestment of dividends, the stock price at $t = 1$ would be $120 + $12 = $132. The price that would prevail without market recognition would be expected to be $110 (the market anticipates $10 in dividends, rather than $12, for reinvestment). Hence, when recognition occurs $\lambda = 1$ and:

$$
\begin{aligned}
E(\tilde{r}_1 | I^*) &= R + \lambda \left(\frac{\tilde{P}_1^* - \tilde{P}_1}{P_0} \right) \\
&= .10 + (1) \left(\frac{132 - 110}{100} \right) \\
&= .32 \text{ or } 32\%.
\end{aligned}
$$

Again, the time required for convergence is seen to be very important. If it took the market three years to converge (suppose the stock paid \$12 in years one and two, but the market only became convinced that this dividend could hold in the long run in period $t = 3$), the return would be just about 18 percent[5]. If the market never came to accept the analyst's judgment (and the yield, in fact, was 12 percent), he or she would still earn a larger return than that expected by the market. On the other hand, if the analyst's view were adopted by the market, his or her return would be far greater still. Thus, having better insight than the market would have given the analyst a 2 percent higher return in our example (12 percent − 10 percent), whereas having insight about patterns of market expectations could produce a 20 percent additional return (i.e., 32 percent − 12 percent).

The movement of stock prices from expected levels $(\tilde{P}_1)$ to those that the analyst thinks are justifiable $(\tilde{P}_1^*)$ is called *convergence*. The principle of convergence has been recognized by conventional authorities for years (see [10]) and the importance of timing in convergence has been given renewed consideration (see [29]). It is clear that rapidity of convergence will greatly influence returns, and the longer the time required for the market to recognize the superior insight of the analyst, the lower will be his or her annual rate of return. In terms of the Security Market Line, convergence takes place as the market realizes that a security is under (or over) priced and the security moves toward the line. In Figure 4.4, two securities are shown. The first, A, is seen to be "under" priced. That is, the return from $A(\mu_A)$ is greater than it should be given A's risk characteristic. Once the market agrees that A is underpriced, demanders will bid up prices until the security reaches the SML at A'. At this point, the rate of return will be μ'_A, which, of course, is lower than μ_A. In the process of reaching equilibrium, the security rises in price, and investors who bought A when its yield was μ_A receive a windfall gain. Similarly, security B is overpriced, given its risk characteristic. Its price will be bid down (expected returns rising from B to B') until it reaches the SML, if the analyst is correct in his or her appraisal. Investors who sold B short would reap a windfall gain as the price of B falls to yield μ'_B.

---

[5] $100 = \sum_{t=1}^{3} X_t (1 + .18)^{-t}$, where $X_1 = X_2 = 12$ and $X_3 = 132$.

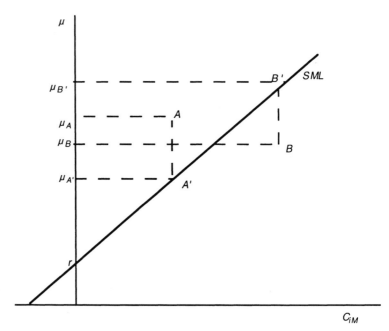

Figure 4.4. Convergence and the Security Market Line.

The investor who wishes to achieve returns above the market must consistently find securities that are under (over) priced, buy (sell short), and hope that the market soon agrees. Efficient market proponents, of course, would argue that no one could repeatedly find securities off the SML.

Expectations play a major role in convergence. Furthermore, it is *changes in expectations*, rather than simply new information, that influence returns. This distinction is important because all new information does not change expectations. As we shall see in Chapter 5, a stock split, for example, does not change the fundamental position of the firm, and it should not be surprising that the long run effect of a split on investor returns is nil. On the other hand, expectations can change *even though new information does not develop.*

Although all three conditions producing disequilibrium can affect future expectations with respect to individual stocks, we will focus our attention on the market price of risk (i.e., the equity premium for stocks as a group) and not the parameters of an individual security. Examination of market factors, rather than unique security characteristics, is more conducive to generalization and comparison to prior theoretical work. Extending a scheme developed by economist J. R. Hicks more than a half century ago [11], we may argue that there exists an elasticity of expectations about equilibrium rates of return that may be defined as follows:

$$\epsilon_\chi = \frac{E[E\tilde{P}_{t+1} - EP_t]}{(EP_t - EP_{t-1})}, \tag{4.25}$$

where

$$EP_t = \mu(X)_t - r_t$$

for the simple one-period forecasting model. ϵ_χ may be negative, zero, positive, positive and unity, or positive and greater than unity. If $\epsilon_\chi \neq 0$, expectations about changes in future returns do depend on changes in historical returns. If $\epsilon > 0$, patterns of historical changes will alter expected future returns in the same direction (and in the same proportion if $\epsilon = 1$). If $\epsilon_\chi > 1$, this historical pattern will "make people feel that they can recognize a trend, so that they try to extrapolate" ([11], pp. 260–262; 281–282), which can produce bubbles. Finally, if $\epsilon_\chi < 0$, people make "the opposite kind of guess, interpreting the change as the culmination point of fluctuation" ([11], p. 205). Although a reasonably strong *a priori* case can be made for $\epsilon_\chi = 0$, (which is implied by the EMH) empirical tests using expectations are very difficult to construct. It may be that other elasticities do exist at times (e.g., when interest rates have been advancing or declining very rapidly), and the analyst must be alert to such things as stock market "whim" and "fancy."

As a concrete example, from the mid 1920s to the mid 1990s, the equity premium averaged 7–9%. In the late 1990s, it was much higher. Were premia of 15% or more ($\epsilon = 0$) the new order of the day or, indeed, the signal of 25–30% returns to come ($\epsilon > 0$ or even $\epsilon \geq 1$). Some investors appeared to believe so. Since March 2000, however, it appears that $\epsilon < 0$ would have been the better guess. Subsequently, most of us have been hoping that prices reflect $\epsilon = 0$ at the historical average values. Although it could not be timed perfectly, almost any shrewd investor should have realized that dot com stocks that had been in business less than a year, that made no money and barely registered revenues, were not worth ten times more than their original offering price, where many such stocks sold in January of 2000. Many dot coms had a "burn rate" of invested capital equal to many times their monthly revenues, and most of these companies were bankrupt by mid 2002!

4.8 Conclusions

So why did the authors not get rich shorting dot coms? Keynes reminds us that "the market can remain irrational longer than you can remain solvent." These stocks were not worth their *offering (IPO) prices* either. Anybody shorting near the offering price (or even at a multiple of the offering price) would have been sold out of his or her position and lost everything long before the peak was reached. The problem in trying to profit from crazy markets is that one can never know how long the insanity will last or, indeed, if it will worsen. Hence, we sat that one out.

Problems

Problem 4.1.

a. Assuming that the market portfolio has $\mu = .12$ and $\sigma = .15$ and the riskless rate of interest is .06, draw the Capital Market Line.

b. How much expected return would be demanded under this system at a σ level of .20?

c. How much would μ increase if a σ level of .25 were adopted?

d. State, in general terms, the relationship between $\Delta\mu$ and $\Delta\sigma$ in this system.

Problem 4.2. Classify the following securities as "aggressive" or "defensive" if $\sigma_X = .10$:

a. $b_1 = 1.2$.

b. $C_{2X} = .02$

c. $b_3 = .7$

d. $C_{4X} = -.001$.

Problem 4.3.

a. Compute C_{1X} and C_{3X} from Problems 4.2.a and 4.2.c above. What assumptions must be made?

b. If $\mu_X = .10$ and $r = .03$, determine μ_i for the four securities in problem 4.5.

Problem 4.4. From the data in Problems 4.2 and 4.3,

a. Plot the four securities on an SML using b_i.

b. Do the same thing using C_{IX}.

Problem 4.5. Rank the three securities described in Section 4.3 by the Treynor and Jensen criteria, assuming a riskless rate of .05. Explain your results.

Problem 4.6. The shares of the Serita Corporation are expected to pay a dividend of $1 per year forever and possess a degree of systematic risk corresponding to a return of .10 on the SML. At what price should Serita shares sell?

Problem 4.7. Suppose that markets are not in equilibrium and the data in

Problem 4.6 only refer to the expectations of a particular group of investors, while another group of investors expects a dividend of $1.50 per year forever and has a SML indicating a .08 return for Serita shares. What may now be said about the price of Serita if the first group of investors became more wealthy while the second group became poorer, *ceteris paribus*? What do you imagine to be the impact of a new issue of shares by Serita?

Problem 4.8. In his review of Jensen's work on mutual fund evaluation, Professor Fama states ([4], p. 413) "...regardless of how mutual fund returns are measured (i.e., net or gross of loading charges and other expenses), the number of funds with large positive deviations of returns from the market line...is *less than* the number that would be expected by chance with 115 funds under the assumption that fund managements have no special talents in predicting returns." (Emphasis added.) Comment on the implications of the above for the evaluation of the efficient market hypothesis.

References

[1] Baumol, W. (1963). "An expected gain−confidence limit criterion for portfolio selection," *Management Science*, October, 174−182.

[2] Blume, M. (1971). " On the assessment of risk," *Journal of Finance*, March, 1−10.

[3] Chen, A., Jen, F. and Zionts, S. (1971). "The optimal portfolio revision policy." *Journal of Business*, January, 51−61.

[4] Fama, E. F. (1970). "Efficient capital markets: A review of theory and empirical work," *Journal of Finance*, May, 383−413.

[5] ___(1970). "Multi-period consumption−investment decisions," *American Economic Review*, March, 163−174.

[6] Fama, E. F. and French, K. R. (1993). "Common risk factors in the returns on stocks and bonds," *Journal of Financial Economics*, February, 3−56.

[7] ___ (1996). "Multi-factor explanations of asset pricing anomalies," *Journal of Finance*, March, 55−83.

[8] Findlay, M. C. and Williams, E. E. (1986). "Better betas didn't help the boat people," *Journal of Portfolio Management*, Fall, 1986, 4−9.

[9] ___ (2000−2001). "A fresh look at the efficient market hypothesis: How the intellectual history of finance encouraged a real 'fraud-on-the-market," *Journal of Post Keynesian Economics*, Vol. 23, No. 2, 181−199.

[10] Graham, B., Dodd, D. and Cottle, S. (1962). *Security Analysis*, 4th ed. New York: McGraw-Hill.

[11] Hicks, J. R. (1946). *Value and Capital*, 2nd ed., London: Oxford University Press.

[12] Jacob, N. (1971). "The measurement of systematic risk for securities and portfolios: Some empirical results," *Journal of Financial and Quantitative Analysis*, March, 815−834..

[13] Jensen, M. C. (1968). "The performance of mutual funds in the period 1945−1964," *Journal of Finance*, 389−416.

[14] ___(1969). "Risk, the pricing of capital assets, and the evaluation of investment portfolios," *Journal of Business*, April, 167−247.

[15] Keynes, J. M. (1936). *The General Theory of Employment, Interest, and Money*. New York: Harcourt, Brace and World.

[16] Levy, R. A. (1971). "Stationarity of beta coefficients," *Financial Analysts Journal*, November-December, 55−63.

[17] Lintner, J. (1965). "Security prices, risk, and maximal gains from diversification," *Journal of Finance*, December, 587−615.

[18]___ (1965). "The valuation of risk assets and the selection of risky investments in stock portfolios and capital budgets," *Review of Economics and Statistics*, February, 13−37.

[19] Mossin, J. (1969). "Optimal multi-period portfolio policies," *Journal of Business*, April, 215−229.

[20] Roll, R. and Ross, S. A. (1980). "An empirical investigation of the arbitrage pricing theory," *Journal of Finance* , December, 1073−1103.

[21] Ross, S. A. (1976). "The arbitrage theory of capital asset pricing," *Journal of Economic Theory*, December, 341−360.

[22] Roy, A. D. (1952). "Safety first and the holding of assets," *Econometrica*, July, 431−449.

[23] Sharpe, W. (1963). "A simplified model for portfolio analysis," *Management Science*, Jan., 277−293.

[24] ___(1966). "Mutual fund performance," *Journal of Business*, January, 119−138.

[25] ___ (1964). "Capital asset prices: A theory of market equilibrium under conditions of risk," *Journal of Finance*, September, 425−442.

[26] ___ (1972). "Risk, market sensitivity and diversification," *Financial Analysis Journal*, January-February 74–79.

[27] Sharp, W. F., Alexander, G. J., and Bailey, J. V. (1999). *Investments*, 6th edition, Upper Saddle River, NJ.: Prentice-Hall.

[28] Smith, K. (1971). *Portfolio Management*. New York: Holt, Rinehart, and Winston.

[29] Taylor, W. M. and Williams, E. E. (1991). "Market microstructure and Post Keynesian theory," *Journal of Post Keynesian Economics*, Winter, 233–247.

[30] Thompson, J. R. and Williams, E. E. (1999–2000). "A post Keynesian analysis of the Black–Scholes option pricing model," *Journal of Post Keynesian Economics*, Vol. 22, No. 2, 251–267.

[31] Treynor, J. (1965). "How to rate management of investment funds," *Harvard Business Review*, 63–75.

[32] Treynor, J. and Mazuy, K. (1966). "Can mutual funds outguess the market?" *Harvard Business Review*, July-August, 131–136.

[33] Whitmore, G. A. (1970). "Third-Degree stochastic dominance," *American Economic Review*, June, 457–459.

[34] Williams, E. E. and Findlay, M. C. (1974). *Investment Analysis*, Englewood Cliffs, NJ.: Prentice-Hall.

Chapter 5

Equilibrium Implying Efficiency: The Neoclassical Fantasy

5.1 Introduction

Chapter 1 discussed some of the basic principles of stock market efficiency. Some of the underpinnings and implications were outlined in Chapters 2 through 4. In this chapter, we will review the evidence that has been accumulated by scholars over the past forty or so years. We shall see that the data supporting the efficient market hypothesis (EMH) were never very strong, and that some of the arguments advanced early on by academic researchers were perhaps more a manifestation of presumptions rather than something normal people might view as proof. We shall conclude that the chain of logic supporting the efficient market position was always based on presumption, and this will validate our implicit assumption in later chapters dealing with market inefficiency. As we develop our arguments, we shall also take time to review some of the intellectual history of financial economics. [1]

5.2 A Formal Restatement of the Hypothesis

Initially, we need to have some common understanding of just what "efficiency" in efficient markets means. As we saw in Chapter 1, all perfect markets are efficient, but a market need not be perfect to be efficient. The usual characteristics of perfect markets (e.g., many buyers and sellers no one of whom can influence price; perfect, costless information available to

---

[1]This chapter borrows heavily from [17] and [18].

everyone; an absence of entry barriers; the existence of an homogeneous product or service; and an absence of taxes and transactions costs) are not found in any market. The issue therefore reduces to whether (and to what extent) efficiency can be observed in actual (imperfect) markets, and the ultimate test of applicability must be empirical in nature and application.

The early formulations of the EMH maintained that market prices fully reflected all available information (i.e., that the market processes information efficiently). An implication of this formulation of the theory was that new information (i.e., "news") would be rapidly and unbiasedly impounded into price. This implied that markets would neither overshoot nor undershoot the "correct" market price. Thus, by the time one heard of "news," it would be too late to profit from trading on it. As we observe in an earlier work ([17], p. 182]):

> For the last 30 years or so, the EMH has been further subdivided (based upon the set of information assumed to be reflected in the price) into *weak, semi-strong*, and *strong-form* efficiency.
>
> With weak-form efficiency, the information set is limited to past share price (and, perhaps, volume) data. The implication is that any information about future price movements contained in past price movements is already reflected in the current price. Technical analysis of the stock market [such as that provided by many brokerage houses and charting services], hence, is a waste of time.
>
> Under semi-strong efficiency, all publicly available information is impounded in stock prices. In this case, fundamental analysis (such as that purportedly undertaken by securities analysts) is useless. From a philosophical perspective, at least regarding public information, any search for a second value (e.g., fundamental value, true value, just price, etc.) to compare with market price in order to determine whether a stock is "cheap" or "dear" is wrongheaded. Either there are no cheap or dear stocks or else they cannot be discerned from public information.
>
> Under strong-form efficiency, all information is considered including private information. If this condition held, even insiders could not trade at a profit. Few economists (even those of staunch neoclassical persuasion) accept this version as being descriptive of any market.

As a result of the above taxonomy, most writers now adopt efficiency at the semi-strong level as the only significant meaning of the term "efficient markets," and we shall follow that adoption in this book.

5.3   Who Are the EMH Investors?

A basic issue in the development of the EMH was how much one could reasonably assume about the integrity of the stock pricing mechanism without making independent investigation or analysis. As we noted in another work ([17], p. 183):

> Clearly, an investor could "free ride" on the market's analysis by conducting no independent analysis and simply take the price being offered. In a semi-strong efficient market, this would be reasonable behavior for an investor who possessed no private information. An investor who placed a constant periodic amount (e.g., no belief in an ability to "time" the market) into the same broadly diversified portfolio like an index fund (e.g., no belief in an ability to pick stocks or managers) would be acting in such a fashion.

Now casual empirical observation reveals that few investors have ever actually acted in this manner. More importantly, those who have so acted have clearly not been the marginal investors setting prices in the market. Further casual empirical observation suggests that ([17], p. 183):

> Most investors act as though they think they can pick "cheap" stocks or that they can pick funds that can pick "cheap" stocks. A fair number act as though they think they can time the market (e.g., switch from stock funds to bond or money market funds). Very few people invest as though they think the price is always "right". Yet this is the implication of the EMH. Hence, the first problem with the presumption that all investors rely on the integrity of the pricing mechanism is that the vast majority demonstrably do not.

Many economists, such as Nobel Prize Winner Milton Friedman, have advocated market solutions as being generally superior to other available approaches. Professor Friedman never argued that market solutions are perfect (because, among other reasons, perfection is generally undefined and invariably unobtainable). Other economists, such as Keynes (see Chapter 1) have pointed out that excesses can occur on the stock exchanges. Below we describe how an application of Friedman's methodological constructs have been applied to the securities markets to "assume away" some of the very legitimate concerns raised by Keynes nearly 70 years ago.

5.4   Some Early History

From the perspective of the twenty-first century, it is possible to identify intellectual developments in economics which would ultimately be incorporated into financial theory and the evolution of the EMH. In his book,

Essays in Positive Economics published in 1953, Friedman [19] introduced at least two concepts that would later appear in the EMH. The first of these was that of the "rational speculator." A "rational speculator" was a person who bought when prices were too low and sold when they were too high. A policy of "buy low/sell high" should by definition lead to profits. Profit-seeking behavior of this kind should, in turn, cause prices to converge at "true" or "fundamental" value. Friedman originally made this argument in the development of a theory of foreign exchange markets to promote his contention that fixed exchange rates should be abandoned in favor of floating rates. Opponents to Friedman's view have argued that speculation might destabilize a floating rate system. He argued in return that destabilizing speculation would be unprofitable and traders following such a strategy would exit the market. Thus, only rational (i.e., profitable) speculation would survive in the market and this sort of speculation would be stabilizing. The EMH theorists applied this notion to the stock market by assumption. It was assumed that all investors (or, in some versions, at least the marginal investors who set the stock price) are "rational speculators" such that if the price of a stock were "too high" ("too low"), everyone would be a seller (buyer) and the price would immediately revert to fair value. The Keynesian notion of an investor buying (selling) when the price was "too high" ("too low") in the expectation of selling to (buying from) an even "Bigger Fool" at a higher (lower) price was eliminated from market behavior by assumption.[2] This was a key step in the EMH's argument that momentum investors do not exist.

Friedman was also a philiosophical positivist. He argued that, since any theory is an abstraction of reality, models should not be judged by the descriptive accuracy of their assumptions but rather by their predictive accuracy. This position was roundly criticized at the time by many economists as being incomplete (e.g., one's confidence in the continued predictive power of a model might depend upon how well its assumptions comported with reality). In subsequent years, however, positivism was used as an intellectual "weapon" to justify ignoring any criticism of the assumptions of a model. The result has been the development of models with assumptions which were often ridiculously absurd and frequently in contradiction to clearly observed facts! Examples of these models are presented later in this chapter.

As we observed in Chapter 3, Harry Markowitz ([38] and [39]) developed the mathematics of portfolio selection in the 1950s. As the originator of this analysis, he obviously had to make some choices with respect to definitions and assumptions. Many of these were employed by those who followed and have remained unchanged for 50 years (and unexamined for probably 40 years). He assumed that security returns could be depicted by a normal distribution. A normal distribution can be fully described by its mean (as

---

[2]Keynes noted that it made no sense to buy "cheap" stocks if one thought they would go even lower in price or sell "dear" stocks if one expected even higher prices.

a measure of central tendency) and variance (as a measure of dispersion) or standard deviation (square root of variance). It was further assumed that the mean (as a measure of expected return) was good but that variance (as a measure of risk) was bad, such that the investment problem involved a trade-off of the one against the other. This was basically the point at which variance of return as *the* measure of risk entered the finance literature.

Markowitz's key insight was that return aggregated but risk did not. In other words, the expected return (mean value) of a portfolio will be a (dollar) weighted average of the returns (means) of the separate individual securities. The risk (variance) of the portfolio will only be such an average in the limiting case where all of the securities are perfectly correlated with each other. In all other cases, it will be less. Thus it could be said that Markowitz provided a mathematical proof of the benefits of diversification. If the returns on separate securities are uncorrelated with each other, a large number of small investments will produce a near riskless portfolio of investments even though the investments are individually risky — the idea being that some will be up when others are down. Markowitz then defined another concept of efficiency as that portfolio which had the highest return (mean) for a given level of risk (which he took to be variance) or, equivalently, the lowest risk for a given return.[3] The set of efficient portfolios was called the efficient frontier, and it was posited that all rational risk adverse investors would choose a portfolio from that set.

Some years later, other writers (e.g., Sharpe [49], Lintner [32], and Mossin [42]) carried this observation one step further as discussed in Chapter 4. They assumed all investors to be essentially the same. That is, all investors were assumed to possess the same information and be able to borrow or lend at the same riskless rate. (Critics of the realism of these assumptions received a citation to Friedman's positivism). It would then follow that everyone would wish to hold the same portfolio on the efficient frontier. They then demonstrated that, in equilibrium, the only portfolio everyone could hold would be a microcosm of the market. This sort of portfolio was known as the *market portfolio* and was often proxied by a broad index like the S&P 500.

The Markowitz methodology for computing the efficient frontier for a realistic number of securities was infeasible at the time he wrote. He suggested some simplifying assumptions. One involved assuming that securities were only correlated with each other through their correlation with a common index. The later writers also employed this notion.

With everybody assumed to hold the market portfolio, the relevant risk of a given security is its contribution to the risk of the market portfolio (i.e., that risk which is not eliminated by diversification). This is measured as the ratio of its risk (security's standard deviation) to the risk of the market portfolio (standard deviation of some proxy index), all multiplied by the

---

[3] In the opinion of the authors of this book, the time indexed cumulative distribution function is a far more appropriate measure of risk.

correlation between the two. The last is also called the security's beta coefficient and the whole theory has been called the Capital Asset Pricing Model (CAPM). The implication of the CAPM is that, in equilibrium, the expected return on any security is the risk-free rate plus the product of its beta coefficient times the difference between the expected return on the market and the risk-free rate. The CAPM is important to our discussion because it was originally identified as the equilibrium of which the EMH was the manifestation. Hence, among other implications, it is often used as a benchmark in tests of the EMH.

5.5   Science and the "Social Sciences"

The rise of the "social sciences" as academic disciplines played an important role in the intellectual history of the EMH. The relationship of such areas of study as sociology, anthropology, psychology, political science, and economics with the "hard" sciences (physics, chemistry, etc.) at most universities was often difficult and occasionally adversarial in the past century. Originally, the social "sciences" were deemed to be rather soft subjects which offered taxonomy rather than "proof" or "hypothesis testing" as their principal means of inquiry. In order to become more academically acceptable, the social sciences adopted the paradigm of the natural sciences as they began to gather "empirical" data and apply statistical methods to analyze these data. As the softest of the social sciences, business was looked upon with disdain by other disciplines which often had ill-concealed doubts about the intellectual merit of business study and whether it really belonged on a university campus. Comparisons with business colleges (i.e., proprietary operations that taught bookkeeping, typing, etc.) were often made. It was perhaps not by chance that when Harvard established its business school, it located it across the Charles River from the rest of the campus.

Although they liked the money that the business schools brought in, most universities still regarded business education as being intellectually bankrupt in the 1950s; and these views were not improved by the findings of the Ford and Carnegie Foundations which funded simultaneous studies of business higher education in the latter 1950s. Known by the names of their authors as the Gordon and Howell Report and the Pearson Report, these studies tended to confirm the worst fears that many business schools were trade schools at best (e.g., offering courses in "Hotel Front Desk Management" and the like) and lacked the analytical rigor to be granting university degrees. Clearly, something had to be done. Equally clearly, the business schools needed the prestige of the university affiliation and the universities needed the money produced by the "B" schools. This was, of course, also the era of the space race and the general glorification of science by Congress and the public at large. Research money was available from sources like the National Science Foundation for science. Many of the basic disciplines which business schools applied (e.g., economics, psychology, etc.) had already declared themselves "sciences" and as mentioned above were beginning to apply the "scientific method" to their respective areas of inquiry. The answer to the "B" school dilemma became apparent: trade your peddler's cart for a smock! The business school disciplines thus became social sciences. Finance became financial economics. Personnel management became behavioral science, etc. Being based upon a tautology (i.e., assets ≡ liabilities + net worth), the transformation was not easy for accounting (imagine arithmetic as an "experimental science"), but academic accountants borrowed heavily from finance and economics to give accounting the aura of "science." Of course, it was never demonstrated anywhere that the scientific paradigm was appropriate for these disciplines; it was simply and conveniently adopted.

Beginning in the 1950s, however, and expanding greatly later, a subtle change took place. Encyclopedic business texts such as Guthmann and Dougall's *Corporate Financial Policy* [22] and Hunt, Williams, and Donaldson's *Basic Business Finance: Text and Cases* [26] were being supplemented by more "rigorous" (read: mathematical) treatments, and journals such as *Management Science* were started. Even the *Journal of Finance*, the eminent academic journal in finance, began to publish an occasional article that was not completely descriptive (e.g., articles with very simple equations and occasional diagrams).[4] By the late 1950s, the economic journals (*American Economic Review, Journal of Political Economy*, etc.) were publishing an increasing number of "scientific" (read: containing equations and/or empirical data) articles. In 1958, Professors Modigliani and Miller (MM) published their now famous paper [41] which demonstrated that, absent taxes, the value of an income stream should not depend upon the number or types of claims upon it (e.g., the debt/equity ratio does not affect the value of the firm). The notion itself was hardly unique. Two decades earlier, J. B. Williams [59] had introduced it as the "Law of Conservation of Investment Value," discussed its implications and discarded it in a paragraph. What was claimed to be unique by MM was that they had demonstrated an equilibrium result by an arbitrage proof (i.e., they were being *scientific*).

For a long time before MM, arbitrage meant the simultaneous purchase and sale of the same commodity at different prices so as to earn an instantaneous profit on no investment at no risk. Without the passage of time, the making of investment, or the incursion of risk, arbitrage could be conducted on a large scale such that a single individual's trading could eliminate a price discrepancy. The principal feature of arbitrage which made it riskless was that the commodity bought constituted good delivery against the simultaneous sale.

---

[4]It is difficult to describe how totally out of place the 1952 Markowitz article appeared relative to the others in that issue of the *Journal of Finance*.

MM's arbitrage proof assumed two identical firms, one of which borrowed and one which did not. Ignoring taxes and assuming individuals and firms borrowed at the same rate, they analyzed the return from buying shares of the levered firm versus borrowing personally to buy shares of the debt free firm. They maintained that, unless the total values of the two firms were the same, an arbitrage profit could be earned. In equilibrium, no arbitrage profits should exist. Hence, they believed they "proved" that corporate debt policy did not affect the value of the firm.

The MM position was debated for years, but several institutional factors were clear. Even assuming two identical firms, delivery of shares of the debt free firm subject to a margin loan would not cover a short sale of shares of the levered firm (and vice versa). Instead, the definition of "arbitrage" has been extended to mean simultaneous long and short positions in statistically "similar" investments which are expected to generate differential returns over time. Now notice that positive investment and the passage of time have been added back to the picture, and this has changed the nature of true arbitrage (which the MM mechanism really was not). If the correlation between the positions is less than perfect, we have also added back risk. Thus, MM began the process of gutting meaning from arbitrage, so that now Wall Street discusses "risk arbitrage," a concept which would have been an oxymoron 40 years ago.

The positive investment "problem" with "arbitrage" proofs of this sort has, over the years, generally been met by another assumption: a perfect short sale mechanism. The latter assumes that proceeds from the short sale are available for investment by the seller. Hence, any security expected to under-perform (even if the expected return were positive) could be profitably sold short when the proceeds were placed in a security of average performance. In real life, of course, many market participants (e.g., mutual funds, pension funds) are prohibited from short selling. As to the others:

- The proceeds of sale are held by the broker as margin. This means that the stock price must decline (not merely not rise fast enough) for a profit to arise.[5] Additionally, most brokers do not permit retail clients to earn interest on short sale proceeds. If one has $100,000 in margin debt and is short $60,000 in stock, interest is charged on the entire $100,000 balance.

- The short seller is responsible for dividends. This means that the price decline must be greater than the yield.

- Additional margin must be posted and only the large players are

---

[5] If the shares are borrowed from a second broker, the proceeds are held by the latter who, in the normal case, will pay a modest rate of interest to the first broker (suggesting that the larger players may be able to earn some low rate on this margin). The problem, according to recent research, is that the rate declines and becomes negative (e.g., cases of 40% have been reported) for shares in limited supply. This seems to be the growing alternative to the unavailability of shares to borrow discussed in the fourth bullet point.

allowed to post interest-bearing securities. For everyone else, the price decline must provide a return on this margin as well before a profit is realized.

- The short-seller may be forced to cover his position before he would wish if the broker can no longer borrow the stock.

- To the extent the link with the corresponding long position is less than perfect, the specter of the unlimited liability of the short position looms.

In a world of the perfect short sale mechanism, any market participant can play the role of rational speculator with respect to an overpriced stock by selling it short. Given the risks and costs of real world short sales, a stock would need to be fairly seriously overpriced to attract short sellers. Hence the first line of defense becomes sales by current shareholders. In a world where people differ, these would tend to be those with the more optimistic views of the security in the first place. Hence, while these "arbitrage" arguments do not require both the perfect short sale and "everybody the same" (often called homogeneity) assumptions, they do tend to require one of them.

5.6   Risk versus Uncertainty

Historically, at least prior to the 1950s, a certain world was deemed to be one in which all future events had a single outcome and both the events and their outcome were known (i.e., probability $= 1$). A world of risk was one in which future events could have multiple outcomes but the outcomes and their probabilities were known (such that the probabilities of the outcomes of an event summed to 1). An uncertain world was one in which neither events nor outcomes were known and assignment of probabilities was a meaningless exercise. Both Frank Knight [31] and Keynes [30] wrote on these issues and the matter would have appeared settled by the 1950s. It would also seem unexceptional to state that we live in an uncertain world, parts of which, at our peril, we may attempt to model as (near) certain or risky.

Then, beginning in the 1950s, and expanding greatly later, a subtle change in the research paradigm of business took place. As we saw in Chapter 3, Markowitz ([38] and [39]) made the simplifying assumption that stock returns were normally distributed and, hence, could be fully specified by their mean and variance and that the correlation between the returns of different stocks could somehow be estimated. The identical firms in the MM arbitrage proof could then (especially by later writers) be described as having return distributions which had the same mean and variance and were perfectly correlated with each other. Hence, a long position in the shares of one firm would be a perfect hedge against a short position in the other, but only in a world of risk. In a truly uncertain world, one could have no confidence about the future co-movement of the prices of shares. As we pointed out elsewhere ([17], pp. 186−187):

> With *arbitrage* defined as simultaneous long and short hold-ings of "equivalent" positions, "equivalent" increasingly came to be viewed in statistical terms (mean, variance and correlation). If historical data existed, they were often employed (as in stocks' beta co-efficients). If no data existed, subjective estimates of re-turn distributions and correlations (e.g., as in Bayesian statis-tics) were made. Any future event was assumed to be subject to a probability distribution that, in turn, was assumed amenable to estimation. More complex models presented distributions of distributions. In any event, the real world was increasingly assumed to be one of risk where the forces of ignorance and darkness were overcome by the assumption of probability dis-tributions.

Language has also been debased in the "looser" use of the word "arbitrage." For many years now in the academic literature this world of probability distributions has been called "uncertainty." What would have been called "risk" fifty years ago is now called "uncertainty." What Knight and Keynes called "uncertainty" has vanished from the lexicon of financial economists. Long Term Capital Management provides an interesting example. When, after only a couple of years of operation, they lost almost all their equity in August 1998, they blamed a "six sigma" event. This would be six standard deviations from the mean of a (presumably) normal distribution, an event with a probability of occurrence of about one in a billion. Thus the LTCM managers claim they made the "right" decisions but were simply "unlucky"!

As we observed above, equilibrium notions began appearing in the fi-nance literature about this time. A system is deemed to be in equilibrium when it has no further tendency to change. Prices are in equilibrium when nobody has a further desire to trade but rather is content to hold what one has at existing prices. The MM proof consisted of showing that, if two firms had different values, further profitable incentives to trade would exist; only when the two were equal would there be no such incentives and hence, an equilibrium result. The EMH relied on similar arguments about exhaust-ing profitable trades on information; that is, no further trading incentive (equilibrium) only occurs when all information is reflected in price.

Thus, consistent with our analysis above, it is clear that the wide-spread use of equilibrium structures in academic finance was part of the shift to the paradigm of science. All sorts of "no profitable trading" structures have been modeled and the resulting pricing relations have been mathe-matically derived. Empirical testing (generally employing regression in the early years and more advanced econometric techniques later) has been done to see if these conditions held (or more precisely, could not be shown not to hold). Numerous opportunities for research grants, reduced teaching loads, publication, and tenure have thus been provided.[6]

As we pointed out elsewhere ([17], p. 187–188):

> In many economic applications, equilibrium is a hypothetical position toward which systems move over time. For example, if analysis suggested that a given market could consume the output of five firms in equilibrium and there were at present ten, we might reasonably forecast an exit of five firms from the market over time. If we were one of the ten, we might use this data to plan our strategy. The usefulness of the paradigm arises in providing a forecast time path or target.

> Unlike real economic variables, stock prices can adjust instantaneously and without cost. Hence there is no time path. To assume that equilibrium structures apply to the stock market is simply to assume that the stock market is always in equilibrium - that is, we are always at the target.

It should be remembered that the EMH is a positivist, empirical hypothesis that prices impound information; it really does not try or even purport to explain how this happens. If one asks for a normative model, EMH advocates have usually told stories of this sort: *The market acts as though it takes all participants' probability distributions of future cash flows, aggregates them somehow, and discounts for time and risk to arrive at current price. With no new information, an investor at a given price would expect to earn the appropriate rate over time. The arrival of new information (which by definition would not be anticipated and, hence, could be viewed as a random event) would cause an immediate recomputation of the above and the setting of a new price.* (Paraphrasing [17], p. 188).

If the above description were, in fact, true, it might well be reasonable to rely on the integrity of the pricing mechanism. It should be remembered, however, that this is not a scientifically testable hypothesis. It is, rather, the totally unsubstantiated normative "story" often told along with the EMH where the latter is merely an assertion of truth. To see the difference, imagine how one might even observe the market aggregating everyone's probability distributions (assuming that these exist) of the future. Imagine further how one might test whether or not the market did so correctly. It stretches the imagination, does it not?

A further argument in favor of the reasonableness of the equilibrium concept is often provided by EMH advocates by use of the device of "temporary" equilibrium. Markets are deemed to be in a "temporary equilibrium" if they are clearing at a price. Of course, this is merely an assertion

---

[6]Given the numerous incentives within academia to play this game, the fact that it is still played (e.g., numerous research projects, journal articles, etc.) should provide no evidence that it has any relevance to the real world.

that markets are serving their basic function of finding a price that equates supply and demand at that instant. It says nothing whatsoever about what information is reflected in price, nor does it provide any clue as to where the next temporary equilibrium may be. Thus, there is no normative story here. Moreover, this process is consistent with any price path, including gross violations of the EMH. Hence, the "temporary" equilibrium approach saves the concept by destroying its usefulness.

5.7   The 1960s and 1970s

In 1961, MM [40] returned to show that, absent taxes, dividend policy did not matter either. Using similar proofs and constructs, they demonstrated that the ex-dividend price decline would offset the value of the dividend (holding investment policy constant). In their demonstration, MM introduced a new concept to the finance literature, *symmetric market rationality* (SMR). This assumes not only that all market participants are Friedman's rational speculator but further that everyone knows or believes this (i.e., everyone in the stock market assumes everyone else is rational). In sum, not only do Bigger Fools not exist but also nobody believes they do. Hence, nobody would buy when the price is "too high" or on momentum and bubbles cannot exist. As MM themselves granted ([35] p. 428):

> Symmetric market rationality cannot be deduced from individual rational behavior in the usual sense since that sense does not imply imputing rationality to others. It may, in fact, imply a choice behavior inconsistent with imputed rationality unless the individual actually believes the market to be symmetrically rational. For if an ordinarily rational investor had good reason to believe that other investors would not behave rationally, then it might well be rational for him to adopt a strategy he would otherwise have rejected as irrational.

In one form or another, this assumption has persisted in models of the capital markets for more than 40 years. In recent years, it is often embedded in "transversality conditions" (which basically do not admit behavior that would cause the models to blowup). If the choice is between preserving the models or depicting reality, the prevalent philosophy will protect the models. There is actually a sick joke here. The discipline of finance used to teach people how to do things, for example, run banks, trade stocks, buy and finance companies, etc. The discipline of economics has been more passive, involving observation, measurement, analysis, and prediction. The new discipline of financial economics appears to involve observing finance practitioners doing things while explaining to them that they are wasting their time!

Despite the increasing irrelevance of academic finance, in the early 1960s Merrill Lynch donated millions of dollars to the University of Chicago to set up the Center for Research in Security Prices (CRSP). The money was used to gather historical price data extending back to the mid-1920s and to maintain current files of such data. The availability of what became known as the "CRSP" tapes obviously invited analysis and resulted in a huge volume of semi-strong EMH dissertations and articles. Among these articles was a publication by Chicago Professor Eugene Fama in 1970 [9]. This review piece has become one of the most significant historical documents in the development of the EMH, and the very name of the EMH and the various forms of it were formulated in this work. Fama re-characterized the prior random walk literature as weak form testing. He reviewed the much newer (in 1970) semi-strong form testing (much of which was going on at Chicago) and outlined what little evidence existed at the time on the strong form. It is interesting to read some of the behind-the-scenes commentary on just how that research was conducted. Niederhoffer, cited in a recent volume by Lo and MacKinlay [34], revealingly tells us that (p. 15):

This theory and the attitude of its adherents found classic expression in one incident I personally observed that deserves memorialization. A team of four of the most respected graduate students in finance had joined forces with two professors, now considered venerable enough to have won or to have been considered for a Nobel prize, but at that time feisty as Hades and insecure as a kid on his first date. This elite group was studying the possible impact of volume on stock price movements, a subject I had researched. As I was coming down the steps from the library on the third floor of Haskell Hall, the main business building, I could see this Group of Six gathered together on a stairway landing, examining some computer output. Their voices wafted up to me, echoing off the stone walls of the building. One of the students was pointing to some output while querying the professors, "Well, what if we really do find something? We'll be up the creek. It won't be consistent with the random walk model." The younger professor replied, "Don't worry, we'll cross that bridge in the unlikely event we come to it."

I could hardly believe my ears. Here were six scientists openly hoping to find no departures from ignorance. I couldn't hold my tongue, and blurted out, "I sure am glad you are all keeping an open mind about your research." I could hardly refrain from grinning as I walked past them. I heard muttered imprecations in response.

Niederhoffer went on to manage a $500 million trading fund which collapsed when he was forced to liquidate to meet margin calls on his short position in S&P future puts (see discussion in Chapter 11 below). Some practitioners now refer to his experience as "pulling a Niederhoffer." He summarizes his thoughts regarding the more recent research in academic finance on market anomalies (see discussion later in this chapter) as follow ([43], p. 270):

> All the anomalies that have appeared in the published literature that, in turn, has formed the basis of the professors' reformulations are mere ephemera designed to lure the unwary followers of professorial authority into ruin.

5.8   The Weak Form of the EMH

We have briefly digressed into the politics of religion above, but it is important to understand not only the chronology of "thought" regarding the EMH but also how and why that thought developed. At this point, we might briefly summarize several interesting elements about weak-form EMH before proceeding to semi-strong form in more detail.

To begin with, all of this appears to have started with a very modest paper by Harry Roberts [46] who was at the time a statistician at the University of Chicago. He assumed stock price changes conformed to a normal distribution, made random selections from such a distribution, added the results to an arbitrary starting price, and noted that the result looked a lot like the Dow Jones Industrial Index over time. From this he cautioned readers that the analysis of stock price "patterns" might be nothing more than the analysis of noise. It should be noted that the statistical process of making independent random drawings from a normal distribution is called a random walk. Let us consider

$$\Delta Y = \frac{\Delta S}{S} = \mu \Delta t + \sigma \epsilon \sqrt{\Delta t}, \qquad (5.1)$$

where ϵ is a normal variate with mean zero and variance 1. Now suppose we somehow "backed out" the growth part $\mu \Delta t$, say, via using something like

$$\Delta Z = \Delta Y - \mu \Delta t. \qquad (5.2)$$

Then, we would have something like

$$\Delta Z = \sigma \epsilon \sqrt{\Delta t}. \qquad (5.3)$$

And this would give us a process which, as time evolved, would get very far from $Z(0)$, but would be equally likely to be at any point below as above it. That is to say,

$$E(Z(t)) = E(Z(0)). \qquad (5.4)$$

This notion of the expected value of a variable being unchanging in time defines, roughly, a *martingale*;[7] and it forms the mathematical formalism

---

[7]A *martingale* is a stochastic process $S(t)$ which has the property that $S(t+\tau|\text{past}) = S(t)$ for all positive τ. The process is a *submartingale* if $S(t+\tau|\text{past}) \geq S(t)$.

behind the "weak form" of the Efficient Market Hypothesis. Suppose that the market is really efficient. We know that technological advances of many various sorts cause the market to grow over time. Incrementally, if all buyers and sellers of stocks have full information and identical utilities for each evaluation of the stock at each future time, at each given instant the market price of a stock would capture the present value of the stock, which would, of course, include the entire profile of future prices of the stock. In such an event, we might attempt to model the price of a stock via (5.3). In such a situation, one might well be tempted to invest in a portfolio of a large number of stocks selected across the market sector in which the investor has the desire to invest. There would be no advantage available via forecasting the value a stock might achieve in, say, one year from today. All such information would have been captured in the present market price of the stock. But then, the present market price must also capture information about the value of the stock six months from now. And, somehow, the time based utilities (assumed to be equal) of all players in the market would have to be included in the price for all time horizons. Or, we could assume that transaction costs were zero and that utilities were replaced simply by dollar costs. That would enable us simply to agree, in the aggregate of market players, on the value of the stock at the next instant of time. Even under these circumstances, there would still be the problem of risk profiling. Expectations are not enough. The wise investor should ask what might be, for example, the 10% lower percentile for which his portfolio will be worth at various slices in future time.

The early empirical work, following Roberts, was produced by the "random walk" hypothesis researchers primarily to attack "technical analysis" (employed by those who believe there are patterns in stock price movements that allow one to forecast future stock prices). This line of reasoning actually goes back to Louis Bachalier's 1900 doctoral dissertation, *Theórie de la Speculation* and was followed up with Holbrook Working's research on commodity prices in the 1920s. Technical analysts have for many years contended that by analyzing only the past price movements of a security, it is possible (or, "they are able" − depending upon the size of the ego of the technical analyst involved) to predict future price movements and thus make greater-than-normal profits. This is achieved by a variety of black boxes, Gestalts, and, on occasion, reasonable time series forecasting. One of the early computer models, that developed in the Wharton School of Business, was supposed to be so good that the institutional investors using it agreed with the federal agencies not to exceed certain purchasing limits. (The feds need not have worried.) The early advocates of the geometric Brownian model for stock price $S(t)$ in (5.1) questioned whether such strategies could work. Cootner ([5], p. 232) put the argument this way:

> If any substantial group of buyers thought prices were too low, their buying power would force up the price. The reverse

would be true for the sellers. Except for appreciation due to earnings retention, the conditional expectation of tomorrow's price, given today's price, is today's price. In such a world, the only price changes that would occur are those that result from new information. Since there is no reason to expect that information to be nonrandom in appearance, the period-to-period price changes of the stock should be random movements, statistically independent of one another.

The early advocates of modeling a stock's progression in time via a geometric Brownian process then concerned themselves with demonstrating that successive price changes were statistically independent of each other, that various mechanical trading rules based upon price changes did not yield profits statistically superior to a simple "buy-and hold" strategy, and that "price changes from transaction to transaction are independent, identically distributed random variables with finite variances" (implying, by the Central-Limit Theorem, that for numerous transactions price changes will be normally distributed). Subsequent writers refined certain parts of the basic random-walk arguments. In the first place, it was suggested that price changes do not follow a true random walk (with an expected value of zero), but rather a submartingale (with an expected value of greater than zero). Thus, the theory can take long-run price trends into account and accept a very modest amount of serial correlation in successive price changes.[8]

The challenge to the technicians became to demonstrate that their rules could earn greater-than-normal profits. It was also contended that price changes are not normally distributed about an expected trend but rather belong to a broad family of stable Paretian distributions of which the normal is only a limiting case. The implication would be that the "fat-tailed" stable distributions may have an infinite variance, such that the usual portfolio approach employing mean and variance cannot be used if prices (and, thus, returns) are so distributed. Fama [6] demonstrated, however, that a similar form of analysis, employing a different dispersion parameter, could be employed. With a bit of work, the stochastic modeling argument could be sustained even if the normal distribution were not the driver of the process. Indeed, the attack on underlying normality was ill placed. It is true that according to the geometric Brownian model, changes of a stock in the range of 5% could occur with miniscule probability. In fact, changes of such magnitude occur frequently. But this is a relatively minor matter. There are other major weaknesses in the plain vanilla geometric Brownian model,

---

[8]There were strong mathematical arguments to indicate that geometric Brownian motion is a natural model for a security. Later, we consider the prices of a stock at times t_1, $t_1 + t_2$, and $t_1 + t_2 + t_3$. Then if we assume $S(t_1 + t_2)/S(t_1)$ to be independent of starting time t_1, and if we assume $S(t_1 + t_2)/S(t_1)$ to be independent of $S(t_1 + t_2 + t_3)/S(t_1 + t_2)$, and if we assume the variance of the stock price is finite for finite time, and if we assume that the price of the stock cannot drop to zero, then, it can be shown that $S(t)$ must follow geometric Brownian motion and have the lognormal distribution indicated.

which we discuss at greater length in Chapter 9. If stocks really did behave like geometric Brownian motion, even if μ and σ had some drift over time, the world of investing would be a kinder gentler place than it is. In such a situation, portfolios really would begin to have the safety of a bank account. The fact is that it is rapid bear changes in the market which must be included as patches if the model is to be at all realistic. Such changes can be included using simulation based approaches.

Interestingly, by the time the geometric Brownian random walk hypothesis appeared as the "weak form" of the EMH, the game was basically over as far as financial researchers were concerned. After 1970, most academics had convinced themselves that there were no exploitable patterns in stock price time series, and little further work was published. Nevertheless, technical analysis has continued to thrive on Wall Street and elsewhere. In recent years, certain academic-types (i.e., "rocket scientists") have been hired on Wall Street and elsewhere to perform a more sophisticated type of technical analysis. Using large computers and very complex concepts (e.g., fractiles, neural networks, chaos theory) they are back seeking exploitable price patterns. It is not clear that the pessimism of the EMH school concerning the inability of enlightened technicians to beat random strategies is justified. But it is probably the case that we need to do better than black box modeling. And now we have the computing power to do better in the aggregation from the micro (where modeling is easier) to the aggregate (where modeling is not very good).

5.9   The Semi-Strong Form of the EMH

We now turn our attention to semi-strong market efficiency. Most semi-strong EMH studies fall into one of two types. The first examine what happens to stock prices when an event occurs. The second consider whether stocks of particular (publicly known) characteristics "beat the market" over time. The latter clearly require a market benchmark. The early examples of the former, done with monthly data, had some benchmark problems; the more recent examples have used daily data so that benchmarks do not matter much. Events studies go back a long way. MacKinlay [37] tells us that perhaps the first study was performed by Professor James Dolley of the University of Texas in 1933 [7]. His work involved taking a sample of 95 stock splits from 1921 to 1931, and seeing what happened to the post split price.

Over the years, studies have become more sophisticated, but all of them have certain common elements. We describe these elements elsewhere ([17], p. 191) as follows:

> An event study begins by picking an "event" (usually an announcement, so that it can be dated) common to enough firms to get a reasonably sized sample (e.g., firms which announced a

dividend increase). The study then takes all recent or otherwise usable examples of the event (which tends to make the sample become the universe) and assembles them in "event time" (the announcement in each case becomes Day 0, the day before becomes Day−1, etc.). The study then takes the frequency distribution of stock returns on Day 0 (or Day 0 and +1) and determines whether the mean of the return distribution is statistically significantly different from zero. If, for example, the event is "good news" and the mean is positive at 5 percent or 1 percent significance levels, the test is said to support the EMH. Tests of this sort are generally regarded to provide the most unambiguous support for the EMH.

Now we must offer a few caveats before we can jump to the conclusion that the aforementioned tests actually "prove" the validity of the EMH. These caveats are also explored in ([17], pp. 191−193).

1. All these tests measure is that "good news" ("bad news") events, on average, are associated with price up-ticks (down-ticks). No study has addressed the issue, on average (let alone in each case), whether the magnitude of the price movement bore any rational relation to the news (a minimum necessary, but not sufficient, condition for the assertion that the price both before and after the news reflected all public information as required by the EMH). Hence, while a failure of such a test would reject the EMH, a non-failure provides little support.

2. What is a "good news" event beyond something to which the market reacts favorably? Hence, what is one testing beyond whether the market is consistent? Some might say that "good news" could be determined by theory. Consider dividend increases: A). Under the old "bird in the hand" view endorsed by financial theorists in the 1950s and earlier, this is good news. B). Under the MM view of the 1960s, it is no news. C). Under more recent theory that a dividend increase signals fewer lucrative investment opportunities for the corporation, it is bad news. In fact, most studies have found that the market tends to view dividend increases as "good news." Since the "bird in the hand" theory was (is) not popular with most believers in efficient markets, they came up with a story that dividend increases are "good news" not because dividends per se were of value but rather because they were signals of higher sustainable levels of earnings. Based upon this last story, dividend increases being "good news" events were deemed consistent with the EMH. Circumstances when the specific price reaction was negative or insignificant were still deemed consistent with the EMH by its advocates as examples of (C) and (B) above. Hence, any price reaction would be deemed consistent with, and supportive of, the EMH.

3. All the tests have been conducted with a strong prior belief in the EMH. As a consequence, contrary results have often been explained away or ignored. The example of dividends cited above is just such as case. But consider earnings: Originally, it was thought that earnings increases would be sufficient to be "good news." However, this did not always produce the "right" results (i.e., significantly positive announcement stock returns). The next step was to define "good news" only as earnings above those which a model would forecast or, later, actual analysts would forecast. This generally produced the "right" results, until recently. In 1999, some firms produced earnings well above forecasts and their stock prices did not move. What happened? EMH theorists have suggested that the answer to this unfortunate dilemma was that the earnings of the aforementioned companies did not exceed the forecasts by as much as the market was expecting! Since it is impossible to ask "the market" a question, who is to say just what the market was expecting. A scientific hypothesis should, at least in principle, be refutable. This does not appear to be the case in "tests" of the EMH, however.

4. As we discussed above, the EMH tests have generally used the mean price reaction. However, the distributions are generally skewed, such that the median would be closer to zero. Further, some (or many) will be of the opposite (i.e., wrong) sign. For example, findings of price declines on "good news" have not been uncommon. What sort of similar "event" is being tested which produces price reactions of differing magnitude and, in some cases, direction? Let us be specific. Based upon these tests, we can say that an unexpected earnings or dividend increase will tend to be accompanied by a price increase, although not always and certainly not of a uniform magnitude. How far does that take us toward being able to say that all public information is impounded into the prices of all stocks all the time?

5. In the "hard" sciences, there is a conventional wisdom at any point in time. If a new theory comes along, a test is proposed. The conventional view plays the role of the null hypothesis and the challenger plays the role of the alternative hypothesis. To be accepted, not only must the alternative be superior, but it must be so much so that its superiority could only occur by chance 1% or 5% of the time (which is where significance levels fit into hypothesis testing). Advocates of the "new finance" (including most EMH theorists) apparently felt that, prior to their arrival, there was no wisdom in finance, conventional or otherwise. Hence, they felt no obligation to run tests by the conventional rules (despite their protests of being "scientific"). Sometimes they have tested their views as the null hypothesis (e.g., "no news" events) such that the higher the significant levels, the greater the chance they would be accepted. The impact would thus be that

the result could not reject the null of efficiency. By that test, one cannot disprove that markets are efficient, which may be a great deal different from a demonstration that they are.

6. Another element of true scientific hypothesis testing is that the null and alternative should be set up as mutually exclusive and exhaustive when possible so that a critical experiment could provide an unambiguous result, and rejection of the null would imply acceptance of the alternative. The EMH advocates do not investigate in this manner either. On those occasions when they have subjected their views to the risk of bearing the burden of an alternative hypothesis, they lighten the load by setting up a "straw-man" as the null. Consider the discussion of "good news" events: What is the null? That the market no longer reacts as it did? Does rejecting that (and thus accepting some continuity of reaction) really imply we have no choice but to accept the EMH?

7. EMH researchers selectively interpreted their own results. Moreover, by the time these results are summarized in textbooks, law review articles, etc., all of the exceptions, reservations, and potential alternative interpretations have conveniently disappeared.

5.10   An Example of "Soft" Results Becoming "Conclusive"

The discussion in the previous section can be illustrated by considering the classic event study conducted by Fama, Fisher, Jensen, and Roll (FFJR)[9] and published in 1969 [11]. This research is still regularly cited in the literature, and it has become a model for events studies. FFJR examined monthly data for the 60 months around the effective date of 940 stock splits from January 1927 through December 1959.

When this work was first published, at least two of the authors of this book were puzzled over how any one could take this paper seriously. It concerned itself with an event that almost the entire spectrum of academia in finance and economics would have deemed to be a non-event: stock splits. (The Dolley work in 1933 looked at the same phenomenon, but this was 36 years earlier and the real economic effect of splits was not so nearly well understood then.) Worse yet, they purported to find an effect!

Next, we found the technical hypothesis testing to be odd. The paper never really addressed the real question that begged to be answered, to wit: why do firms split their shares in the first place? The only story they cite, which is the only one we have ever heard, is a version of managerial belief

---

[9]At least three of whom we suspect to have been members of the Group of Six referred to in the quotation from Lo and MacKinlay [34] earlier in this chapter.

in the optimal range of a price per share, a story often repeated in the brokerage folklore (but never taken seriously by "rational" economists). The stated purpose of their research is provided as follows ([35], pp. 186−187):

> More specifically, this study will attempt to examine evidence on two related questions: (1) is there normally some "unusual" behavior in the rates of return on a split security in the months surrounding the split? And (2) if splits are associated with "unusual" behavior of security returns, to what extent can this be accounted for by relationships between splits and changes in other more fundamental variables?

One searches in vain for a potentially falsifiable statement here. Furthermore, no formal hypothesis testing framework is even set up at the outset of the research. Indeed, given that we do not know what motivates firms to split their shares in the first place, it would be rather difficult to hypothesize a market reaction to such. This tendency to avoid a formal framework became a prime characteristic of most EMH work that followed FFJR.

Continuing on with the FFJR work, despite having no formal hypothesis to test, they do generate data and conclusions. Among those conclusions is the following ([35], pp. 197−201).

> We suggest the following explanation for this behavior of the average residuals. When a split is announced or anticipated, the market interprets this (and correctly so) as greatly improving the probability that dividends will soon be substantially increased. . . .If, as Lintner suggests, firms are reluctant to reduce dividends, then a split, which implies an increased expected dividend, is a signal to the market that the company's directors are confident that future earnings will be sufficient to maintain dividend payments at a higher level. If the market agrees with the judgments of the directors, then it is possible that the price increases in the months immediately preceding a split are due to altering expectations concerning the future earnings potential of the firm (and thus of its shares) rather than to any intrinsic effects of the split itself.

By the time this work is analyzed and reinterpreted at the textbook level, these "suggested explanations" have become solid conclusions! For example, Lorie and Brealey in their readings book introduction to the FFJR article state: ". . . the authors conclude that, although splits do not themselves affect the aggregate value of the shares, they do convey information about the company's prospects" ([35], p. 103). The textbook interpretation notwithstanding, it must be remembered that all FFJR did was present data and tell a story. One story, however, does not necessarily preclude another. As we have previously observed ([17], p. 195):

Consider first the basic FFJR story: Splits signal a (71.5 percent) prior of a dividend [increase] which signals a maintainable earnings increase. This would have greater appeal if splits and dividends were signaled by different agents within the firm. In fact, both require the action of the board of directors. This, in turn, implies a theory of why firms split their shares: To signal a high *a priori* belief of an almost immediate dividend increase. This begs the question again of why firms do not avoid the expense of a split and simply accelerate the dividend announcement. It becomes entirely inoperative in the instance of a simultaneous split and dividend announcement (a very frequent occurrence). It would seem that a rather bizarre implied theory of motivation for splits underlies the FFJR story.

Alternate stories, more complex stories, in which splits, dividends, and combinations provided different signals, could also be told. The difficulty remains, however, of dismissing questions of motivation on the one hand and interpreting results to imply, at least in a positivist sense, theories of motivation on the other. Now consider the augmented FFJR story: The stock price rose in the first place because of the enhanced future earnings expectations. One could envisage a dividend information effect here, but to generate a separate split effect would seem to require a theory regarding additional information from the split itself. In place of the FFJR story, suppose that:

1. Stock prices are determined by Harry Roberts' generator;

2. At least some managers believe in an optimal trading range and will split stock for this reason;

3. This range has a lower bound as well, and managers will be loathe to split if there is a chance the price will subsequently fall into a cat and dog range; and

4. Split announcements can and will aborted if the price starts falling.

Over any arbitrary interval, some stocks will have positive residuals. In up markets especially, this will translate into abnormal share price increases. Some of these stocks will split and, given four above, all of these will have essentially unbroken patterns of price increase immediately prior to the split. This is sufficient to generate the basic FFJR results.

They set $T = 0$ at the effective date; they attribute the positive residuals in the prior month or two to announcement effects of the split. They also report a random sample of 52 of their splits having a median of 44.5 days between the announcement and effective dates (median announcement data $= T - 1.5$ months). Suppose we have two stocks in our sample. X announces at $T = -3$ and Y at $T = 0$. If we aggregate and average, what do we get: a residual increasing up to T = 0 but at a decreasing rate as we get close. This would also seem sufficient to explain the FFJR results.

An alternative story has assumed that there is *no* split effect (not simply one which might only be traded upon with inside information).[10] We can introduce such an effect however. Strangely enough, to get a split effect it seems that we must invoke an optimal trading range notion. If such a range does exist, one would expect to see a split effect as the price moved into the optimal range. If, on the other hand, management possesses inside information (a relaxation of the price generating assumption (1) above) and behaves as though it thought there were an optimal trading range, then splits would have the potential to possess information content per se for reasons exactly analogous to the information content of dividends arguments. (As we are avoiding technical issues, we might also simply note that the dividend effects in FFJR seem to be somewhat controversial and may be quite sensitive to sample and variable definition.)

5.11   Other Studies

Our example of the FFJR study is just one of many events studies used to "prove" the EMH when, in fact, no such proof has been offered at all. We shall continue to explicate the empirical foundations of EMH research and provide many more such examples below.

If FFJR is deemed the classic finance events study, the counter-part in accounting would be Ball and Brown [1], who looked at unexpected earnings. Over time, they found a direct (and possibly significant) correlation between unexpected earnings and the residual. They were not able to demonstrate a clear-cut event effect (i.e., residual discontinuity), but this may be attributable to the use of monthly data. They did, however, purport to demonstrate the competitive context of accounting information, a notion which has taken on some significance in that literature.

In their study, they partitioned their sample into positive and negative unexpected earnings groups at T = 0. For the entire sample over the prior 12 months, they then found that the average residual was 25 percent of its mean absolute value; this was concluded to be the percentage of information that persists. They then constructed an arbitrage portfolio on unexpected earnings, found that its residual was 50 percent of the sample average, and thus concluded that half of the persisting information was attributable to the annual report. As they readily acknowledged, these classifications were somewhat crude.

By the time these conclusions reached the text level, however, the interpretation is much more elegant and conclusive ([6], p. 325):

---

[10]See FFJR (in [35], p. 206). This is also consistent with FFJR's finding that the mean absolute deviation of the residuals was twice their mean.

. . . returns appear to adjust gradually until, by the time
of the annual report, almost all the adjustment has occurred.
Most of the information contained in the annual report is anticipated by the market before the annual report is released. In
fact, anticipation is so accurate that the actual income number
does not appear to cause any unusual jumps in the API in the
announcement month. Most of the content of the annual report (about 85% to 90%) is captured by more timely sources of
information. Apparently market prices adjust continuously to
new information, as it becomes publicly available throughout
the year. The annual report has little new information to add.

One begins to wonder if this entire literature is being reinterpreted subject to a *pre hoc* fallacy. All that Ball and Brown reliably demonstrated
was a positive correlation between accounting numbers and stock prices.
(Arguably they did not even demonstrate that accountants are not useless.
The positive correlation would be consistent with a world where accounting
numbers had symbolic, but not productive economic, value.) Adding that
correlation to the Harry Roberts' generator could, in principle, produce all
of the above results as artifacts of the grouping procedures at $T = 0$. No
causation is implied nor, if it is found to exist, is its potential direction
unambiguous. (At least isolated examples can be found of big baths being
taken *because* prices declined and of earnings being manipulated to validate the expectations underlying price increases.) In sum, the competitive
context story is manufactured of whole cloth.

If the competitive context story effectively precludes the market from
being fooled by (non-fraudulent) accounting manipulations, its absence
opens another possibility. In a study involving accounting changes with
cash flow implications, Sunder [53] found that firms changing to the Last-
In-First-Out (LIFO) accounting method for inventory costing exhibited no
announcement effect (although they had a 5% positive residual over the
prior year). For those switching to the First-In-First-Out (FIFO) accounting method for inventory costing, he found no announcement effect, no
consistent change for 4−5 months, a positive spike at about month +3,
and then a fairly systematic decline in the residuals from month +5−13.
The LIFO result would seem to require a foreknowledge story to be consistent with market efficiency. It is not clear how the FIFO results can be
reconciled, because a short position between months +3 and +10 would
appear to earn an excess return of 20%.

In another famous study, Kaplan and Roll [29] examined accounting
changes without cash flow implications. In a sample of firms switching
to a flow through treatment of the Investment Tax Credit, they found no
announcement effect, a tendency for the residuals to rise over the next
quarter, and then fall back over the following quarter. In so doing, they
presented 80% confidence intervals about the mean residual which did not
lie exclusively on any one side of the abscissa. A textbook interpretation

of this result is that ([6], p. 322):

> ...even if one considers the positive residuals following the reported accounting change of higher eps to be significant, it appears that people were only temporarily fooled because cumulative residuals return to zero approximately 25 weeks after the announcement.

In addition to being consistent with the textbook's rather flexible concept of market efficiency, it should be noted that these results are also consistent with the market being fooled once on the initial announcement, being fooled again when the results are not sustained by the subsequent quarterly report, and never realizing what actually happened.

Kaplan and Roll also looked at firms switching to straight-line depreciation for reporting purposes. There is no announcement effect, but it does appear that the residuals might be rising around that time, although prior to it they decline modestly and afterward they decline sharply. We can find nothing here inconsistent with the assertion that the market is fooled by the accounting change but that the more fundamental problems of these companies soon return to dominate the price. Beyond this, the consistent price declines in the post-change period (implying a 7% return to a short position) raise questions regarding the consistency of this study with market efficiency.

More recently, the EMH has become the maintained, rather than tested, hypothesis in event studies. In other words, the "event" is increasingly classified *ex-post* (rather than *ex-ante*) as "good news" or "bad news" based upon the price reaction. If the result is not especially intuitively appealing, the "stories" begin (e.g., as in the case of dividend increases discussed above).

Finally, it should be noted that there have been so many events studies published (and Ph.D. dissertation generated) that few of them receive more than footnote coverage in the textbooks. Nevertheless, as a whole, they are taken to "prove" the EMH. Sharpe's best selling textbook provides the generally accepted conclusion (in a "Summary of Efficient Markets") as follows ([50], p. 103):

> Tests of market efficiency demonstrate that U.S. security markets are highly efficient, impounding relevant information about investment values into security prices quickly and accurately. Investors cannot easily expect to earn abnormal profits trading on publicly available information, particularly once the costs of research and transactions are considered.

Sharpe does go on to say ([50], p.103):

> Nevertheless, numerous pockets of unexplained abnormal returns have been identified, guaranteeing that the debate over the degree of market efficiency will continue.

And it is the subject of these "unexplained abnormal returns" to which we turn in the next section.

5.12 Intertemporal Analyses

The other major classification of semi-strong EMH studies look at "beating the market" overtime. These clearly require a benchmark (i.e., what market performance is to be "beaten"). They thus become the test of what in the literature has been called [11] a joint hypothesis: that the EMH is true and that the benchmark is correct. A failure of the test could be ascribed to either or both of the joint hypotheses.

The common choice of benchmark has been the CAPM. This assumes that excess return (over the risk free rate) is solely a function of beta (i.e., the contribution to the risk of the market portfolio). It is equivalent to an "arbitrage" holding which is long (short) the stock(s) and short (long) a portfolio with the same beta (as a complete measure of the relevant risk priced in the market).

The last two decades have seen an explosion of studies which find abnormal profits from various strategies employing public information. Firms with small capitalizations (found by multiplying number of shares by price at the beginning of the period, which are rather simple bits of public information) have been found to outperform other shares by substantial amounts over the last half-century (see [2]). Firms having low price-earnings ratios have also been found to outperform the market (see [3]). Shares in general have been found to earn abnormal returns in January [47]. More recent research has even found that different returns can be earned based upon the day of the week (or even time of day) when orders are entered (see [54] and [55]).

These studies, generally classified as the "capital market anomalies" literature, have highlighted fundamental questions regarding the extent to which semi-strong efficiency is valid in any market in general, or on average. Introducing a collection of early studies, Jensen [28] stated, "[V]iewed as a whole, these pieces of evidence [of anomalous valuations] begin to stack up in a manner which make a much stronger case for the necessity to carefully review both our acceptance of the efficient market theory and our methodological procedures." More recently, Reinganum [44] has made a similar point. The near universal interpretation is to blame the CAPM alone (holding EMH harmless — although the implication would be that

---

[11] The text discussion follows the evolution of the literature. A moment's reflection, however, should indicate that a frictionless market in equilibrium would be efficient virtually by definition. (1) Hence, if the benchmark is an equilibrium model, there is but a single hypothesis: that the market conforms to the benchmark. (2) If the benchmark is not (or may not be) an equilibrium model (e.g., Fama–French), the question becomes how a frictionless market can have fully informed pricing and not be in equilibrium.

the price adjusts to provide an appropriate return which is now unknown).[12] As a result, the literature has rejected the CAPM for this reason, although it is prominent in all the textbooks and widely taught. The most common excuse for the latter is that employers expect students of good programs to know it. Second place probably goes to the claim it is intellectually rigorous (e.g., as in high school Latin). A distant third is that the semester is long and the professor does not know anything else.

As we entered the 1990s, advocates of the EMH were, reluctantly, prepared to disavow the CAPM (except in the classroom) in order to preserve their belief in the EMH (in the name of science, of course). While this meant that failures of the intertemporal testing could not be blamed on the EMH, it did nothing to advance the EMH and raised serious questions about why such testing should continue (and be funded and published). Published event studies were by now into the hundreds and researchers were running out of events. Funded empirical research was threatened.

Once again demonstrating that no problem is so great that an assumption cannot be found to eliminate it, Professors Fama and French published a series of papers beginning in 1992 (see [8] and later [9]). They (FF) showed that beta did not explain the cross section of stock returns in recent years and that certain of the prior results of the "anomalies" research did explain differential returns. They were able to collapse the latter into two measures: firm capitalization (i.e., price times the number of shares outstanding), where small firms had higher returns, and the ratio of book value to market value per share, where stocks with higher ratios (i.e., roughly equivalent to value stocks vs growth stocks) had higher returns. They then show that a revised benchmark model based upon these two factors plus beta makes the excess returns to anomalies go away, which, they contend, supports the EMH.

An immediate response to this procedure is to note that they have added the excess return from the inefficiencies back to the market benchmark. They then show that, defining the market this way, the strategies no longer "beat the market". Hence, they have appeared to save the EMH by destroying its meaning. FF claim their procedure is justified because the two factors are proxies for risk and that a higher return should be earned for greater risk. Recall that for at least forty years, risk had been defined as return variation, either for individual stocks (as in Markowitz's variance) or as a stock's contribution to a portfolio's risk (e.g., beta). All of the prior tests controlled for risk defined that way. These two factors are different from, and in addition to, that risk. No intuitive explanation is forthcoming as to why a small and/or value firm is "riskier" than a large and/or growth firm with the same variance and/or beta.[13]

---

[12]Apparently advocates now trust the dealer so much they do not even bother to look at their cards.

[13]FF point to potential "financial distress" costs for these firms. For the prior 20 years, however, EMH advocates had pointed to a Chicago dissertation based on railroad

Recently, Professor Haugen [24] wrote a book featuring FF[14]. His frustration on the issue is rather apparent ([24], pp. 64–65):

> In spite of the fact that the FF risk-adjustment process seems to be meaningless and self-fulfilling, the procedure seems to have developed a wide following among their academic disciples. Many disciples are now compelled to use it so as to "properly" risk adjust the returns in their studies "verifying" the efficient market.
>
> Sometimes FF find they must "risk adjust" using more than two factors. You see, it depends on how many anomalies they are trying to "get rid of." If they want to eliminate the small firm premium, for example, they add a factor equal to the monthly difference in the returns to small and large firms. They then find that there's no significant difference between the returns to portfolios of small and large stocks in market environments in which the performance of both is the same. Jegadeesh and Titman find evidence of intermediate momentum in stock returns. To eliminate this anomaly, rank stocks by their performance in the previous year, and create a factor equal to the difference in the monthly returns to stocks that did well and the stocks that did poorly in the previous year. Then you will find surely no significant difference in the returns to these two kinds of stocks in market environments in which their performance is equal.

In sum, it would appear that, like event studies, intertemporal analyses post FF have adopted the EMH as the maintained hypothesis. That is, since "everybody knows" markets are efficient and only risk is priced, given that these factors appear to be priced it must follow that they are proxying risk.

Based on empirical results, that is, looking at portfolios of real stocks over time, it appeared that the advocates of random stock buying strategies were correct. The managed portfolio performances, in the aggregate, were not necessarily worse than those of the index funds (not surprising, if a managed fund was doing systematically worse than "normal," then the fund manager could become an overnight genius by simply shorting all

---

bankruptcies from the 1930s for the proposition that such costs were small, indeed too small to balance the tax savings from having a large amount of debt in the capital structure. For this reason, Merton Miller, in his American Finance Association Presidential Address, referred to the costs of distress as the rabbit in "horse and rabbit" stew.

[14]Viewing FF as an equilibrium benchmark (case 1, in footnote 11 of this chapter), the question becomes why the two new factors, previously anomalies, are now risk (e.g., where is the small firm argument in the utility function?). When attacked on this front (e.g., Haugen below), FF often retreat into case 2: that their benchmark is appropriate to determine if new anomalies are being discovered or just manifestations of known anomalies. The latter, of course, begs the question of what fully informed price means if it is coexisting with free lunches.

his positions). However, the higher transaction costs plus mutual fund management fees associated with the managed funds made the index funds a better investment. The EMH afficionados then argued that their point had been proved: the market was acting efficiently, taking into account all information, including forecasting, more or less instantaneously. Of course, this is rather like saying that since nobody in the 1930s had a vaccine for polio that no such vaccine could exist. The empirical studies prove that the existing managed funds examined did not do better, in the aggregate of all such funds, than the index funds. The studies did not prove that building such models was impossible. Indeed, they did not even prove that some of the existing mutual fund managers had not developed good ways to beat the market. There exist some funds which have consistently beaten the average of the market. Unfortunately, that would be true even if all the managed funds were using random choice. Every year, we see funds which have always beaten the market, have their first year in which they did not beat the market.

The most current state of academic opinion regarding efficient markets is summarized by Lo and MacKinlay ([34], pp. 6–7).

> There is an old joke, widely told among economists, about an economist strolling down the street with a companion when they come upon a $100 bill lying on the ground. As the companion reaches down to pick it up, the economist says (sic) "Don't bother — if it were a real $100 bill, someone would have already picked it up." This humorous example of economic logic gone awry strikes dangerously close to home for students of the Efficient Markets Hypothesis, one of the most important controversial and well-studied propositions in all the social sciences.
> . . . . . .

> What can we conclude about the Efficient Markets Hypothesis? Amazingly, there is still no consensus among financial economists. Despite the many advances in the statistical analysis, databases, and theoretical models surrounding the Efficient Markets Hypothesis, the main effect that the large number of empirical studies have had on this debate is to harden the resolve of the proponents on each side.
> One of the reasons for this state of affairs is the fact that the Efficient Markets Hypothesis, by itself, is not a well-defined and empirically refutable hypothesis. To make it operational, one must specify additional structure, e.g., investors' preferences, information structure, business conditions, etc. But then a test of the Efficient Markets Hypothesis becomes a test of several auxiliary hypotheses as well, and a rejection of such a joint hypothesis tells us little about which aspect of the joint hypothesis

is inconsistent with the data. Are stock prices too volatile because markets are inefficient, or is it due to risk aversion, or dividend smoothing? All three inferences are consistent with the data. Moreover, new statistical tests designed to distinguish among them will no doubt require auxiliary hypotheses of their own which, in turn, may be questioned.

More importantly, tests of the Efficient Markets Hypothesis may not be the most informative means of gauging the efficiency of a given market. What is often of more consequence is the *relative* efficiency of a particular market, relative to other markets, e.g., futures vs. spot markets, auction vs. dealer markets, etc. The advantages of the concept of relative efficiency, as opposed to the all-or-nothing notion of absolute efficiency, are easy to spot by way of analogy. Physical systems are often given an efficiency rating based on the relative proportion of energy or fuel converted to useful work. Therefore, a piston engine may be rated at 60% efficiency, meaning that on average 60% of the energy contained in the engine's fuel is used to turn the crankshaft, with the remaining 40% lost to other forms of work, e.g., heat, light, noise, etc. Few engineers would ever consider performing a statistical test to determine whether or not a given engine is perfectly efficient — such an engine exists only in the idealized frictionless world of the imagination. But measuring relative efficiency — relative to a frictionless ideal — is commonplace. Indeed, we have come to expect such measurements for many household products: air conditioners, hot water heaters, refrigerators, etc. Therefore, from a practical point of view, and in light of Grossman and Stiglitz, the Efficient Markets Hypothesis is an idealization that is economically unrealizable, but which serves as a useful benchmark for measuring relative efficiency.[15]

5.13 More Evidence That Markets Are Not Efficient

It might seem to the reader that all states of the world other than those in which price impounded all public information would reflect market inefficiency (i.e., all states other than those where the EMH held). By virtue

---

[15] It is interesting to note, of course, that engineers can quite easily measure the relative efficiency of an engine. The relative efficiency of the stock market remains in the "world of the imagination" of financial economists. There are not presently available quantitative measures of relative stock market efficiency, and it is not clear what such a measurement would mean. If the stock market were "60% efficient," how do we determine the idealized 100%, and where does the other 40% go. Such concepts certainly make a great deal more sense in the real world of the engineer than in the hazily conceived one of the "social scientist."

of the advocates' appropriation of the null hypothesis, this is generally not the case. They tend to define inefficient markets as those in which one or more strategies can be demonstrated to beat the market. Failing such a demonstration, the EMH is presumed to hold (e.g., all ties go to the dealer in this casino and virtually all games end in ties). Combined with their strong belief in the EMH, the result is that no test or showing would rebut the EMH in their view.

In the first place, anyone with a winning "system" would have every incentive to conceal it. In the case of event studies, so long as the market reaction is defined as the efficient reaction, no rejection of the EMH is possible. When the anomalies tests did purport to beat the market, the advocates first discredited the benchmark and later built the anomalies into the benchmark, called it risk, and defied anybody to beat *that*.

Consider all possible states of market information processing. A subset of this would be covered by the EMH. Clearly what the advocates call inefficiency (where systems work to beat the market) is only a subset of the remainder. Call the rest non-efficiency, which tends to be ignored (much like true uncertainty) in tests and discussions. This would include Keynes' views on the market, non-rational bubbles (EMH advocates will only consider rational bubbles, which they can show not to exist in equilibrium), momentum trading, and a host of other possible market environments.

The concept of non-efficiency also tends to level the playing field in terms of burdens of proof. The way the advocates set up the game, the opponent must demonstrate a workable system to beat the market over long periods (e.g., decades), clearly an extreme position. Otherwise, the EMH wins by default—an equally extreme position. This is a game where the end zone seats sell at a premium because no action ever takes place mid-field. With the alternative of non-efficiency, ties no longer go to the EMH and its advocates must finally start proving rather than assuming.

The concept would also appear to comport with most investors' views of the market. Even those who think they have winning strategies believe them to be specific to market conditions, and thus, transitory. They certainly do not impute rationality to all other market participants. Indeed, those who have "systems" generally plan to profit from the irrationality of their fellow market participants. They may believe that markets converge to fundamental value over time, but certainly not at every trade.

The distinction can also be made in terms of price adjustment to information. With the EMH, adjustment is rapid and unbiased. With inefficiency, we know where the price is going but there is time to transact at a profit. With non-efficiency, who knows? The price may arrive at rational value eventually, but nobody knows when that will happen or how far it may diverge in the meantime. This view is consistent with Warren Buffett's concept of "Mr. Market." The latter is kind enough every trading day to offer a price at which one could buy or sell. At times, this price may seem much too low or high, but one is under no obligation to transact and should

not do so unless the price seems favorable. In other words, the function of
the market is not to be believed (as in the EMH), but rather, when possible, to be taken advantage of. At last report, Warren Buffett had made
more money for himself and his investors than all the EMH researchers put
together and raised to a power! Furthermore, when investments do not
work out, he does not go broke and plead a "six sigma event"(see Long
Term Capital Management discussion above). The lesson is the investor
can never know whether he or she was right only that one emerged from
the casino with more money than one entered.

More evidence suggestive of non-efficiency follows. Because the condition does not have the regularity of the other two, the tests are not as neat.
Since they are not assuming their conclusions, they are also not as conclusive. Recall the normative "story" (e.g., rational expectations) which
accompanies the EMH. New information causes everyone to reformulate
his or her future cash flow probability distributions, which the market aggregates and discounts back to a new price. On a normal day or week,
the amount of new information entering the market, relative to the level
of existing information, should be rather low. Hence the change in price,
relative to the level of the price, should be rather small. The latter concept
is price volatility (e.g., variance or standard deviation of return). Furthermore, with little information entering the market, most of the volume would
reflect investors buying with excess funds or selling to pay their bills. Such
liquidity traders are presumed not to be trading on information and the
volume they generate is assumed to be fairly uniform over time. In sum, in
the EMH type world, we would expect usually to see price bouncing around
an upward trend (reflecting retention of earnings) on uniform volume.

The academic literature on these issues goes back some two decades
and has recently been surveyed at length in another of Professor Haugen's
books (see [23], especially pp. 1–57). Thus, only a brief summary shall be
presented here.

1. Stock Prices Are Too Volatile.

   The basic tests here construct a perfect foresight price. To do this,
   go back many years and compute the price by discounting the actual
   future dividends at the existing interest rates. Reiterate the process
   coming forward in time. Compare the changes in the price computed
   with changes in the actual market price. By rules of logic, the volatility in the actual price should be less than or, in the limit, equal to
   the variability in the perfect foresight price. In fact, the actual price
   is found to have far greater variability. (The advocate solution to this
   problem is to assume required returns are time-varying. The result
   becomes so seriously underdetermined, that is, far more unknowns
   than equations, that it becomes consistent with any data and, hence,
   the EMH cannot be rejected by this version of the model.)

2. The Volatility Itself Is Too Volatile (Unstable).

These tests are based upon the notion that, while individual stocks may react to many different events, the overall market should only be impacted by major events such that one might expect its volatility to be relatively stable over time. Stable volatility might be illustrated by the Dow Jones Averages going up or down 100 points a day, as contrasted to a situation where they go up or down 50 points for a time and then shift to going up or down 500. For the latter to occur from rational valuation, a fundamental shift in the risk of the economy must have occurred.

Haugen ([23] p. 8) reports results averaging seven shifts per year for the last century and asks how it is conceivable that the risk of the economy could be shifting that frequently? Increases (decreases) in volatility tend to be associated with decreases (increases) in stock prices — which would be a rational reaction to increased risk. Volatility increased substantially before the Crash of October 1987, which therefore fits the pattern. However, in the vast majority of cases including 1987, there are no news reports of events which would indicate increased risk to the economy. In sum, the increase in volatility itself did not appear to be a reaction (rational or otherwise) to any reported event.

These results raise the further question as to the impact of economic and financial conditions on the stock market. One study looked at the unanticipated changes in numerous key indicators and could only explain 18% of the changes in the market index; 82% of the variability was caused by unknown factors. Pearl Harbor and war against Japan (December 8th and 9th, 1941) only drove the market down 7.6%; the Crash of '87 drove it down over 20% for no known reason. The evidence begins to accumulate that something is driving the market beyond a rational reaction to information.

3. Risk Stops at the Closing Bell.

Another useful statistical insight is that, if price changes are random, variance is proportional to time (i.e., the variance of returns over two day intervals should be twice the variance over one-day intervals). This implies that the variance of returns measured Friday close to Monday close should be 3 (by days) or almost 4 (by hours) times that of weekday returns. In fact, it is virtually the same. The algebraic solution suggests the variance is 70 times greater when the exchange is open than when it is closed. This perhaps provides support for Keynes' proposal that the stock market should be allowed to open only once a year!

The initial advocate reaction was that weekends are slow news days. Similar tests were run on exchange holidays and the period in 1968 when the NYSE shut down on Wednesdays to catch up on paperwork.

The tests continue to show that the implied variance is much lower when the exchanges are closed. This differential also varies by firm size. For the largest firms, variance is 12 times greater when the exchange is open, while for the smallest it is 50 times greater. Finally, this effect also works in the other direction. If the shares trade more hours in the day (either because the exchange lengthens its hours or because of multiple − including global − listings), variance rises.

It should come as no surprise that the advocates' answer takes the form of an assumption: Information is only gathered while the market is open so that it can be traded upon immediately (i.e., the Private Information Hypothesis). On the West Coast, this would imply that, no later than 1:30 PM, analysts, fund managers, etc. hang up on company officials, adjourn meetings, unplug their computers, and join the brokers and traders in the bar. Haugen ([23], pp. 30−33) devotes pages to having fun with this one.

5.14 Conclusions

The conclusion that emerges is that the vast majority of stock price volatility is not the result of a rational reaction to economy-wide, industry, or firm specific news. In other words, we do not have information causing price, which is the foundation of the EMH. What we appear to have is: Price causing price. The greed and fear of heterogeneous investors, worrying what the "fools" may do next and how they may react to (often trivial) "information," whether there will be a chair for themselves when the music stops (do we hear the last strain?), etc., is what causes the bulk of the volatility. (Haugen makes the interesting observation that while real estate did not tank in the October 1987 Crash, REITs did!) This is the second, third, and higher levels of Keynes' newspaper picture contest − all of which is assumed away by the EMH.

Consider the newspaper picture contest. Under the homogeneity assumptions everyone has the same tastes, desires, and attitudes. Under symmetric market rationality they ascribe this sameness to each other. Hence, everyone· picks the same pictures (the market portfolio) and the contest ends without a unique winner. Even if the prize covered postage for all participants, the game would probably fail for lack of interest. Under rational expectations assumptions (one name for the normative "story"), the contest (market) would have known its outcome in advance and never run itself!

The EMH has effectively adopted Keynes' casino analogy for the stock market, but in reverse. Investors are assumed to be the house, playing against "nature." The game is assumed to have a positive expected return (the casino's cut), known odds, and independent realizations over time. Results of one spin of the wheel provide no information about where the ball will drop on the next spin. This is the fantasy land that neoclassical financial economic research has brought us.

References

[1] Ball, R. and Brown, P. (1968). "An empirical evaluation of accounting income numbers," *Journal of Accounting Research*, Autumn, 159–178.

[2] Banz, R. (1981). "The relationship between return and market value of common stocks," *Journal of Financial Economics*. Vol. 9, No. 1, March, 3–18.

[3] Basu, S. (1977). "Investment performance of common stocks in relation to their price-earnings ratios: a test of the efficient market hypothesis," *Journal of Finance*, Vol. 32, No. 3, June, 663–682.

[4] Bernstein, P. L. (1997). "How long can you run and where are you running," *Journal of Post Keynesian Economics*, Vol. 20, No. 2, Winter , 183-189.

[5] Cootner, P. (ed.) (1964). *The Random Character of Stock Market Prices*. Cambridge, MA: MIT Press.

[6] Copeland and Weston. (1983). *Financial Theory and Corporate Policy*, 2nd edition, Reading, MA: Addison-Wesley.

[7] Dolley, J. C.(1933). "Characteristics and procedure of common stock split-ups," *Harvard Business Review*, April, 316–326.

[8] Fama, E. F. (1965). "Portfolio analysis in a stable paretian market," *Management Science*, January, 404–419.

[9] _____ (1970). "Efficient capital markets: a review of theory and empirical work," *Journal of Finance*, May, 383–423.

[10] _____(1991). "Efficient capital markets II," *Journal of Finance*, December, 1575–1617.

[11] Fama, E. F., Fisher, L., Jensen, M. and Roll, R. (1969). "The adjustment of stock prices to new information," *International Economic Review*, February, 1–21.

[12] Fama, E. F. and French, K. (1992). "The cross-section of expected stock returns," *Journal of Finance*, June, 427–465.

[13] _____ (1996). "Multifactor explanations of asset pricing anomalies," *Journal of Finance*, March, 55–58.

[14] Findlay, M. C. and Williams, E .E. (1980). "A positivist evaluation of the new finance," *Financial Management*, Summer, 7–18.

[15] _____ (1985). "A Post Keynesian view of modern financial economics: In search of alternative paradigms," *Journal of Business Finance and Accounting*, Spring, 1–18.

[16] _____ (1986). "Better betas didn't help the boat people," *Journal of Portfolio Management*, Fall, 4–9.

[17] _____ (2000),"A fresh look at the efficient market hypothesis: How the intellectual history of finance encouraged a real 'fraud on the market,'" *Journal of Post Keynesian Economics*, Winter, 181–199.

[18] Findlay, M. C., Thompson, J. R. and Williams, E. E. (2003), "Why We All Held Our Breath When the Market Reopened," *Journal of Portfolio Management*, Summer, forthcoming.

[19] Friedman, M. (1953). *Essays in Positive Economics*. Chicago: University of Chicago Press.

[20] Glassman, J. and Hassett, K. (1999). "Stock prices are still far too low," *Wall Street Journal*, March 17, (op.ed.).

[21] Graham, B. and Dodd, D. (1934). *Security Analysis*, New York: McGraw-Hill.

[22] Guthmann, H. G. and Dougall, H. E. (1962). *Corporate Financial Policy*, 4th edition, Englewood Cliffs: Prentice-Hall.

[23] Haugen, R. A. (1999). *Beast of Wall Street: How Stock Volatility Devours Our Wealth*. Englewood Cliffs, NJ: Prentice Hall.

[24]_____ (1999). *The New Finance: The Case Against Efficient Markets*, second edition, Englewood Cliffs, NJ.:Prentice-Hall.

[25] (1999). "Ho-Hum, another earnings surprise," *Business Week*, May 24, 83–84.

[26] Hunt, P., Williams, C. M., and Donaldson, G. (1966). *Basic Business Finance: Text and Cases*, 3rd edition, Homewood, IL: Richard D. Irwin, Inc.

[27] Ikenberry, D., Lakonishok, J. and Vermaelen, T. (1995), "Market underreaction to open market share repurchases," *Journal of Financial Economics*, October-November, 181–208.

[28] Jensen, M. (1978). "Some anomalous evidence regarding market efficiency," *Journal of Financial Economics*, 6, 95–101.

[29] Kaplan, R., and Roll, R. (1972). "Investor evaluation of accounting information: some empirical evidence," *Journal of Business*, April.

[30] Keynes, J. M. (1921). *A Treatise on Probability*. London: Macmillan.

[31] Knight, F. (1921). *Risk, Uncertainty and Profit*, New York: Harper and Row.

[32] Lintner, J. (1965). "The valuation of risk assets and the selection of risky investments in stock portfolios and capital budgets," *Review of Economics and Statistics*, February, 13–37.

[33] Lo, A. W. and MacKinlay, A. C. (1988). "Stock market prices do not follow random walks: evidence from a simple specification test," *Review of Financial Studies*, Spring, 41–66.

[34] _____ (1999). *A Non-Random Walk Down Wall Street*, Princeton, NJ: Princeton University Press.

[35] Lorie, J. and Brealey, R., eds. (1972). *Modern Developments in Investment Management*. New York: Praeger Publishing.

[36] Mackay, Charles (1869). *Memoirs of Extraordinary Popular Delusions and the Madness of Crowds*, London: George Routledge and Sons.

[37] MacKinlay, A. C. (1997). "Event studies in economics and finance," *Journal of Economic Literature*, Vol. 35, No. 1, March, 13–39.

[38] Markowitz, H.(1952). "Portfolio selection," *Journal of Finance*, March, 77–91.

[39] _____(1959). *Portfolio Selection*, New York: John Wiley & Sons.

[40] Miller, M. and Modigliani, F. (1961). "Dividend policy, growth, and the valuation of shares," *Journal of Business*, October, 411–433.

[41] Modigliani, F. and Miller, M. (1958). "The cost of capital, corporation finance, and the theory of investment," *American Economic Review*, June, 261–297.

[42] Mossin, J. (1966). "Equilibrium in a capital asset market," *Econometrica*, October , 768–783.

[43] Niederhoffer, V. (1997). *The Education of a Speculator*, New York: John Wiley & Sons.

[44] Reinganum, M. (1981). "Abnormal returns in small firm portfolios," *Financial Analysts Journal*, No. 2.

[45] _____ (1990). "The collapse of the efficient market hypothesis," in Fabozzi and Fabozzi, *Current Topics in Investment Management*, New York: Harper & Row, 33–47.

[46] Roberts, H. (1959). "Stock market 'patterns' and financial analysis: Methodological suggestions," *Journal of Finance*, March, 1–10.

[47] Roll, R. (1983). "Vas ist das?" *Journal of Portfolio Management*, Vol. 4.

[48] Samuelson, P. (1965). "Proof that properly anticipated prices fluctuate randomly," *Industrial Management Review*, Vol. 6, 41−49.

[49] Sharpe, W., "Capital Asset Prices: A theory of market equilibrium under conditions of risk" (1964). *Journal of Finance*, September, 425−442.

[50] Sharp, W., Alexander, G., and Bailey, J. (1999). *Investments*, 6th edition, Upper Saddle River, NJ: Prentice Hall.

[51] Simmonds, A., Sagat, K., and Ronen, J. (1992-3). "Dealing with anomalies, confusions and contradiction in fraud on the market securities class actions," *Kentucky Law Journal*, Vol. 81, No. 1.

[52] Smith, E. L. (1924). *Common Stocks as Long Term Investments*. New York: Macmillan.

[53] Sunder, S. (1973). "Relationships between accounting changes and stock prices: Problems of measurement and some empirical evidence," *Empirical Research in Accounting: Selected Studies.*

[54] Thaler, R. H. (1987). "Anomalies: The January effect," *Journal of Economic Perspective*, 197−201.

[55] _____(1987). "Anomalies: seasonal movements in securities prices II—weekend, holiday, turn of the month, and intraday effects," *Journal of Economic Perspective*, Fall, 169−177.

[56] Thompson, J. R. and Williams, E. E. (1999−2000). "A Post Keynesian analysis of the Black−Scholes option pricing model," *Journal of Post Keynesian Economics*, Winter, 247−263.

[57] Williams, E. E. (1969). "A Note on accounting practice, investor rationality and capital resource allocation," *Financial Analysts Journal*, July-August, 37− 40.

[58]_____and Findlay, M. C.(1974). *Investment Analysis*. Englewood Cliffs, NJ: Prentice-Hall.

[59] Williams, J. B. (1964). *The Theory of Investment Value*, Amsterdam: North Holland. First published 1938.

Chapter 6

More Realistic Paradigms for Investment

6.1 Introduction

To the economist objects are valuable only if they are scarce and provide utility. Scarcity is an important criterion of value, since free goods have no value. Air, for example, is quite useful, but because there is an abundance of air, it is worthless. Similarly, there is a scarcity of alchemists; but, since the utility of their services is questionable, these services have little if any value.

Although defining value is a relatively simple matter, determining value is much more difficult. In a market economy, there is a tendency to value goods and services at their market prices. Nevertheless, because of differences in tastes, the "value" to one individual (i.e., the utility) of a certain commodity may be less than or greater than the "value" of the identical commodity to someone else. Indeed, the very rationale for sales transactions is the fact that a particular commodity has greater utility for one individual than for another. The price being offered to the seller is greater than the value he places on the commodity being sold. Thus, he sells. The price paid by the buyer is less than the value he places on the commodity. Thus, he buys.

Economists have wrestled with questions of valuation for years, and the theoretical pronouncements on the subject are voluminous. In this book we are not concerned with the esoteric properties of value. We shall be content with examining the techniques and methodologies of valuation available to the investor. Our assumption will be that economic value is determined by utility. However, as we discovered in Chapter 2, the expected utility U is equal to the expected dollar value of a game payoff only in the special case of a linear utility function. Risk averse investors will require greater dollar returns (i.e., they will only pay a sum less than the expected value of a payoff distribution) to compensate for the fact that there are no certain payoffs. Even the short-term U.S. Treasury bill (a direct obligation of the United States Government, with the nation's "full faith and credit" behind it) has some risk of non-payment, although its repayment terms (typically 90, 180, 270, or 360 days) makes the risk so small as to be considered "riskless" by most investors.

Common stocks and derivatives are altogether a different story. Here, the payoffs are never certain, either in dollar amount or in time of payment. Dollar returns from common stocks are derived from two sources: (1) the dividends paid by the firm, and (2) the price appreciation of the security. The ability of a firm to pay dividends, of course, depends upon its earning power. A firm that generates only a negligible return on its asset investment will obviously be in no position to pay dividends. Nevertheless, even firms that obtain very good returns on investment sometimes elect not to pay large dividends. Such firms may be able to benefit stockholders more by retaining earnings and reinvesting in profitable assets. This procedure will produce higher earnings in the future and improve the long-run dividend paying potential of the firm. Furthermore, such a policy may allow stockholders in high income-tax brackets a substantial tax saving. Cash dividends are normally taxed at ordinary income rates, whereas capital gains are given preferential treatment. When a firm retains earnings for reinvestment in profitable projects, it improves its long-run earnings stream. This improvement should be reflected in the price of the firm's shares. Thus, by retaining earnings, the firm may, in effect, raise the price of its stock. Investors would then have a capital gain rather than the ordinary income that they would get if a dividend were paid.

6.2 Growth

J. B. Williams [10] worked out the math of asset valuation in the 1930s. The basic idea was, like any asset, the price of a share can be viewed as the present value of the future stream of cash to which the owner is entitled. For stock, this stream is dividends (defined to include corporate share repurchases, liquidating dividends, etc.). Hence the basic valuation equation is:

$$P(0) = \frac{D(0)}{(1+r)^0} + \frac{D(1)}{(1+r)^1} + \frac{D(2)}{(1+r)^2} + \ldots \qquad (6.1)$$

$$= \sum_{t=0}^{\infty} \frac{D(t)}{(1+r)^t},$$

where

$$P(0) = \text{Price (Value) today}$$

$$D(t) = \text{Dividend } t \text{ years in the future}$$
$$r = \text{Market's (investor's) required rate of return.}$$

The choice of interpretation of the variables relates to how the value equations are employed. If we insert the current price as $P(0)$ and future dividend estimates and then solve for r, the latter becomes the implied return on the stock. If we add the trappings of equilibrium, it becomes the market's required return (presumably for firms of the risk class—whatever that may mean given the current state of the literature). If, instead, we insert what we feel to be an appropriate r and solve for $P(0)$, the latter becomes a measure of fundamental or intrinsic value. This value can be compared to price in a search for cheap or dear stocks (i.e., the exercise deemed wrongheaded by the semistrong EMH). Finally, in formulations such as equation (6.3) below, we can insert all the other values and solve for μ. The latter would be the growth rate in dividends required to earn a return of r on a market price of $P(0)$.

For finite holding periods, (6.1) may be stated as

$$P(0) = \sum_{t=0}^{n} \frac{D(t)}{(1+r)^t} + \frac{P(n)}{(1+r)^n}$$

where the best estimate at $t(0)$ of $P(n)$ is

$$P(n) \approx \sum_{t=n+1}^{\infty} \frac{D(t)}{(1+i)^{t-n}}.$$

Frequently, it will be a best guess to assume that the dividends increase according to the formula

$$D(t) = D(1+\mu)^t. \tag{6.2}$$

Then, equation (6.1) becomes [1]

$$P(0) = D \sum_{t=0}^{\infty} \left(\frac{(1+\mu)}{(1+r)}\right)^t = D\frac{1+r}{r-\mu}. \tag{6.3}$$

The continuous time version of (6.3) is easily seen to be

$$P(0) = D \int_0^{\infty} e^{(\mu-r)t}dt = D\frac{1}{r-\mu}. \tag{6.4}$$

In both (6.3) and (6.4), it is essential that r be greater than μ. When $r \leq \mu$, $P(0)$ is infinite in value.

---

[1] Here we are using the fact that a series of the form $1 + p + p^2 + p^3 + \ldots = 1/(1-p)$ if p is greater than 0 and less than 1.

It is important to note that the discounting valuation models in (6.3) and (6.4) deal with dividends rather than earnings. In recent years, similar discounting models have been applied to value the firm as a whole using *free cash flow* in place of dividends. Roughly speaking, *free cash flow* equals earnings after taxes plus depreciation minus capital expenditures and does represent funds available to pay down debt or distribute to shareholders. Assuming the funds do find their way to investors, the approach is legitimate. We thus arrive at valuation models based upon earnings. If the firm retains b proportion of earnings $E(t)$ every year, then

$$D(0) = [(1 - b)]E(0)$$

and if both grow at μ, (6.4) may be restated as

$$P(0) = \frac{(1 - b)E(0)}{r - \mu}. \tag{6.5}$$

In terms of price to earnings multiples (P/E), we may simplify the notation by using m to signify this ratio. Hence,

$$P = mE(0), \text{ and} \tag{6.6}$$

$$m = \frac{1 - b}{r - \mu}. \tag{6.7}$$

Finally, if the firm earns on average x on retained earnings, it will retain $bE(0)$ and earn x, such that the growth in earnings caused by retention will be

$$xbE(0)/E(0) = xb = \mu. \tag{6.8}$$

With these building blocks, we can now consider the interesting case of "profitless" growth. This is the situation where earnings retained only earn at the same rate shareholders can obtain elsewhere in the market (i.e., $x = r$). Substituting into (6.7), we have

$$m = \frac{1 - b}{r - br} = \frac{1 - b}{r(1 - b)} = \frac{1}{r}. \tag{6.9}$$

In this special case, it does not matter how much is paid out because the same return is earned both inside and outside the firm; and the P/E multiple (m) is the inverse of the required return (e.g., if investors require 12.5%, $m = 8$). This also illustrates that, the more of earnings retained, the greater μ must be for the shareholders to break even vis-à-vis having been paid the funds as dividends.

Now consider the firm that pays no dividends currently. Shares that are never expected to generate payments are worthless. Hence, there must be some expectation of future payments. The standard approach is to go forward to the point where payments are expected to begin and apply one of the above equations to get price at that time. This price is then discounted to the present to obtain the current price. The wrong thing to do is to discount earnings. This approach double (and more) counts because this year's earnings reinvested also appear in future earnings.

There are several implications of this case. Until dividends begin, the entire return will be in the form of expected price appreciation at the rate r; once they begin, expected appreciation will fall to r minus the dividend yield. Over this period, earnings will grow at μ. If $x = r$ (and with $b = 1$), $xb = \mu$ will also equal r. Earnings, as well as price, will grow at r. However, $m = 1/r$ in this case. It is only when $x > r$ that $\mu > r$ and earnings grow at a faster rate than price. This can only occur over a finite period (or else, as discussed above, $P \to \infty$) and it is only during such periods that, rationally, $m > 1/r$ (i.e., big multiples can be justified because of rapidly growing earnings only over finite periods).

As we have demonstrated, rational valuation models do accommodate growth. There is nothing magical about it, and it is clearly not the case that no price is too great to pay for it. Growth can be "profitless" (e.g., increasing retentions for run of the mill investments). The only growth worth a premium multiple is that reflecting access to lucrative investments (i.e., $x > r$) and even that can only be expected to occur for a few firms over finite time intervals. Nevertheless, much of the foolishness in the market relates to growth.

6.3 Rational Valuation and Growth

One implication of the J. B. Williams model developed in equation (6.1) (and explicated with the work of Professors Modigliani and Miller in Chapter 5) is that the value of the whole should equal the sum of the values of the parts (often called the "Value Additivity" Principle). During the 1960s, companies in unrelated industries were combined into conglomerates, which for a time traded at a premium to the values of the components. This upset EMH advocates greatly, because it made no sense that the market would pay a premium for a portfolio of companies (acquired at premiums) when the same portfolio could have been assembled on the market at no premium. True to their tradition, however, advocates attempted to find explanations consistent with the EMH.

The best explanation appears to be a "growth fallacy" inconsistent with the EMH. Suppose a conglomerate is selling at 12 times earnings and acquires a (non-growth) company selling at 6 times earnings. For every share the purchaser issues, it gets twice its current earnings in return. Hence, its earnings grow over time by such acquisitions even if the earnings of each acquisition are not growing. This illustrates the problem with paying a premium for "growth" irrespective of its source. Unfortunately, the game also works in reverse. A fall in multiple would mean that such acquisitions could no longer profitably be made. Hence, "growth" would decline and the price could fall further. (At various times, it has also been popular to believe that the whole was worth less than the sum of its parts. Waves of spin-offs, carve-outs, etc. followed.)

During the early 1970s, high-growth stocks were variously called "one decision stocks" (i.e., buy them, put them in your safe deposit box, and never worry about them again) and the "nifty fifty" (there were about 50 of them including then popular stocks such as IBM, Xerox, and Polaroid). This was another era when "no price was too high for growth." Tables were published to show how many years of continual earnings growth would be required before the current price would produce a reasonable multiple (10−15 years was not uncommon). The stock market collapse which began in 1974 ended the "growth at any price" strategy, at least for a while. It reappeared in the 1990s in the form of "high tech" growth stocks which went on to collapse in the 2000−2001 crash! (As an aside, it is interesting to note that one of the three stocks mentioned above, Polaroid, has gone bankrupt, and another, Xerox, is doing very poorly today. This suggests how ridiculous the notion of "one decision stocks" really was. We shall return to this notion at the end of this chapter.)

The regular reappearance of "growth" as a means to justify higher and higher stock prices raises the question of what a reasonable estimate of the finite period of continued extraordinary earnings growth might be. In other words, how many years of such growth might one reasonably price in arriving at a multiple of current earnings (i.e., how much should $m > 1/r$). The literature on this issue goes back almost 40 years (see [5]). It consistently suggests that earnings growth rates are strongly mean-reverting within 3−5 years. Hence, companies selling at prices capitalizing extraordinary growth beyond five years should be few in number and have compelling stories. The number of such companies that periodically appear as "growth stocks" in the aforementioned boom/bust cycles would seem to exceed any reasonable figure by far, raising further doubts about the EMH.

6.4   Momentum and Growth

Companies are said to have earnings momentum if they report quarterly earnings increases or, if the industry in which the company is classified has a seasonal component, increases over the year earlier quarter. The key issue is not that earnings are generally rising, but that the record be unbroken (even if the increase is only a few pennies). In rational valuation, the issues are exactly reversed: the overall, long term growth rate is very important but a few pennies here or there on a quarterly basis are irrelevant. Price momentum involves buying stocks which have risen in the expectation of a continual rise. As one fund manager put it, "I would rather buy a stock that makes new highs every day than one making new lows." Investing on earnings momentum will generate price momentum. As illustrated by the Internet stocks prior to March, 2000, price momentum investing can occur with stocks that have *no* earnings.

Price momentum investing violates the rational speculator assumptions of the EMH if it occurs with any regularity. Although data are hard to come by, surveys and observations suggest that substantial numbers of both professional and individual investors in the bull market that ended in March, 2000, traded on momentum (see [5] and [7]). Even the SEC has expressed concern about corporate accounting which may be employed to maintain earnings momentum. Clearly, if instead of pushing price back to value, the market is extrapolating it further from value until momentum stops (at which point it may crash below value), the EMH is in shambles. The situation is so serious that EMH advocates have advanced, from their apparently limitless supply, an assumption to deal with it! If the initial price reaction to news is too small, momentum trading could carry the price to true value and, thus, in principle, be stabilizing (see [1] and [8]). Unfortunately, this story has problems. In the first place, if the initial reaction is too small, this is per se inefficiency (recall that the reaction to news is supposed to be rapid and unbiased in the EMH). Hence, the advocates once again save the EMH by destroying it. Furthermore, if the price reaction to the news is wrong, how will price reacting to price (i.e., momentum trading) know when to stop at true value? As discussed in Chapter 5 above, this could certainly increase volatility.

6.5   An Application

One approach to making the above analysis concrete which is regularly advocated in investment texts involves preparing probability distributions for all the unknowns. If the analyst has prepared such distributions, he or she may use them to determine deviations around the expected valuation. Such an approach will be explored in this section, although the reader should be forewarned that calculating expected values (and deviations around them) may lead to meaningless results. It is tempting to compute means, variances and covariances of returns in this fashion (rather than relying on historical price movements to make such calculations) in order to implement the portfolio-theoretics outlined in Chapter 3. A better method will be suggested in the next section.

Suppose an analyst has specified a probability distribution for the sales of the Ace Co. He has also determined a conditional distribution of costs given each level of sales. Finally, he has calculated an overall distribution of net income and dividend payments (in millions of dollars).

Sales		Total Expenses	Net Income
100(.3)		70 (.2)	30(.06)
		80(.6)	20(.18)
		90(.2)	10(.06)
90(.5)		65(.1)	25(.05)
		75(.7)	15(.35)
		85(.2)	5(.10)
80(.2)		60(.2)	20(.04)
		70(.5)	10(.10)
		80(.3)	0(.06)

If the firm has 10 million shares outstanding, the data may be grouped into a distribution of earnings per share (EPS):

EPS	Probability
$3.00	.06
$2.50	.05
$2.00	.22
$1.50	.35
$1.00	.16
$.50	.10
$.00	.06
	1.00

From this, the analyst may construct a dividend payment distribution. Suppose he believes that management will pay a maximum dividend of two dollars but will distribute less if earnings are low. Within the range of EPS from one dollar is two dollars, he believes a .75 payout will take place. Below earnings of one dollar, no dividend will be paid. The dividend per share (DPS) distribution would be

DPS	Probability
$2.00	.11
$1.50	.22
1.12\frac{1}{2}$	.35
$.75	.16
$.00	.16
	1.00

The expected value of EPS and DPS and their standard deviations may be computed. They equal:

	EPS	DPS
$\mu =$	$1.50	$1.06
$\sigma =$	.72	.58

The analyst may repeat the above process for each year in the anticipated holding period of the security. Suppose he obtained the following:

Year	EPS	DPS
1	$\mu = \$1.50$	$\$1.06$
	$\sigma = .72$	.58
2	$\mu = 1.80$	1.20
	$\sigma = .85$	.64
3	$\mu = 2.00$	1.40
	$\sigma = .95$	.76

Thus, he would have expected values and deviations for EPS and DPS over the three-year holding period. He could next determine a distribution of multiples to get the expected price in three years. Assume he forecasts the following:

P/E Multiple	Probability
10×	.6
12×	.4

This distribution could be combined with the earnings distribution for year three to obtain a distribution of prices. Ideally, each multiple should be combined with each EPS to get the distribution. If the EPS distribution were

EPS	Probability
$3.50	.2
2.00	.6
.50	.2

and if it were assumed that the level of multiple was independent from the forecasted EPS (an assumption that may not hold), the price distribution would be:

Multiple	×	EPS	=	Price
		$3.50(.2)		$35.00(.12)
10× (.6)		2.00(.6)		20.00(.36)
		.50 (.2)		5.00(.12)
		3.50 (.2)		42.00(.08)
12× (.4)		2.00(.6)		24.00(.24)
		.50(.2)		6.00(.08)

Grouping the data:

Price	Probability
$42.00	.08
35.00	.12
24.00	.24
20.00	.36
6.00	.08
5.00	.12
	1.00

The expected price and its standard deviation would be: $\mu = \$21.60$, $\sigma = \$10.50$.

In the example presented above, all the separate return distributions were assumed to be independent. Most likely, they will not be. Generally, if a firm does poorly in the early years of an investment holding period, it will continue to do poorly. Thus, the dividend distributions of later years may well depend upon the outcomes in previous years. Furthermore, the EPS distribution in the terminal year will almost always be correlated with the dividend distribution of that year. Finally, the value of the P/E multiple in the terminal year will most likely be influenced by the earnings and dividend performance in earlier years. Thus, the analyst should attempt to construct conditional probabilities, where appropriate, for all the distributions involved.

Suppose an analyst has projected the following distributions for a stock that is to be held for two years as portrayed in Figure 6.1:

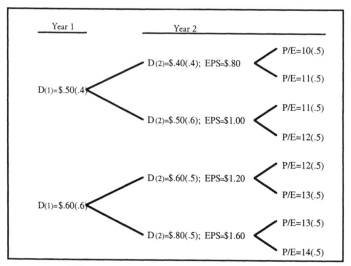

Figure 6.1. Two-Year Profile of Stock.

Eight outcomes (with associated probabilities) would be possible:

Combination	$D(1)$	$D(2)$	$P(n)$	Prob
a.	$.50	$.40	$8.00	.08
b.	.50	.40	8.80	.08
c.	.50	.50	11.00	.12
d.	.50	.50	12.00	.12
e.	.60	.60	14.40	.15
f.	.60	.60	15.60	.15
g.	.60	.80	20.80	.15
h.	.60	.80	22.40	.15
				1.00

The yield r from each combination (a–h) can be computed, given the current market price of the security. Basically, the notion is that of compound interest. We assume the dividend is reinvested in the stock and grows at i. Thus, for a two-year horizon, we have:

$$P(0) = \frac{D(1)}{1+r} + \frac{D(2) + P(2)}{(1+r)^2}.$$

If $P(0)$ is assumed to be \$8, the results may be placed in a probability distribution.

Combination	r	Probability
a.	5.6%	.08
b.	10.4	.08
c.	23.1	.12
d.	28.2	.12
e.	40.7	.15
f.	46.1	.15
g.	68.1	.15
h.	74.1	.15

The expected yield from this distribution is 41.8 percent with a standard deviation of 22.6 percent.[2]

For situations where more outcomes are designated (and where the time span is longer) many combinations are possible. In these instances the analyst must rely on a computer to do the computational work. We shall do this in the next section when we develop fully a simulation approach to valuation. At this point, however, we might simply observe that once the analyst has specified DPS and terminal price distributions for each year, he may compute an overall return-risk measure for the equity under investigation. If it is assumed that all distributions are independent, the following equations may be employed:

$$V(\mu) = \sum_{t=0}^{n} \frac{\bar{D}(t)}{(1+r)^t} + \frac{\bar{P}(n)}{(1+r)^n} \tag{6.10}$$

and

$$\sigma = \sqrt{\sum_{t=0}^{n} \frac{\sigma_D^2(t)}{(1+r)^{2t}} + \frac{\sigma_P^2(t)}{(1+r)^{2n}}}. \tag{6.11}$$

Here:

- $V(\mu)$ is the present value of the dividend stream and terminal selling price.

---

[2] Returns of this magnitude should strike the analyst as being "high," perhaps causing him or her to re-work the assumptions.

- $\bar{D}(t)$ is the expected dividend in period t.

- $\bar{P}(n)$ is the expected terminal price at period n.

- r is the required or market rate of return from the investment.

- σ is the overall standard deviation around the present value.

- $\sigma_D(t)$ is the standard deviation of dividends in period t.

- $\sigma_P(t)$ is the standard deviation around the expected market price at period n.

Referring to the earlier example of the Ace Co., we may solve for $V(\mu)$ and σ. Suppose an investor wished to determine the probability of earning at least 10 percent on the purchase of the stock. The present value of the projected return stream would be

$$V(\mu) = \frac{1.06}{1.10} + \frac{1.20}{(1.10)^2} + \frac{1.40}{(1.10)^3} + \frac{21.60}{(1.10)^3} = \$19.13.$$

If the current market price of the stock were \$19.13, the investor would expect to obtain the 10 percent return. If the current market price were lower, he would expect a larger return. If it were higher, a lower return would be expected.

The standard deviation of this distribution would be

$$\begin{aligned} \sigma &= \frac{(.58)^2}{(1+.10)^2} + \frac{(.64)^2}{(1+.10)^4} + \frac{(.76)^2}{(1.+.10)^6} + \frac{(10.50)^2}{1.+.10)^6} \\ &= \$7.94. \end{aligned}$$

If the stock were selling at \$11.19 (i.e., \$19.13 − \$7.94) and the underlying distributions were close to normal, then the investor could determine the probability that he would earn less than 10 percent:

$$P(X|X \le V(\mu) - 1\sigma) \approx .16.$$

An examination of the normal tables in the Appendix B.1 shows that the probability a normal variate lies below one standard deviation from its mean is around .16.)

Now, at the opposite extreme from having completely independent flows is the case of perfectly correlated returns. In this instance, any deviation from the mean value of a flow in any particular period would be matched by deviations in all other periods in exactly the same manner. The present value of the dividend stream and terminal selling price would be the same as in the independent case. However, the overall standard deviation around the present value would be given by

$$\sigma = \sum_{t=0}^{n} \frac{\sigma_D(t)}{(1+r)^t} + \frac{\sigma_P(n)}{(1+r)^n}. \tag{6.12}$$

If the flows in the above example were perfectly correlated, the σ of the distribution would be

$$\sigma = \frac{.58}{(1+.10)} + \frac{.64}{(1+.10)^2} + \frac{.76}{(1+10)^3} + \frac{10.50}{(1.+.10)^3}$$

$$= \$9.52.$$

Not surprisingly, the overall σ is greater when returns are perfectly correlated than when returns are independent.

Naturally, in the real world, it will be unusual for returns to be either perfectly correlated or independent. A more likely situation would show something in between. We can emulate things by a mixture of the two with parts of the return stream being independent and other parts being perfectly correlated. In the mixed case, the expected value remains the same, but the overall standard deviation becomes something more complex (see [4]):

$$\sigma = \sqrt{\left(\sum_{t=0}^{n}\frac{\sigma_D^2(t)}{(1+r)^{2t}} + \frac{\sigma_P^2(n)}{(1+r)^{2n}}\right) + \sum_{k=1}^{m}\left(\sum_{t=0}^{n}\frac{\sigma_D^k(t)}{(1+r)^t} + \frac{\sigma_P^k(n)}{(1+r)^n}\right)^2},$$

(6.13)

where $\sigma_D(t)$ and $\sigma_P(n)$ are the dividend-price standard deviations for the independent portion of the flow returns; and $\sigma_D^k(t)$ and $\sigma_P^k(n)$ are the dividend-price standard deviations for stream k of a perfectly correlated portion of the flow returns.

Reconsidering our example, suppose it can be determined that the dividend stream is perfectly correlated. That is, poor corporate performance in period $t = 1$ that results in a lower dividend payment than the expected value would also mean poor performance in period $t = 2$ and $t = 3$. Because the stock price at period $t = 3$ depends on future dividends (beyond $t = 3$), and these dividends would also be influenced by the poor performance of earlier years, at least part of the variation in stock price for $t = 3$ would be perfectly correlated with the dividend-stream variations. Nevertheless, there may be other factors influencing stock prices (such as the overall level of interest rates or simply random events) so that part of the variation in stock price is also independent. Assume that we can determine that about 80 percent of the stock price variation in period $t = 3$ is traceable to the riskiness of the dividend stream, and 20 percent results from other (independent) factors. Given this information, we could reformulate equation (6.13) to get the overall deviation for the investment:

$$\sigma = \sqrt{\frac{\sigma_P^2(3)}{(1+.10)^6} + \left(\sum_{t=1}^{3}\frac{\sigma_D(t)}{(1+.10)^t} + \frac{\sigma_P(3)}{(1+.10)^3}\right)^2}$$

$$= \sqrt{(2.10)^2(.564) + [(.56)(.909) + (.64)(.826) + (.76)(.751) + (8.40)(.751)]^2}$$

$$= \$8.09.$$

We note that the overall standard deviation of the mixed case lies between that of the independent and perfectly correlated cases.

The above formulations can also be used to generate a distribution of rates of return. The following steps should be employed:

1. Compute $V(\mu)$ and σ for an initial value of r.

2. Determine the number of standard deviations that would equalize the present price of the stock and V_μ (i.e., solve the equation $V(\mu)+Z\sigma = P(0)$ for Z).

3. Find $P(X|X \leq V(\mu) + Z\sigma)$. This will be the probability that $V(\mu)$ will be less than the present price of the stock at the given r and is, in turn, the probability that the rate of return will be less than r.

4. Plot the probability determined above on a cumulative probability distribution.

5. Iterate steps 1–4 above varying r in order to generate a complete cumulative probability distribution.

6. Convert the cumulative distribution to a discrete probability distribution of rates of return.

7. For a two-parameter description, the fiftieth percentile of the cumulative probability distribution will provide an estimate of the mean (expected) rate of return. Dividing the interquartile range by 1.35 will provide an estimate of σ.

When the analyst has specified distributions with many more possible outcomes than we have dealt with above, "back of the envelope" calculations like the above must be replaced with computer intensive methods. Generally, that will involve the kind of simulation techniques extensively employed in this book. Here is a suggested approach for the situation where the variables are assumed to be independent of each other (see [2] and [3]).

1. Determine a probability distribution for each variable.

2. Transform any continuous distributions into discrete distributions with each discrete interval representing, say, 1 percent of the distribution.

3. Determine the midpoint of each interval in each discrete distribution.

4. Assign a number 1 to n (where n represents the number of intervals into which each distribution has been divided) to each interval. For example, if a distribution were divided into 100 intervals, the first percentile would be assigned the number 1, the second would be assigned 2, and the hundredth, 100.

5. The distribution of returns should also be divided into intervals (say, whole percentages) and counters defined (for example, .5 percent$\leq$ r_1 < 1.5 percent; 1.5 percent $\leq r_2$ < 2.5 percent, and so on) which are initialized at zero (that is $r_1 = r_2 = r_3 = \ldots r_n = 0$).

6. Generate a two digit random number for the first independent distribution (which will be designated $D(0)$).

7. Determine the interval on the distribution that was assigned a number corresponding to the random number generated and select the midpoint.

8. Repeat steps 6 and 7 for the remaining distributions such that there is one observation for each variable.

9. Compute the rate of return for this set of observations. For the example given, this would require inputing the simulated variables to the formula:

$$P(0) = \sum_{t=1}^{3} \frac{D(t)}{(1+r)^t} + \frac{P(3)}{(1+r)^3}$$

and solving for r.

10. The r computed in step 9 would then be assigned to the appropriate interval in the return distribution and the counter for that interval would be incremented by one. For example, if the first computation yielded an $r = 5.7$ percent, then r_6 would be incremented and, at the end of the first iteration, $r_{1-5} = r_{7-n} = 0$ and $r_6 = 1$.

11. Repeat steps 6−10 a large number of (N) times in order to generate a frequency distribution of r's.

12. Each counter can then be normalized to provide a probability distribution [for example, $r_1/N = p(r|.5$ percent $\leq r < 1.5$ percent$)$].

13. From this distribution, μ_i, σ_i, or any other desired statistic may be computed. The larger N is, the better will be the resulting estimate.

6.6   The "Risk Profile" Approach to Stock Selection

The previous section outlined a method for computing means (expected values), variances, and covariances for specific variables associated with analyzing the "value" (or expected return) of a stock. Although we shall improve upon this method below, it should be clear from the above that the value of a stock depends upon a number of fundamental economic and accounting variables. As a starting point, we shall assume that it is the goal of the analyst (investor) to find stocks which have existing market prices below (or above in the case of short-sales) the determined values. By definition, these stocks will only exist in inefficient markets.

Now one way to approach the portfolio (stock) selection problem would be to simply examine historical returns, variances around those returns, and calculate all implied covariances. The optimal portfolio would thus be one that is efficient (no greater returns given a level of risk or no lower risk given a level of returns) and consistent with the risk preferences (utility) of the investor. This is the sort of analysis typically recommended in books that assume market efficiency. The cardinal rules are that all portfolios should be diversified, and all stocks are efficiently priced. Often, it is concluded that the investor is best served by holding some index (such as the S&P 500 or the Wilshire 5000) since professional portfolio managers cannot offer returns in excess of those earned by the index anyway. The calculations suggested in Section 6.5 above may be perfectly consistent with this procedure except returns (variances, covariances) are determined directly by forecast rather than by using historical price (and dividend) data. Of course, inefficiency assumptions could also be made indirectly by observing actual market prices which were lower (or greater) than those implied by the calculations.

In this section, we shall suggest what we consider to be a better method. This method assumes that portfolios are diversified but that individual stocks may be mis-priced in inefficient markets. Basically, the approach assumes that we can forecast future cash flows to be received by investors and that, using the riskless rate of interest as a discount rate, we can determine a histogram of possible gains (losses) depending upon assumptions we make about the variables underlying the future cash flows. To us, this is the real measure of risk. What is the probability that we shall lose money in the future if we make a risky investment rather than a riskless one?

A simple example for one possible outcome would be the following: Suppose XYZ Corp. common stock is selling for $10 per share, and an analysis of the future revenues, expenses, etc. of XYZ for the next five years (our contemplated holding period) is made. From these, we may determine the likely dividend that XYZ would pay in each of the years (annual flows will be assumed although dividends are usually paid quarterly). These data, plus other expectations, will set in place a given stock price for XYZ in five years so we can calculate a future value price for XYZ. Let us assume that we expect XYZ to pay dividends as follows at the end of each of the next five years:

Year 1	$.50
Year 2	$.55
Year 3	$.60
Year 4	$.65
Year 5	$.70

We also expect XYZ stock to be selling for $14 in five years, so we would receive that amount as well if we sold the stock then. Ignoring taxes, and assuming a riskless rate of interest of 5%, we would determine the future value of all cash flows to be received from XYZ as

$$(.50)(1.05)^4 + (.55)(1.05)^3 + (.60)(1.05)^2 + (.65)(1.05)^1 + .70 + 14.00 = \$17.29.$$

The riskless alternative would produce: $(\$10.00)(1.05)^5 = \$12.76.$[3]

Now we should make the same set of calculations for every possible set of outcomes we can determine (perhaps many thousand). Each specific calculation would be deemed a certain result (hence the use of the riskless rate of interest with which to compound) but there may be several sets of calculations that give the same dollar result. Thus, we may draw a histogram of all such outcomes. This histogram provides us with a "Risk Profile" to see whether our investment is very risky (many outcomes below $12.76), risky (only a few outcomes below $12.76), or not risky at all (no outcomes below $12.76).

Of course, this simple outcome produces an obvious result: Since the stock price alone is expected to be above $12.76, we know without further calculations that this is a "non-risk" result. Other possibilities might not be so clear. Assume the following dividend expectations:

Year 1	$.050
Year 2	$.055
Year 3	$.060
Year 4	$.065
Year 5	$.070

Further assume, the stock price rises to only $11.00. The compounded value of the cash flows would be: $.061 + $.064 + $.066 + $.068 + $.070 + $11.00 = $11.33, which would show a negative return for taking the risk.

6.7 The "Risk Profile" Approach After-Taxes and Transactions Costs

The EMH generally assumes a lack of taxes and transactions costs. In the real world, however, these costs most definitely exist. Federal (plus state) income taxes vary from about 15% to well over 40% on ordinary income (dividends, interest, etc.). Long term capital gains are generally taxed at 20% (in the state of California, the rate is an additional 9.4%,

---

[3]Unfortunately, there may exist no riskless medium to preserve purchasing power (after taxes and transactions costs). Even the CPI-adjusted Treasuries tax the adjustment as well as the real return. What is the alternative (in an opportunity cost framework)? Consume all your wealth today and die? Perhaps the after-tax 5-year Treasury strip (or a high grade 5-year municipal bond) would be an appropriate choice. These instruments typically yield well under the so-called "riskless" rate.

which results in an overall rate as of this writing of 29.4%!) Transactions costs have come down greatly in recent years to as low as a few dollars per transaction. Generally, these costs now average about $.03 per share, or less than .1% of a total transaction.

Let us review the calculations in the first example of Section 6.6, assuming a tax rate (federal and state) of 50% on ordinary income (dividends, interest, etc.) and 20% on long term capital gains (holding period of over one year). Let us also assume transactions costs of .1% on purchases and sales of stock. The revised calculations in the example for dividends would be: Dividend × (1 − tax rate) or,

Year 1	$.50 (.50) = $.250
Year 2	$.55 (.50) = $.275
Year 3	$.60 (.50) = $.300
Year 4	$.65 (.50) = $.325
Year 5	$.70 (.50) = $.350.

With transactions costs at .1%, a share would cost: ($10.00)(1.001)= $10.01. It would be sold for: ($14.00)(.999) = $13.99. The capital gains tax on the profit would be: (.2)($3.98) = $.796. Thus, the after-tax proceeds would be: $13.99 − $.796 = $13.194. Assuming minimal transactions costs for the "riskless" security, it would still be taxed as ordinary income, producing a yield of: (5%)(.5) = 2.5%. After five years, the $10 would have grown to $(10.00)1.025)^5 = $11.31 after-tax in the riskless instrument. This would compare with the following in the stock investment:

$$(.25)(1.025)^4 + (.275)(1.025)^3 + (.30)(1.025)^2 + (.325)(1.025) + (.35) + (13.19) = \$14.76,$$

and we would still find the outcome ($14.76 vs. $11.31) to be profitable. Now each outcome contemplated above in Section 6.6 should be adjusted for taxes and transactions costs just as we have done in this example and a new "Risk Profile" histogram should be determined.. Of course, the "Risk Profile" would shift back (more risk) due to the addition of taxes and transactions costs, but this might be more than offset by the unfavorable tax treatment of the "riskless" alternative.

6.8   Realistic Capital Market Theory

As a final note to realistic paradigms for investment, the reader should appreciate that the elements of capital market theory presented in Chapter 4 are not completely destroyed by the critique in Chapter 5 (especially as they relate to the portfolio theory material developed in Chapter 3). We may modify some of the diagrams (and theoretical analysis) to make them more attuned to the real world. Moreover, we shall provide a framework for analysis in the chapters to follow that will make it possible for investors and securities analysts to get a practical handle (with the use of models) on the elusive but not entirely fanciful arena of investing one's wealth.

The authors believe that security analysis as performed in Chapters 7–10 below is a necessary task for every investor. There is no other way that one can make determinations of risk and reward. Even if it were impossible to achieve above-average rewards, forecasts of expected risk-reward relationships are required for all but the most simple-minded decision algorithms. It is also true that portfolio analysis is a required activity if one is to decide in a rational framework which securities should be purchased in order to maximize utility. Although some aspects of portfolio analysis in the current state of the art are rather abstract, we maintain that operational decisions can be made using the tools we will suggest. Utility analysis is an important aspect of portfolio construction. The investor must be able to determine his or her personal risk preferences in order to make optimal selections. Finally, an understanding of the constructs of capital market theory as explored in Chapter 4 is essential if one is to grasp clearly the significance of many of the issues that we have raised above.

Notwithstanding the pitfalls of capital market theory as explicated earlier in Chapter 5, we would argue that it is still possible for the investor to generate efficient portfolios, drawn with no more than the constrained proportions of securities which may be chosen (EF in Figure 6.2).

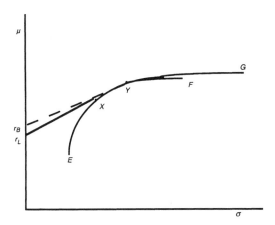

Figure 6.2. Constrained Individual CML.

Because virtually anyone can buy Treasury bills (or, failing that, put money into a savings account or bank certificate of deposit), the lending line ($r_L - X$) can be determined. To the extent the investor may also borrow, a borrowing line may also be determined ($r_B - Y - G$). Borrowing is generally more expensive than lending. Also, costs tend to rise with the amount borrowed, and total borrowing is subject to some limit. Thus $Y - G$ is a curve, and $r_B - Y - G$ becomes discontinuous at G. The rather cumbersome looking $r_L - X - Y - G$ is the individual's revised Capital Market Line (CML), which is a more realistic depiction than the equilibrium figure for the entire market provided in Chapter 4.

6.9 Conclusions

This chapter presents several realistic paradigms that may be useful to the real-world investor. We shall develop these into practical models in the remaining chapters of this book. Since these models will generally not assume that the Efficient Market Hypothesis reflects real-world conditions, we shall summarize our conclusions about market efficiency below:

1. For many years it was believed by financial economists (and more than a few practitioners) that simply because most institutional investors could not achieve long-term aggregate returns greater that those generated by the market averages (S&P 500, DJIA, etc.) that no one could do. This, in fact, led to the development of "index" funds which do not try to "beat the market" but only purport to mimic it. It is our view that there are strong reasons for believing that the activities of expert participants are not properly motivated. In this regard, some of the astute observations of John Maynard Keynes are applicable ([15] pp. 154−155).

> It might have been supposed that competition between expert professionals, possessing judgment and knowledge beyond that of the average private investor, would correct the vagaries of the ignorant individual left to himself. It happens, however, that the energies and skill of the professional investor and speculator are mainly occupied otherwise. For most of these persons are, in fact, largely concerned, not with making superior long-term forecasts of the probable yield of an investment over its whole life, but with foreseeing changes in the conventional basis of valuation a short time ahead of the general public. They are concerned, not with what an investment is really worth to a man who buys it "for keeps," but with what the market will value it at, under the influence of mass psychology, three months or a year hence. Moreover, this behavior is not the outcome of a wrongheaded propensity. It is an inevitable result of an investment market organized along the lines described. For it is not sensible to pay 25 for an investment of which you believe the prospective yield to justify a value of 30, if you also believe that the market will value it at 20 three months hence.

The lessons that Keynes drew from his study of security markets do not make pleasant reading for proponents of efficient market hypotheses. Particularly, those who believe that well informed professional investors make use of all available data to select securities and thereby eliminate the possibility of above average returns are ill at ease with Keynesian conclusions. And yet, even today, Keynes' arguments ring true to many.

2. It must follow that above average return opportunities can be created if some market participants are constrained to accept below average

returns. Legal and institutional constraints, transactions costs, differing expectations, imperfect information, and so on may make possible above average returns for some investors. We might conclude from this that it is indeed possible for the investor to achieve greater than "average" returns. Although institutional investors may not be able to achieve these results, others can. Of course, we must assume that an individual has sufficient funds for investment to justify the costs (in terms of time and money) of doing adequate research and to obtain adequate diversification. There are many investors who do have the sums required, and those who do not probably should not be investing in the first place. Individuals who master the material in this book should be well prepared to begin the task of obtaining above average profits, although the authors tend to accept part of the conventional wisdom that luck may play an important role. Furthermore, we would emphasize the importance of being willing to take a bearish stand when conditions warrant. There is compelling evidence that many investors are unwilling to get out of the "game," and this undoubtedly accounts for the poor performance of many individuals in the market. Market hubris and exuberance at the end of the bull market in 1999 and early 2000 undoubtedly caused many investors to overstay their welcome, especially in high tech stocks that were bid up to totally unreasonable levels by any standard.

3. We would argue that *"average" returns may in fact be better than average.* To the extent that institutional portfolio constraints cause superfluous diversification and suboptimal investment policies, other investors may benefit. Thus, when institutions are confined to purchasing an inordinate amount of low-yield-low-absolute-risk securities, bargains may develop in the range just above the constrained cutoff. This will make it possible for individuals to get even greater returns at given levels of risk than they otherwise would be able to obtain. Thus, even though apparent average returns would accrue to investors (differing only according to risk-class preferences), these average returns would tend to be higher than those that would prevail if institutional investors were not constrained.

4. It may be observed that some *institutional investors pursue suboptimal investment policies by choice and thus create continuing opportunities for individual investors who do have optimal policies.* Institutions are not motivated by the same goals as the individual entrepreneur seeking his or her own profit. Hence, the "invisible hand" cannot be counted on to clear the capital markets at "efficient" prices any more than in the goods or factor markets. In order to demonstrate that they are working for their stockholders, mutual fund managers try to maintain fully invested positions. Although short-run bearishness may develop from time to time, long-run periods of being completely liquid (holding only cash or Treasury Bills) would invite large scale share redemption. Because management fees are based on the size of the fund, there is a definite reluctance to follow any

consistent strategy that would reduce the volume of the fund's assets (even though long run performance might be significantly better). Individual investors are not constrained to an effective fully invested position, and it would seem logical that one could outperform most of the funds simply by acting rationally. Thus, if an individual who had superior predictive insight foresaw bearish conditions, he could profit by maintaining a bearish position (being out of the market) until his opinion changed.

5. Agency problems with financial institutions go well beyond mutual funds being fully invested. Pension managers are often assigned a style (e.g., small cap value) or sector (e.g., large cap technology) for investment, given a sum to invest, and tracked by the corresponding style or sector benchmark. The primary goal is to track the benchmark, with a secondary goal of beating it (but not by so much as to upset the tracking). Since shortfalls can cause the "sum" to be removed (and associated fees lost), a fair amount of "herding" results.

6. We would suggest that *the risk-return measures used in most of the efficient market studies (and in capital market theory) are not necessarily the same measures as those used by investors in ex ante decision making.* These measures are essentially *ex post* approximations of what investors may (or may not) have expected. Furthermore, the usual risk-return measures are based on annual holding period returns and may in fact overstate the true risk borne by the investor's portfolio if one's holding period is longer. We shall explore this issue in depth in the chapters to follow.

Problems

Problem 6.1. Supergrow, Ltd., is expected to pay no dividends for ten years. From year eleven and thereafter, the firm is expected to pay three dollars per share annually.

    a. If investors require a 12 percent return for holding Supergrow, what price pattern would the stock evidence?

    b. What pattern would prevail if the firm paid a dividend of one dollar for years one to five, two dollars for six to ten, and three dollars thereafter?

    c. What pattern would prevail if the firm paid three dollars throughout?

Problem 6.2. The Uni Corp. currently has the ability to pay a dividend of two dollars per year forever. By retaining all earnings for ten years, however, a dividend of ten dollars per year forever could be paid beginning in year eleven. If Uni shareholders require a 15 percent return, what would be the current price of the stock under each of the policies?

Problem 6.3. A firm pays a dividend of one dollar per share. The market price of the firm's shares is 20, and the market expects dividends to grow by 10 percent annually. What return would investors obtain from purchasing this security, assuming a constant future dividend payout ration and P/E multiple?

Problem 6.4. Equatoroid Camera, Ltd., is a growth company. The firm is primarily financed by retained earnings, pays 50 percent of earnings in taxes, and earns 25 percent on its asset investment. If the firm retains 80 percent of its earnings, what is its rate of growth?

Problem 6.5. Suppose Equatoroid (above) finances with debt as well as equity. The firms tries to maintain a debt-equity ratio of .4. What would its growth rate be if it paid bondholders a rate of 6 percent?

Problem 6.6. An investor is in the 40 percent marginal income tax bracket. He pays capital gains taxes at a rate of 20 percent. He is considering the purchase of a stock that now sells at $100 per share. The stock has an EPS of $4, which should grow by 10 percent compounded annually. The stock pays a dividend of $1, which should be increased to $1.20 after two years and to $1.50 for the fifth year of holding. The investor believes the current P/E multiple will prevail in five years when he plans to sell the stock. If the investor wants to earn at least 6 percent on his investment *after taxes*, should he purchase the shares?

Problem 6.7. An analyst has prepared the following sets of probability distributions for a stock (sales and costs in millions):

	Years 1−2	Year 3
Sales	25(.4)	30(.3)
	30(.6)	40(.7)
Costs	.6 of Sales (.5)	.6 of Sales (.5)
	.7 of Sales (.5)	.7 of Sales(.5)

a. Prepare net income and EPS distributions for the three years. The firm has 10 million shares outstanding.

b. The analyst expects the firm to pay out one third of earnings in dividends. Determine the expected dividend and its standard deviation for years 1−3.

c. The following P/E multiple distribution is anticipated by the analyst for year 3.

P/E Multiple	Probability
16X	.4
18X	.6

Determine the distribution of market prices for year 3.

d. Compute the expected market price and standard deviation for year 3.

e. Suppose an investor wished to earn 20 percent on this stock. What maximum price would he pay today, given the expected values of the dividend and terminal price distributions?

f. Assume the stock currently sells at the price to yield 20 percent. Determine the standard deviation of the present-value price distribution if the returns are assumed to be independent. Determine overall σ assuming the returns are perfectly correlated.

g. Assume the stock sells at $16. What is the probability that a return below 20 percent will be earned?

Problem 6.8. An investor expects to earn the following:

Year	Dividend	Terminal Selling Price
1	$\mu = \$10, \sigma = \$1$	
2	$\mu = 10, \sigma = 1$	
3	$\mu = 10, \sigma = 1$	
4	$\mu = 10, \sigma = 1$	
5	$\mu = 10, \sigma = 1$	$\mu = \$100, \sigma = \$10$

a. For $r = 8$ percent, what is the present value of the flows, assuming the distributions are independent?

b. What is the standard deviation of the distribution?

c. If the current price of the security is $94, what is the probability that the investor will earn less than 8 percent?

Problem 6.9. An analyst has simulated the future performance of a security. He has obtained a distribution with $\mu(r) = 12$ percent and $\sigma = 4$ percent.

a. If the distribution is approximately normal, what is the probability that the investment will lose money?

b. What is the probability that the return will be 12 percent or greater?

c. What is the probability that the return will will exceed 20 percent?

d. Suppose the distribution of returns for the security were positively skewed. Might this influence the analysts's judgment about the riskiness of the security? What if it were negatively skewed? Can you suggest a better measure of risk in these instances?

Problem 6.10. An analyst has prepared the following flow diagram for dividends, terminal EPS, and terminal multiple for the Random Corp. common shares:

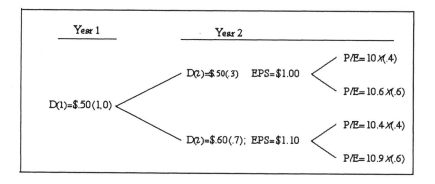

a. What possible outcomes (and associated probabilities) are present?

b. What is the yield for each outcome if the current price of the stock is $10?

c. Specify the probability distribution of yields. What is the mean of the distribution? The variance? The standard deviation?

d. Describe the symmetry of the distribution

References

[1] Chan, Jegadeesh, and Lakanishok, (1996). "Momentum strategies," *Journal of Finance*, 1681–1713.

[2] Hertz, D. B. (1964). "Risk analysis in capital investment," *Harvard Business Review*, January—February, 95—106.

[3] _____(1968)."Investment policies that pay off," *Harvard Business Review*, January—February, 96—108.

[4] Hillier, F. S. (1963). "The derivation of probabilistic information for the evaluation of risky investments," *Management Science*, April, 443—457.

[5] Little, I. M. D. (1962). "Higgledy piggledy growth," *Bulletin of the Oxford Institute of Statistics*, November, 387—412.

[6] "Momentum," (1999). *Wall Street Journal*, March 15, C1.

[7] "Red Flag on Wall Street: Momentum Investors," (1997).*Wall Street Journal*, February 24, C1.

[8] Wermers, R. (1999). "Mutual fund herding and the impact of stock prices," *Journal of Finance*, April.

[9] Williams, E. E. and Findlay, M. C. (1974). *Investment Analysis*. Englewood Cliffs, NJ: Prentice-Hall.

[10] Williams, J. B. (1938). *The Theory of Investment Value* Amsterdam: North Holland.

Chapter 7

Security Analysis

7.1 Introduction

In order to obtain the "outcomes" discussed in prior chapters, it will be necessary to perform an extensive analysis of the security from which the cash flows are expected to be generated. Years ago, security analysis was the mainstay of the investment analysts. Graham and Dodd's classic work *Security Analysis* [5] was devoted to providing insights on how an individual stock should be scrutinized to determined its *intrinsic value*. The principles originally espoused there 70 years ago are still employed by analysts in one fashion or another today, and the title "securities analyst" is still bestowed on individuals who seek to determine those values. (Interestingly, the title "statistician" was given to those performing similar functions prior to the 1930s.) In this book, we will assume an individual analyst (investor) has reviewed in detail the financial statements of, and other publicly available information about, the company whose stock he or she wishes to appraise. Of course, this presumes that the reader has an elementary knowledge of accounting terminology and the meaning of financial statements. We will ultimately seek to generate numerous "outcome" possibilities from this information in order to build a "Risk Profile" for the security in question.

7.2 Financial Statement Analysis

The art of financial analysis has depended for years on the computation of ratios of data secured from the published financial statements of corporations (e.g., balance sheets, income statements, cash flow statements, etc.). The earliest attempts at evaluating a firm and its securities were mainly qualitative assessments and shrewd guesses. These early attempts were succeeded by the computation of a growing battery of comparative financial statistics. These comparisons were of two kinds: comparisons over time and comparisons among different variables. The first form yielded simple but important information. As an example, suppose ABC Corp. earned $1.25 per share last year and $1.50 this year. The computation of a simple ratio [(This Year EPS − Last year EPS)/(Last Year EPS)] would show that EPS was up by 20%. Although such a comparison is extremely crude in that it gives no clue as to *why* EPS increased or whether such an increase will be sustained in the future, it is nonetheless a very significant first step in analyzing the financial position of the company. The second form of comparison is one among variables. Suppose ABC had sales this year of $150 million and net income of $15 million. A comparison of these data would show that net income was 10 percent of sales. This information by itself may not be terribly useful, although if it were observed that such relationship had held fairly steadily over time, perhaps a meaningful conclusion could be drawn.

Ratio comparisons may be made whenever it makes sense to examine the relationship between two or more financial numbers. There is nothing sacrosanct about any given set of ratios, and different comparisons may be necessary when conditions differ. Nevertheless, there is some advantage in attempting to standardize the computation of ratios whenever possible. Hence, over the years, financial analysts have developed a core of ratios that may be used in determining such things as the current financial condition of the firm, the trend in the firm's financial condition, the financial condition of the firm in relationship to the position of other firms, etc. In the paragraphs to follow, we shall explore this core set of ratios.

Figures 7.1−7.8 contain most of the basic ratios that the financial analyst may require. A number of interest groups, including internal management, may find ratios a useful guide for policy making. We shall focus here, however, on the use made of them by investors. Hence, we shall examine each category of ratios as they may apply to the evaluation of the securities of the firm.

Ratio	Method of Computation	Interpretation
a. Current ratio	Current assets / Current liabilities	Ability to meet current debts with current assets
b. Cash ratio	Cash+Short-term securities / Current liabilities	Ability to meet current debts with cash on hand
c. Quick (acid test) ratio	Cash+Short-term securities + Receivables / Current liabilities	Ability to meet current debts with more liquid current assets
d. Basic defensive interval	Cash+Short-term securities + Receivables / Daily operating expenditures	How long the firm could meet cash obligations from liquid assets
e. Working capital to total assets	Current assets − Current liabilities / Total assets	Liquidity of total assets and working capital position

Figure 7.1. Liquidity Ratios.

Short-term creditors, bondholders, and stockholders are all concerned with liquidity of the firm and ratios may be calculated to determine liquidity (see Figure 7.1). Liquidity is important to short-term creditors because payment of their claims may depend on the cash-conversion ability of the firm. Bondholders are concerned because deteriorating liquidity may impair the security of the firm's debt. If the firm's cashable assets (cash, accounts receivable, and in some cases, inventory) are low in comparison with liabilities soon to come due, then the payment of bond interest may be jeopardized. Even if the firm has adequate fixed assets, the conversion of these assets into cash in order to meet liabilities may be expensive (the assets may have to be liquidated by the firm at prices below their economic values) and time consuming. Stockholders are also concerned with corporate liquidity because dividend payments, like interest payments, may be jeopardized if the firm is short of cashable assets. Furthermore, the long-run earning power of the firm may be impaired if it becomes necessary to liquidate fixed assets to meet current liabilities.

The maintenance of profit margins is necessary if the firm is to be able to service its debt obligations, pay dividends, and increase earnings. The ratios in Figure 7.2 are designed to test the ability of the firm to control costs and keep profit margins intact. The long-run profitability of the firm is clearly far more important to bondholders and stockholders than the asset security of the firm's debt and equity. Because asset liquidation is only a final resort relied upon by the unprofitable firm, profit-margin indicators assume a more important role for all the firm's investors.

Ratio	Method of Computation	Interpretation
a. Gross profit margin	$\dfrac{\text{Net sales} - \text{Cost of goods sold (CGS)}}{\text{Net sales}}$	Gross profit per dollar of sales
b. Operating income ratio	$\dfrac{\text{Net sales} - \text{CGS} - \text{Selling \& adm expenses (S\&A)}}{\text{Net sales}}$	Operating profit before interest and taxes per dollar of sales
c. Operating ratio	$\dfrac{\text{Operating expenses (CGS} + \text{S\&A expenses)}}{\text{Net sales}}$	Operating expenses per dollar of sales
d. Net profit margin	$\dfrac{\text{Net income}}{\text{Net sales}}$	Net income per dollar of sales

Figure 7.2. Profit Ratios.

Turnover ratios (Figure 7.3) indicate the ability of the firm to generate revenues from its asset investment. Even if profit margins are high, the firm may be unprofitable if a large investment in assets is required to generate a meager sales volume. In addition to identifying the revenue-generating capacity of the firm, turnover ratios may also serve to portray the liquidity of the firm (along with the ratios in Figure 7.1). In particular, the working capital ratios (b through f in Figure 7.3) may give clues to the cash cycle of the firm, that is, its ability to convert receivables and inventories into cash.

Profit ratios and turnover ratios lie behind a most important group of comparisons—those describing return on investment (Figure 7.4). Indeed, the earning-power-of-total-investment ratio is simply the EBIT ratio times the total-asset-turnover ratio. Because a firm's return on investment is a function of both its ability to produce cheaply (margins) and sell in quantity (turnover), the rates of return on investment ratios are good summaries of the overall earning position of the firm.

Ratio	Method of Computation	Interpretation
a. Total asset turnover	$\dfrac{\text{Net sales}}{\text{Total assets}}$	Ability of invested capital to produce gross revenue
b. Receivables turnover	$\dfrac{\text{Net credit sales}}{\text{Average receivables}}$	Quality and liquidity of receivables
c. Average collection period	$\dfrac{\text{Average receivables} \times 365 \text{ (or 360)}}{\text{Net credit sales}}$	Period required to collect average receivables
d. Inventory turnover	$\dfrac{\text{Cost of goods sold}}{\text{Average inventory}}$	Liquidity of inventory and tendency to overstock
e. Average day's inventory	$\dfrac{\text{Average inventory} \times 365 \text{ (or 360)}}{\text{Cost of goods sold}}$	Holding period of average inventory
f. Working capital turnover	$\dfrac{\text{Net sales}}{\text{Current assets} - \text{Current liabilities}}$	Indication of the cash cycle of the firm

Figure 7.3. Turnover Ratios.

Ratio	Method of Computation	Interpretation
a. Net earning-power ratio	$\dfrac{\text{Net income}}{\text{Total assets}}$	Net earning power of invested capital
b. Earning power of total investments	$\dfrac{\text{EBIT}}{\text{Total assets}}$ also (EBIT ratio) $\times$ (Total asset turnover)	Ability of invested capital to produce income for all investors (bond and stock holders) Eliminates "leverage" effect from net profit-total-assets ratio
c. Net profit to common equity	$\dfrac{\text{Net income}}{\text{Total common equity}}$	Net earning power of common capital

Figure 7.4. Return on Investment Ratios.

Ratio	Method of Computation	Interpretation
a. Total debt-to-equity ratio	$\dfrac{\text{Current liabilities} + \text{Long-term debt (L.T.D.)}}{\text{Total common equity}}$	Total amount of debt leverage per dollar of common equity
b. Total debt to total capital or Total debt to total assets	$\dfrac{\text{Current liabilities} + \text{L.T.D.}}{\text{C.L.} + \text{L.T.D.} + \text{Pfd. Stk.} + \text{Total common equity}}$ $\dfrac{\text{Current liabilities} + \text{L.T.D.}}{\text{Total assets}}$	Debt financing per dollar of total finance Asset security of debt sources of finance
c. Long-term-debt-to-equity ratio	$\dfrac{\text{L.T.D.}}{\text{Total common equity}}$	Long-term debt leverage p dollar of common equity
d. Tangible-assets debt coverage	$\dfrac{\text{Total assets} - \text{Intangibles} - \text{Current liabilities}}{\text{L.T.D.}}$	Asset security of long-term debt sources of finance
e. Total-debt-plus-preferred-to-equity-ratio	$\dfrac{\text{Current lib.} + \text{L.T.D.} + \text{Pfd. Stk.}}{\text{Total common equity}}$	Debt and preferred levera, per dollar of common equi
f. Times-interest-earned ratio	$\dfrac{\text{EBIT}}{\text{Long-term debt interest}}$	Income security of long-te debt
g. Cash-flow-times-interest-earned ratio	$\dfrac{\text{EBIT} + \text{Depreciation}}{\text{Long-term debt interest}}$	Short-run ability to meet interest payments
h. Average of interest and sinking fund payments	$\dfrac{\text{EBIT} + \text{Depreciation}}{\text{Interest} + \text{Sinking fund}}$ $\dfrac{\text{EBIT} + \text{Depreciation}}{\text{Interest} + \text{Sinking fund} - (\text{Tax rate})(\text{Depreciation})}$ $\overline{}$ divided by $1 - \text{Tax rate}$	Coverage of interest and sinking-fund payments when depreciation exceeds sinking-fund payments Coverage of interest and sinking-fund payments wh depreciation is less than sinking-fund paymen
i. Times-interest-earned plus preferred dividends	$\dfrac{\text{EBIT}}{\text{Long-term interest} + \dfrac{\text{Preferred dividends}}{1 - \text{Tax rate}}}$	Income security of the preferred stock

Figure 7.5. Leverage and Capital Structure Ratios.

The use of debt by the enterprise to improve earnings for shareholders is demonstrated by the ratios in Figure 7.5 (leverage and capital structure).

These ratios serve further to test the asset security of debt sources and the risk accruing to bondholders and stockholders from financial leverage. The income security of bondholders and preferred-stock holders is also indicated by these ratios.

The asset-relation ratios (Figure 7.6) may be used to supplement some of the previously outlined groups. The composition of assets into earning and liquid parts is suggested by ratio "a" in Figure 7.6. Ratios "b" and "c" are useful supplements to the liquidity and turnover ratios of Figure 7.1 and Figure 7.3. Ratio "d" is an indicator of the regularity and uniformity of depreciation allowances and may imply the necessity of adjusting the firm's depreciation accounts. Ratio "e" provides information about the average life of plant and equipment. It can also be used to determine whether reported earnings have been overstated (understated) when book depreciation has been inadequate (excessive).

Ratio	Method of Computation	Interpretation
a. Plant and equipment to total assets	$\dfrac{\text{Net plant} + \text{Net equipment}}{\text{Total assets}}$	Proportion of operating-earning assets to total assets
b. Inventory to total assets	$\dfrac{\text{Average inventory}}{\text{Total assets}}$	Size of inventory and tendency to overstock
c. Receivables to total assets	$\dfrac{\text{Average receivables}}{\text{Total assets}}$	Size of receivables and credit policy
d. Annual depreciation to plant and equipment	$\dfrac{\text{Annual depreciation}}{\text{Gross plant and equipment}}$	Regularity and uniformity of depreciation allowances
e. Approximate average asset life	$\dfrac{\text{Net plant and equipment}}{\text{Normalized depreciation}}$	Average life of plant and equipment

Figure 7.6. Asset Relation Ratios.

The security of common stock may be determined from the ratios in Figure 7.7. The asset security, income security, dividend security, and leverage risk to shareholders are indicated by the ratios in this group.

The ratio of the market price of the common stock to its book value (Ratio "a" in Figure 7.7) is often identified as Tobin's "q" ratio, although a more refined analysis would attempt to adjust the firm's assets and liabilities to market values (which primarily impacts fixed assets) before computing book value. Keynes [7] noted that, during the Depression, existing capital could be acquired on the stock exchanges from the purchase of existing companies below its cost of reproduction (i.e., $q < 1$), which impeded real investment and hence, recovery. The ratio of market price to book value is also employed as a crude screen for growth (high ratio) versus value (low ratio) stocks. It also is the basis for one of the capital market anomalies (low is better) and serves as the basis for one of the factors in the Fama-French [4] benchmark model. The other anomaly in Fama-French is based upon the market price of the common stock times the number of shares outstanding (the market capitalization, where small is better).

Ratio	Method of Computation	Interpretation
a. Book value per share of common stock	Total common equity / No. shares outstanding	Asset security of the common stock
b. Net tangible assets per share	Total common equity—Intangible assets / No. shares outstanding	Tangible asset security of the common stock
c. Leverage and capital—structure ratios	(see Figure 7.5)	Leverage risk of the common stock
d. EPS ratio	Net income available for common / No. shares outstanding	Earnings per share of common stock
e. DPS ratio	Dividends paid on common / No. shares outstanding	Dividends per share of common stock
f. Cash-flow-per-share ratio	Net income+Depreciation / No. shares outstanding	Cash earnings per share retainable for investment in assets
g. Pay-out-ratio	Dividends paid on common / Net income available for common	Dividend security of the common stock and the dividend policy of the corporation

Figure 7.7. Common Stock Ratios.

The final group of ratios (Figure 7.8) describes the yields of the securities of the firm. As such, these ratios are among the more important class of ratios. The percentage rates of return to bondholders, preferred stock holders, and common stock holders are computed with these ratios. As such, they are critically important for yield-oriented securities (bonds, preferred stock, and certain types of common stock such as those of public utilities). Many common stocks today pay little or no dividends so ratio "c" may not be very meaningful for them. Ratio "d," however, is probably the most frequently computed ratio of all for common stocks.

Ratio	Method of Computation	Interpretation
a. Net yield to maturity	To be read from bond tables or computer files	Expected annual rate of income retur on funds invested in bonds
b. Current yield of preferred	Annual cash dividends on pfd. / Market price of pfd.	Expected annual rate of income retur of funds invested in preferred
c. Current yield of common	Annual cash dividends on common / Market price of common	Expected annual rate of dividend inco return of funds invested in common
d. Price-earnings ratio	Market price of common / Dollars earned per share of common	Price of common relative to earnings

Figure 7.8. Yield and Price Ratios.

Another presentation of income statement and balance sheet ratios is on a common size basis. This form breaks down each category of income, expense, asset, liability, and so on into percentages. Thus, all revenue and cost items are expressed as percentages of net sales. All asset items are expressed as percentages of total assets, and all liability and net worth items are expressed as percentages of total liabilities and net worth. This form of computation allows the analyst to make immediate comparisons of all items in the income statements and balance sheets of a firm over time and among firms at a point in time. Consider the income statement and balance sheet for the Ungar Corp. given in Figure 7.9. Common-sized statements for each report are indicated in the parentheses next to each figure.

Ungar Corp. Common-Sized Statements		
Net sales	$200,000	100%
Cost of goods sold	50,000	25%
Gross income	$150,000	75%
Selling expense	20,000	10%
Administrative expense	30,000	15%
Earnings before interest and taxes	$60,000	30%
Taxes	30,000	15%
Net Income	$30,000	15%
Current assets		
Cash	$50,000	10%
Receivables	40,000	8%
Inventory	60,000	12%
Total current assets	$150,000	30%
Fixed asset	$350,000	70%
Total assets	$500,000	100%
Current liabilities	$100,000	20%
Bonds	200,000	40%
Net worth	200,000	40%
Total liabilities & net worth	$500,000	100%

Figure 7.9. Common-Sized Statements.

7.3 Ameritape, Inc.

We shall now apply the above analysis by reviewing the operations of Ameritape Inc., a fictitious entity which will be our example in the material to follow. Ameritape Inc. (the "Company") is the largest and most technically advanced producer of pressure-sensitive labels and associated products in the United States. Its sales volume in the fiscal year ended December 2002 was $5,438,195,000. The Company's product line includes:

Panel fronts	Property tags	Calibration labels
Nameplates	Parking stickers	Wiring diagrams
Trim strips	Packaging labels	ICC warning labels
Serial number labels	Security tags	Instruction labels
Warning labels	Shipping labels	Membrane switches
Decals	Lapel badges	

Virtually all of Ameritape's customers are located in the United States, and no one customer accounts for more than 5% of revenues. Customers include Dell, Texas Instruments, General Electric, General Motors, plus numerous small manufacturing companies. The customer base is diverse. The Company has an estimated 20% market share and is considered to be a high quality producer. There are 20 other major and many minor competitors in the United States and abroad. E. E. Wilson founded Ameritape in 1956 (originally, the American Tape Company), and has been its CEO and Chairman of the Board since it was founded. The Company went public in 1969 at $1.40 per share (adjusted for splits). He is also the president and one of the 10 directors of the Company.

The typical Ameritape manufacturing plant facility occupies 55,000 square feet. The plant facility is generally divided as follows:

Area:	Sq. Feet:
General and Administration	2,300
Sales	6,300
Art	2,000
Warehouse	8,400
Screen Printing	14,500
Die Cutting	5,000
Flexography	16,500

Plants are constructed to provide for additional capacity and improvements in productivity and quality control. Buildings usually have a controlled air conditioning/heating system for the manufacturing areas that reduces dust, material expansion/contraction and spoilage. In addition, buildings are designed to centralize many of the production management and support departments, and allow for the implementation of job jacket/history filing systems that improve productivity of the manufacturing process. Facilities are designed for $50 million to $70 million manufacturing capacity, with additional capital investments in manufacturing equipment. (Taken from the Form 10K for the period ending December 31, 2002).

The typical plant has approximately 300 printing presses and silk screening machines plus special die cutting and slitting machines. Two basic printing processes are used by the Company: Silk-screening and flexographic printing. Each process accounts for roughly 50% of the Company's volume. Ameritape maintains its own machine shops for equipment repair and customization. The normal work force is 800–950 people per plant. All company personnel are non-union. All production people are trained, many "from scratch," by the Company and work five eight-hour days. The Company has recently instituted night shifts to accommodate growth. A bonus system exists for all employees and takes into account attendance. There are managers for each of four plant departments (accounting, silk-screening, flexography, and sales). Their salaries are in the mid $50,000 range. (Taken from the Form 10K for the period ending December 31, 2002).

Orders for each plant are processed on computer terminals which contain data on every customer order received over the last three years. New and repeat orders can therefore be immediately verified. Similarly, work in process is entered into the computer system on line so that the order department can give current job status reports to inquiring customers who may visit a secure web site via a password. There are generally about 700 orders in process at any one time. Most aspects of operations are computerized, including order entry, job tracking, machine control and general ledger. Many of the computer systems are unique to the Company which developed most of its own software. After an order is received, normal production steps are as follows: Graphic design, typesetting, photo lab preparation, die design, printing, lamination, adhesive application, backing and 100% inspection. The Company produces everything (except raw materials) needed to ship finished goods to its customers. (Taken from the Form 10K for the period ending December 31, 2002).

7.4 The Auditor's Opinion

After a general review of the nature of business, history, and background of a company, most analysts turn to the financial statements. As we observed in Chapter 1, the SEC (in conjunction with the Financial Accounting Standards Board, or FASB) establishes accounting policy for public corporations (and other entities) in the United States. Accounting policy is codified under Generally Accepted Accounting Principles (GAAP), which are predicated upon historical transactions in the affairs of a company. As one authority has put it ([9], pp. 11−12):

> GAAP uses historical cost and accrual accounting to measure economic activity. Historical cost accounting registers events based on the economic value observed when a transaction is completed. A transaction is an exchange that transfers value between unrelated parties. (Within a company, a transaction records transfers of value between parts of the firm.) These values are carried forward in a firm's financial statements, unchanged unless a future transaction affects the item. Thus, even if subsequent events increase the market price of an item in inventory, its book value is not adjusted to reflect market conditions.

With the collapse of Arthur Andersen and the recent furor over the lack of accountability of accountants, it is even more important for analysts (and investors) to understand what is in (and what is *not* in) financial statements. Thus, one must know quite a bit about how financial statements are put together by accountants to understand them. An important place to start is the Report of the Independent Auditors (see Figure 7.10).

To the Stockholders and Board of Directors of Ameritape, Inc.

In our opinion, the consolidated financial statements listed present fairly, in all material respects, the financial position of Ameritape, Inc. at December 31, 2002 and 2001, and the results of its operations and its cash flows for each of the three years in the period ended December 31, 2002, in conformity with accounting principles generally accepted in the United States of America. In addition, in our opinion, the financial statement schedule listed in the accompanying index presents fairly, in all material respects, the information set forth therein when read in conjunction with the related consolidated financial statements. These financial statements and financial statement schedule are the responsibility of the Company's management; our responsibility is to express an opinion on these financial statements and financial statement schedule based on our audits. We conducted our audits of these statements in accordance with auditing standards generally accepted in the United States of America, which require that we plan and perform the audit to obtain reasonable assurance about whether the financial statements are free of material misstatement. An audit includes examining, on a test basis, evidence supporting the amounts and disclosures in the financial statements, assessing the accounting principles used and significant estimates made by management, and evaluating the overall financial statement presentation. We believe that our audits provide a reasonable basis of our opinion.

Arthur Brown, LLP

Houston, Texas

March 29, 2003

Figure 7.10. Ameritape, Inc. Report of Independent Auditors.

It will be observed that this Report holds the Company's management responsible for the financial statements of the company. The role of the auditor is to express an "opinion" on these statements based on the audit which was conducted in accordance with generally accepted auditing standards (GAAS). The key for the analyst is that the auditor finds the statements to be "in conformity with accounting principles generally accepted in the United States of America." This means that the numbers are prepared in accordance with GAAP; and when this is done, the auditors are said to provide a "clean" or "unqualified" opinion. If the auditors cannot offer a "clean" opinion, they will "qualify" their Report. A "qualified" opinion should generally be taken as a "red flag" that there may be material problems with the record keeping, accounting practices, or even possible solvency of the company. Section 7.10 contains further reflections on accounting numbers and accountants.

It should be noted that even a "clean" option does not necessarily mean that the figures presented are always useful. As one accounting professor has observed ([9], p.28):

The purpose of GAAP is to ensure that a firm's financial statements faithfully represent its operating results and economic circumstances. Remember that GAAP is transaction oriented. Financial statements record events when an exchange occurs, but other things that affect the value of a firm happen without exchanges. For example, an increase in the know-how of employees adds to value; so does creation or improvement in the good will of customers toward the firm. Yet neither phenomenon is an event that GAAP records. GAAP in general does not attempt to value these intangible assets.

Despite the fact that financial statements prepared according to GAAP may not tell us everything we need to know about the economic activities of a company, the statements are a very useful place to begin to understand these activities. Of course, the shrewd analyst (investor) will always keep in mind the old accountage adage, "Profit is an opinion, but cash is a fact!"

7.5   The Historical Record

A review of Ameritape, Inc.'s most recent Annual Report and Form 10K for the year ending December 31, 2002 produces the information found in Figures 7.11 through 7.15. An examination of Figure 7.11 suggests that the Company has been growing over the past five years. Net sales have increased by almost $1.1 billion (24.9%), or by a compounded annual rate of about 5.7%. This figure is calculated by solving the following for g:

$$(\$4,355,301)(1+g)^4 = \$5,438,195.$$

This rather crude compounding calculation assumes continuous growth by examining only the beginning and ending sales figures, whereas the actual numbers fluctuated a bit around that smooth compounding curve. More precise methods of calculation for more volatile patterns will be discussed later in Chapter 8.

AMERITAPE, INC. STATEMENT OF INCOME SUMMARY FOR THE YEAR ENDED DECEMBER 31 (IN THOUSANDS EXCEPT PER SHARE NUMBERS)					
	2002	2001	2000	1999	1998
Net Sales	$5,438,195	$4,912,094	$4,525,059	$4,042,539	$4,355,301
Cost Goods Sold	2,818,009	2,687,589	2,655,245	2,195,033	2,314,249
Gross Profit	2,620,186	2,224,505	1,869,814	1,847,506	2,041,052
Operating Expenses	1,447,394	1,268,102	1,218,342	1,067,850	979,229
Operating Income	1,172,792	956,403	651,472	779,656	1,061,823
Other Income (Expense)	27,595	27,889	25,529	(68,943)	(126,372)
Income Before Tax	1,200,307	984,292	677,001	710,713	935,451
Income Tax	408,132	334,659	230,180	241,642	318,053
Net Income	$792,255	$649,633	$446,821	$469,071	$617,398
Average Number Of Shares Outstanding	640,001	640,001	639,811	629,877	623,241
Basic and Diluted Earnings Per Share	$1.24	$1.02	$.70	$.74	$.99
Dividends Per Share	$.40	$.36	$.36	$.36	$.36

Figure 7.11. Ameritape, Inc., Statement of Income 1998–2002.

It should be noted that sales declined in 1999, and began to recover the following year. Further review of Figure 7.11 also indicates that Ameritape's cost of goods sold (CGS) as a percentage of sales has been fairly stable. Calculations over the period 1998–2002 are as follows:

Year	CGS	Net Sales	Percentage
1998	$2,314,249	$4,355,301	53%
1999	2,195,033	4,042,539	54%
2000	2,655,245	4,525,059	59%
2001	2,687,589	4,912,094	55%
2002	2,818,009	5,438,195	52%

With the exception of the year 2000, the cost of goods sold ratio has been fairly stable. It is worth noting that the Annual Report for the year 2001 (not included here) advises the stockholders that prices were cut during 2000 in order to stimulate demand (which had declined the prior year). Since sales increased in 2000 (despite continuing recessed aggregate economic activity) this strategy appears to have been successful. The 2001 Annual Report goes on to say, however, that expenses rose despite price reductions. This had a negative impact on operating margins. The good news, however, is that expenses were contained during 2001 and 2002, resulting in improved profit margins (reduced CGS as a percentage of sales).

At this point, the reader will have observed that we have begun the process of calculating some of the ratios provided in Figures 7.1 through 7.8; in particular the profit ratios of Figure 7.2 may be computed for the years 1998–2002 from the data in Figure 7.11. These are calculated as follows:

	1998	1999	2000	2001	2002
Gross Margin	.47	.46	.41	.45	.48
Operating Income Ratio	.24	.19	.14	.19	.22
Operating Ratio	.76	.81	.86	.81	.78
Net Margin	.14	.12	.10	.13	.15

These calculations confirm that profit rates declined from 1998 to 2000, and then recovered in 2001 and 2002.

Since many of the other ratios require balance sheet information, we need to add the information in Figure 7.12 to the analysis. Once this is done, we may calculate most of the ratios in Figures 7.1 through 7.8. Some of these computations (and an interpretation of their meaning for Ameritape) are provided after we review the notes to the financial statements, in the next chapter, and in Chapter 10.

AMERITAPE, INC.
BALANCE SHEET SUMMARY
AS OF DECEMBER 31, 2002 and 2001
(IN THOUSANDS)

	2002	2001
ASSETS		
CURRENT ASSETS		
Cash and temporary investments (Note 2)	$807,362	$318,429
Accounts receivable, trade, less allowance for		
losses (2002, $58,153;2001, $52,498)	797,038	719,532
Inventories	151,567	131,836
Deferred income taxes (Note 7)	42,600	43,900
Prepaid expenses and other	99,116	43,104
TOTAL CURRENT ASSETS	1,897,683	1,256,801
MARKETABLE SECURITIES, at lower of		
cost or market (Note 3)	21,076	37,037
PROPERTY AND EQUIPMENT, at cost,		
less accumulated depreciation (Note 4)	713,401	653,316
OTHER ASSETS (Note 10)	23,656	25,092
TOTAL ASSETS	$2,655,816	$1,972,246
LIABILITIES		
CURRENT LIABILITIES		
Accounts payable, trade	$186,971	$206,993
Other payables	142,340	-
Other accrued expenses	136,481	183,792
Current portion of long-term debt (Note 5)	5,294	6,618
Current portion of capital lease obligations	15,990	-
TOTAL CURRENT LIABILITIES	487,076	397,403
LONG-TERM DEBT, less current portion	63,194	17,325
above (Note 5)		
CAPITAL LEASE OBLIGATIONS,		
less current portion above (Note 9)	11,773	-
TOTAL LIABILITIES	562,043	414,728
STOCKHOLDERS' EQUITY		
CAPITAL CONTRIBUTED		
Common stock (no par), authorized		
1,000,000,000 shares, 640,001,226 shares		
issued and outstanding as of 2002 and 2001	10,000	10,000
RETAINED EARNINGS	2,083,773	1,547,518
TOTAL STOCKHOLDERS' EQUITY	2,093,777	1,557,518
TOTAL LIABILLITIES		
AND STOCKHOLDERS' EQUITY	$2,655,816	$1,972,246

The accompanying Notes to Financial Statements are an integral part of this statement.

Figure 7.12. Ameritape, Inc. Balance Sheet as of December 31, 2002 and 2001.

AMERITAPE, INC.
STATEMENT OF INCOME AND RETAINED EARNINGS

For the Three Months and the Years Ended
December, 2002 and 2001
(IN THOUSANDS)

	2002		2001	
	Three Months	Year	Three Months	Year
NET SALES	$1,445,232	$5,438,195	$1,234,952	$4,912,094
COST OF GOODS SOLD	733,015	2,818,009	687,319	2,687,589
GROSS PROFIT	712,217	2,620,186	547,633	2,224,505
OPERATING EXPENSES	426,577	1,447,394	470,566	1,268,102
OPERATING INCOME	285,640	1,172,792	77,067	956,403
OTHER INCOME	(1,251)	27,595	(189)	27,889
INCOME BEFORE INCOME TAXES	284,389	1,200,387	76,868	984,292
INCOME TAXES (Note 7)	86,149	408,132	(12,478)	334,659
NET INCOME (LOSS)	198,240	792,255	64,400	649,633
RETAINED EARNINGS Balance beginning of the period		1,547,518		1,128,285
Dividends		256,000		230,400
TOTAL RETAINED EARNINGS		$2,083,773		$1,547,518
Basic and Diluted Earnings per Share	$.31	$1.24	$.10	$1.02

The accompanying Notes to Financial Statements are an integral part of this statement.

Figure 7.13. Ameritape, Inc. Statement of Income and Retained Earnings as of December 2002 and 2001.

7.6   Notes to the Financial Statement

Typically, financial statements carry detailed notes which often contain very important information. Below we present these notes for Ameritape, Inc., for the three months and the years ended December 31, 2002 and 2001.

1. Summary of significant accounting policies.

The significant accounting policies followed by the Company are summarized as follows: Inventory valuation — Inventories are stated at the lower of cost or market. Cost has been determined on the first-in, first-out basis. Market is based upon realizable value less allowance for selling and distribution expenses and normal gross profit. Inventories consist of raw materials used in production. Depreciation — Depreciation of machinery and equipment is computed by the straight-line and declining-balance methods over their estimated useful lives of 3–10 years. Leasehold improvements are amortized by the straight-line method over the shorter of the life of the related asset or the life of the lease.

2. Cash and temporary investments.

Cash and temporary investments consists of the following (in thousands):

	2002	2001
Checking accounts (overdraft)	$139,376	$(117,355)
Savings account	–	129,024
Time deposits	–	200,000
Money Market Fund	424,427	106,464
Money Market Account	243,219	–
Cash fund	340	296
Total cash and temporary investments	$807,362	$318,429

The carrying value of cash and temporary investments approximate their fair market value.

3. Marketable securities.

Marketable securities consist of treasury notes which are valued, in aggregate, at the lower of cost or market. The cost and market value were not materially different at the valuation date and are as follows (in thousands):

2002	2001
$21,076	$37,037

4. Property and equipment.

Property and equipment consist of the following (in thousands):

Cost		
Machinery and equipment	$2,024,071	$1,765,376
Leasehold improvements	120,247	114,678
Total cost	2,144,318	1,880,054
Accumulated depreciation		
and amortization	1,430,917	1,226,738
Net property and equipment	$713,401	$653,316

5. Long-term debt.

Long-term debt consists of the following (in thousands):

	2002	2001
8% note, due August 2014 secured by equipment	$68,488	$12,695
6.65% note due June 2002, unsecured	-	11,248
Total long-term debt	68,488	23,943
Less current portion	5,294	6,618
Non-current portion	$63,194	$17,325

6. Officers' life insurance.

The Company is the owner and beneficiary of life insurance policies carried on the life of the officers of the Company bearing face value in the amounts of $236 million. The net cash surrender value of officers' life insurance consists of the following (in thousands):

	2002	2001
Cash surrender value	$150,730	$137,910
Less loans secured by the cash surrender value	141,862	133,569
Net cash surrender value	$8,868	$4,341

7. Income taxes.

Income tax expense differs from amounts currently payable because certain expenses are reported in the income statement in periods which differ from those in which they are subject to taxation. The principal differences in timing between the income statement and taxable income involve (a) net capital loss carryforwards, and (b) accelerated cost recovery provisions in excess of book depreciation. The differences between income tax expense and taxes currently payable are reflected in deferred tax accounts in the balance sheet.

8. Employees' benefit plans.

The Company has a defined benefit pension plan which covers all full-time employees who meet certain eligibility requirements as to age and length of service. The accounting and funding policy of the plan is to provide amounts sufficient to meet normal cost plus amortization of prior service costs as computed by independent actuaries. The total pension expense for the years ended December 31, 2002 and 2001 was approximately $42 million and $36 million respectively. The approximate benefit and asset information for the plan is as follows actuarial present value of accumulated benefits (in thousands):

Vested	$200,000
Non-vested	—
Net assets available for benefits	$200,000

The above actuarial values are based upon a minimum 6% rate of return. The vested benefits and net assets available for benefits are determined as of December 31, 2002. In addition, the Company has a profit sharing plan for all full-time employees who meet certain eligibility requirements as to age and length of service. Profit sharing contributions charged to operations were approximately $100 million for 2002 and $60 million for 2001. The plan conforms to all ERISA requirements in effect.

9. Lease commitments.

Future minimum rental commitments on non-cancellable leases are approximately $238 million for real estate for the year ending December 31, 2003. The existing non-cancellable leases will expire during fiscal year 2003. Most leases have an option to renew for various periods. The total expense on non-cancellable operating leases for the years ended December 31, 2002 and 2001 was approximately $166 million and $106 million respectively. During 2002, the Company entered into various equipment lease arrangements which have been capitalized. The following is a schedule of future minimum lease payments of such capitalized leases as of December 31, 2002 (in thousands):

Years ending December 31,	Amounts
2003	$6,000
2004	6,000
2005	6,000
2006	6,180
Total minimum lease payments	24,180
Less amount representing interest	12,407
Present value of minimum lease payments	$ 11,773

10. Other Assets (in thousands):

	2002	2001
Cash surrender value—officer's life insurance	$8,868	$4,341
Other deposits	14,788	20,751
TOTAL OTHER ASSETS	$23,656	$25,092

7.7   The Most Recent Year and Ratio Calculations

Figures 7.14 and 7.15 provide detail for determining the cost of goods sold and the operating expenses summarized in Figure 7.11. As observed earlier, the gross margin improved from .45 in 2001 to .48 in 2002.

AMERITAPE, INC.				
FINANCIAL STATEMENT DETAIL				
For the Three Months and the Years Ended December 31, 2002 and 2001 (IN THOUSANDS)				
	2002		2001	
	Three Months	Year	Three Months	Yea
MANUFACTURING LABOR				
Labor - data label	$25,103	$70,064	$ 9,518	$45,98⬦
Labor - flexo	65,771	263,099	71,926	291,91⬦
Labor - art	20,601	73,348	22,184	84,81⬦
Labor - silkscreen	205,562	807,710	196,311	771,90⬦
Labor - temporary	−	672	3,350	4,00⬦
Total manufacturing labor	317,037	1,214,893	303,289	1,198,62⬦
MANUFACTURING MATERIALS				
Flexo and tape stock	151,512	597,007	134,543	538,58⬦
Flexo supplies and ink	11,933	45,260	9,992	36,28⬦
Flexo tools and dies	14,959	56,390	11,660	40,82⬦
Silkscreen stock	104,711	399,474	109,227	385,80⬦
Silkscreen supplies and ink	31,323	121,249	29,300	122,13⬦
Silkscreen tools and dies	29,275	100,621	23,655	88,05⬦
Data label stock	2,883	5,279	1,538	6,13⬦
Data label supplies and ink	843	3,581	1,043	3,68⬦
Data label tools and dies	1,673	8,171	1,890	6,75⬦
Art department - miscellaneous	481	2,410	947	8,27⬦
Art supplies	6,195	23,169	5,923	18,73⬦
Miscellaneous	11,078	(3,853)	(1,317)	(6,531
Total manufacturing materials	344,710	1,358,758	328,401	1,248,72⬦
OTHER MANUFACTURING COSTS				
Maintenance supplies and repairs	8,672	34,544	9,102	34,34⬦
Power	10,285	43,836	9,744	42,57⬦
Depreciation	48,563	147,764	30,897	140,34⬦
Rentals	569	7,041	2,996	12,42⬦
Shipping supplies	3,179	11,173	2,890	10,55⬦
Total other manufacturing costs	71,268	244,358	55,629	240,24⬦
TOTAL COST OF GOODS SOLD	$733,015	$2,818,009	$687,319	$2,687,58⬦

Figure 7.14. Ameritape, Inc. Cost of Goods Sold, 2002 and 2001.

Calculated in reverse, we see that the ratio of cost of goods sold to sales declined from .55 ($2,687,598/$4,912,094) to .52 ($2,818,009/$5,438,195). On the other hand, the operating ratio fell from .81 in 2001 to .78 in 2002. This is due to the fact that operating (selling general, and administrative expenses) rose in relation to net sales from .26 ($1,268,102/$4,912,094) to .27 ($1,447,394/$5,438,195).

AMERITAPE, INC. OPERATING EXPENSES For the Three Months and the Years Ended December 31, 2002 and 2001 (IN THOUSANDS)				
	2002 Three Months	Year	2001 Three Months	Year
SELLING EXPENSES				
Entertainment	$33,849	$102,339	$42,783	$94,891
Travel	7,366	20,458	4,593	22,946
Depreciation	3,736	27,272	8,892	16,000
General	8,716	13,620	(396)	8,265
Total selling expenses	53,667	163,689	55,872	142,102
GENERAL AND ADMINISTRATIVE EXPENSES				
Gardening	340	1,870	510	2,040
Depreciation	14,147	51,289	12,796	68,849
Amortization	166	663	166	663
Dues and donations	1,324	7,647	4,448	5,875
Insurance	26,388	67,006	24,828	41,346
Legal and accounting	8,526	37,240	12,530	36,823
Office salaries	95,176	311,013	89,341	298,234
Office supplies	2,250	8,342	2,370	7,632
Office computer supplies and service	3,877	17,542	10,040	21,619
Office auto leasing	1,095	4,380	1,095	7,071
Office misc.	2,683	7,487	1,895	6,129
Taxes - payroll	30,981	135,005	23,137	114,812
Taxes - property	3,965	15,590	3,863	14,413
Freight, postage and mailing	18,580	62,722	16,636	54,455
Rent	41,516	165,919	25,875	105,655
Bad debt expense	20,406	51,258	40,929	76,165
Telephone	4,534	18,218	4,734	13,421
Uniforms and cleaning supplies	1,902	5,601	1,755	5,015
Heat, light and water	2,732	9,167	3,248	9,190
Janitorial	1,679	7,958	2,735	8,647
Building maintenance and repairs	7,794	28,156	3,891	23,916
Safety and first aid	742	2,706	769	2,425
Employees' benefits	650	2,337	−	1,656
Employees' pension plan	11,974	41,974	35,695	35,695
Employees' profit sharing plan	25,000	100,000	51,154	60,000
Employees' health plan	30,000	80,000	25,000	70,000
General	14,483	42,615	15,254	34,254
Total general and administrative expenses	372,910	1,283,705	414,694	1,126,000
TOTAL OPERATING EXPENSES	$426,577	$1,447,394	$470,566	$1,268,10

Figure 7.15. Ameritape, Inc. Operating Expenses for 2002 and 2001.

7.8   Other Information

In addition to the Form 10K and Annual Report, the Company's "Proxy Statement" also contains valuable information. The "Proxy" is usually sent as part of the "Annual Meeting Notice" which describes when and where the annual meeting of the corporation will take place. Since most shareholders do not attend these meetings in person, the "Proxy" allows them to vote on significant matters that may come before the meeting. The "Notice" for Ameritape is contained in Figure 7.16.

Ameritape, Inc.
4929 Allen Speedway Drive
Houston, Texas

Proxy Statement and
2003 Annual Meeting
Notice

NOTICE OF ANNUAL MEETING OF SHAREHOLDERS
TO BE HELD MAY 10, 2003

TO OUR SHAREHOLDERS:

The Annual Meeting of Shareholders of Ameritape, Inc. will be held in the Company Auditorium on May 10, 2003, at 10:00 AM Houston time for the following purposes:

(1) To elect three directors as members of the class of directors to serve until the third succeeding Annual Meeting of Shareholders and until their successors have been elected and qualified.

(2) To act on such other business that may properly come before the Annual Meeting or any adjournment(s) thereof.

The transfer books of the Company will not be closed, but only holders of Common Stock of record at the close of business on March 22, 2003, will be entitled to notice of and to vote at the Annual Meeting. A majority of the outstanding stock entitled to vote is required for a quorum.

Management sincerely desires your presence at the Annual Meeting. However, so that we may be sure that your vote will be included, please sign and date the enclosed proxy and return it promptly in the enclosed stamped envelope. If you attend the Annual Meeting, you may revoke your proxy and vote in person.

By Order of the Board of Directors,

James Smith, Secretary

Houston, Texas
April 13, 2003

Figure 7.16. Notice of Annual Meeting of Ameritape, Inc.

A market analysis of the pressure-sensitive label and associated products market has also been performed. The results are summarized as follows:

1. The total market size for all Ameritape products in the United States is estimated to be about $25,000,000,000.

2. The market has been growing at a rate equal to the growth rate in real G.D.P. (for the U.S.).

3. Ameritape has gradually increased its market share from about 10% in 1980 to present levels.

4. Ameritape is recognized as the leader in terms of quality and turn-around time, although it is perceived as "high priced."

5. Although technical innovation may require Ameritape to alter production technology over the next decade, such is unwarranted presently.

6. The Company is regularly inspected by OSHA and has been determined to be in compliance with all OSHA rules and standards.

7. Due to increasing workmen's compensation claims, the company has initiated a major safety program. The safety experience in 1998−2002 was such that a large increase in workman's compensation premiums should be experienced for at least the next five years.

8. Various suppliers have reported that the Company has no problem with accommodating any orders. Price increases of 5% are expected next year.

9. Customers report that the Company is a reliable, quality printer, although it charges premium prices and does not offer direct delivery of product.

7.9   Projections and Evaluation

At this point, we shall review Ameritapes's sales, operating income, EPS, and DPS number for the period 1991−2002 (Figure 7.17). Notice that the data include sales and profit figures over an entire business cycle. Coming out of a recession in 1991, sales, operating income, and earnings grow until 1999, when another downturn begins (slightly in advance of the 2000−2001 "mini" recession the U.S.A.). Absolute levels of sales decline in 1999, while operating income and net income (EPS) continue to fall through the year 2000. Sales and profits resume growing after 2000. The sales and profit growth patterns are consistent with that of the overall cycle of the U.S. economy. This should not be surprising since practically all of Ameritape's customers are industrial concerns which also evidence revenue and profit cyclicality.

Year	Sales (S)	Oper. Inc. (OI)	OI/S	EPS	DPS
1991	$3,087,493	$611,323	.20	$.56	$.20
1992	3,272,743	641,457	.20	.60	.20
1993	3,436,379	673,531	.20	.63	.24
1994	3,642,563	713,942	.20	.67	.24
1995	3,533,286	642,548	.20	.61	.24
1996	3,781,402	813,001	.20	.77	.24
1997	4,137,536	861,781	.20	.88	.24
1998	4,355,301	1,061,823	.19	.99	.36
1999	4,042,539	779,656	.19	.74	.36
2000	4,525,059	651,472	.14	.70	.36
2001	4,912,094	956,403	.19	1.02	.36
2002	5,438,195	1,172,792	.22	1.24	.40

Figure 7.17. Sales, Operating Income, EPS, and DPS
1991−2002 (in Thousands Except Per Share Numbers).

Quarterly EPS data over the past three years may be examined (see Figure 7.18).

Year	1st. Quarter	2nd. Quarter	3rd. Quarter	4th. Quarter
2000	$.20	$.22	$.20	$.08
2001	$.30	$.32	$.30	$.10
2002	$.30	$.34	$.29	$.31

Figure 7.18. Quarterly EPS 2000−2002.

Notice that there is some seasonality in the EPS numbers (except for 2002). This is the case even though sales are fairly non-seasonal over the year. The fourth quarter is typically a "catch-up" quarter where expenses are particularly heavy due to the nature of certain annual contract payments that are made on December 31 of the year. The company generally smoothes its dividend payout such that $.09 per quarter was paid in 2001 and 2000. The dividend was raised to $.10 per quarter in 2002. This is a typical pattern for companies since it is generally assumed that "cuts" in dividends are negatively viewed in the market place. Thus, the board of directors would like to have some reasonable degree of confidence that a dividend payout can be maintained before any increases are approved.

Now given the above data, we might attempt to value the stock of Ameritape, Inc. We might start by reiterating equation (6.1):

$$P(0) = \frac{D(0)}{(1+r)^0} + \frac{D(1)}{(1+r)^1} + \frac{D(2)}{(1+r)^2} + \dots \qquad (7.1)$$
$$= \sum_{t=0}^{\infty} \frac{D(t)}{(1+r)^t},$$

where

$$P(0) \quad = \quad \text{Price (Value) today}$$
$$D(t) \quad = \quad \text{Dividend } t \text{ years in the future}$$
$$r \quad = \quad \text{Market's (investor's) required rate of return.}$$

Notice that the focus of this equation is on determining P(0). To get a value, it will be necessary to project future dividends. This will probably require a projection of future earnings. At this point, a review of all of the ratios outlined in Figures 7.1−7.8 is in order. It may be necessary to forecast a number of variables depending upon the stability of the ratios calculated. Spread sheets are usually set up to make the computations. Also, for actual publicly-held companies, many of the ratios are calculated as part of data bases which may be purchased. Some of these are even available on line from brokers and other sources. Since Ameritape, Inc. is a fictitious company, we will have to compute some of the numbers to get a feel for what is going on. First, notice from Figure 7.17 that the sales have increased almost every year since 1991. They were down by a small amount (3.0%) in 1995, and by a larger figure (7.2%) in 1999. On the other hand, sales were up by over 10% in two years (2000 and 2002) and were up by at least 5% in nine of the eleven years for which we can make comparisons. The average annual sales increase was 5.4%, and the compounded growth rate (using just the beginning and ending years) was: $S(11) = S(0)(1+g)^{11}$, or $(5,438,195) = (3,087,483)S(0)(1 + g)^{11}$. Solving, we find $g = .053$, or about 5.3%. Given the relative stability of the sales pattern, we might be safe in projecting sales growth of about 5% per year in the future, but more sophisticated forecasting approaches will be presented in the next chapter.

Second, we also notice in Figure 7.17 that the operating income to sales ratio over 1991−2002 has been fairly stable (except for the year 2000, which was discussed above). The ratio centers around .20. Unless, there is reason to suspect otherwise, this might also be a good number to use for projecting purposes. Finally, we see the EPS ratio pretty much follows the pattern of the operating income to sales ratio. This will generally be the case unless there are major changes in the financial structure of the company (perhaps causing larger or smaller interest expenses), interest rates fluctuate wildly (again causing interest expense to vary), or the firm issues a substantial number of new shares over the period. Although we do not have financial statements in their entirety going back to 1991, it is safe to say that none of these events has happened. Nevertheless, the average and compounded growth rates in EPS are somewhat greater than that of sales. (The compounded growth rate of EPS is 7.2%). Part of the explanation of the difference is that the operating income to sales ratio was .20 in the base year of 1991 and .22 in the last year (2001). The compounded growth rate of operating income is 6.1%, so about half of the difference between the sales growth rate of 5.3% and the EPS growth rate of 7.2% is due to the higher operating margin. The rest is probably due to slight financial structure differences. In any event, our job will be to forecast future EPS and DPS, say, for at least five years hence. Before we can "value" Ameritape, we must pause to consider methods of accomplishing this task which we shall do in the next chapter.

7.10   Accounting Numbers and Corporate Accountability

As this book is written, some very important changes are being made to both the way accounting numbers are prepared and the way public corporations are held accountable. Sweeping legislation in the form of the Sarbanes–Oxley Act of 2002 (the "Act") has been passed by Congress and signed by President Bush to address the corporate and securities issues that have arisen in the past two years (and that have occupied much of our focus in this book). Although a comprehensive analysis of this legislation is beyond our scope here, a summary of some of the Act's key features is provided as follows:

1. A Public Company Accounting Oversight Board with investigative and disciplinary powers over auditing firms is to be appointed by the S.E.C. in consultation with the Chairman of the Federal Reserve Board and the Secretary of the Treasury. This Board will supplement the basically "self-policed" accounting regulatory bodies that exist presently but will have the power to set auditing standards and investigate and punish accountants.

2. Auditors will be limited in their non-audit functions and such "outside" activities as consulting will become unlawful.

3. Auditors will report to the audit committee of public corporations.

4. Corporate audit committees shall have greatly enhanced powers (and responsibilities).

5. Chief Executive Officers (CEOs) and Chief Financial Officers (CFOs) will be required to "certify" all annual and quarterly reports filed with the S.E.C. This certification must state that the signing officer has reviewed the report, that based on the officer's knowledge the report does not contain any untrue statement of a material fact or omit to state a material fact, and that based on the officer's knowledge the financial statements and other financial information in the report fairly presents the financial condition and results of operations of the reporting company.

6. Numerous other requirements affecting the governance and ethical standards of public companies.

Problems

Problem 7.1. Below are income statements and balance sheets for the Wynn Corp. for the years 2001 and 2002.

Wynn Corp. Consolidated Statement of Income (thousands)	2001	2002
Net sales	$114,868	$170,356
Cost of goods sold	108,086	154,064
Operating depreciation	168	187
Gross income	$6,614	$16,075
Selling and administrative expenses	760	2,540
Net operating income	$5,854	$13,535
Other income	35	474
Earnings before interest and taxes	$5,889	$14,000
Income taxes	2,600	6,720
Net income available for common	$3,229	$7,289
Common dividends	732	1,465
Balance carried to retained earnings	$2,497	$5,824
Add retained earnings beginning period	7,115	9,612
Retained earnings end of period	$9,612	$15,436

Wynn Corp. Consolidated Balance Sheet as of December 31 (thousands)	2001	2002
Current assets		
Cash	$3,625	$4,644
Receivables	13,896	20,468
Inventories	34,320	59,341
Total current assets	$51,951	$84,442
Net plant and equipment	$2,228	$3,084
Investment in subsidiaries	$1,253	$1,315
Total assets	$55,432	$88,841
Current liabilities		
Payables	15,000	27,000
Accruals	1,572	2,938
Reserves	2,498	5,851
Total current liabilities	$28,910	$56,496
Fixed liabilities	$-0-	$-0-
Net worth		
Common stock (1,000,000 shares)	$14,657	$14,657
Additional paid in capital	2,252	2,252
Retained earnings	9,612	15,436
Total net worth	$26,521	$32,345
Total liabilities and worth	$55,432	$88,841

a. Prepare common size statements for the Wynn Corp.

b. Compute all the ratios indicated in Chapter 7 for the Wynn Corp. for 2001 and 2002. The following information will be required:
(1) Daily operating expenditures—about $500,000 (*Note:* this may be determined by dividing annual operating expenditures by the number of days in the year. If operating expenses are expected to be higher in the next year than in the last one for which financial data are published, a budget for that year must be obtained. Also, the simple division of the number of days in the year into annual operating expenditures assumes that cash flows out evenly during the year. For seasonal businesses, this may not be a realistic assumption.)
(2) Market price of the stock as of the end of 2002—$146 per share.

c. From the financial statements you have, try to decide what type of firm Wynn is. The asset mix of the enterprise should be of use in making this determination.

d. Analyze the position of Wynn. Would you consider it a profitable operation? Does the firm appear to be a risky venture?

Problem 7.2. Hampstead Ltd. posted the financial statements for 2002

Net sales	$100,000	Current assets	$50,000
Cost of goods sold	50,000	Plant (net)	200,000
Gross income	$50,000	Copyrights	10,000
Other expenses	40,000	Total assets	$260,000
Net income before tax	$10,000		
		Liabilities	$100,000
Income tax	0	Net worth	160,000
Net income	$10,000		
		Total liabilities and worth	$260,000

a. Prepare an adjusted income statement and balance sheet for Hampstead. Cost of goods sold includes depreciation charges of $10 million based on straight-line depreciation. For tax purposes, $20 million was deducted based on accelerated depreciation. The auditors feel the straight-line figure is an adequate figure for real economic depreciation. Copyrights are valued at their acquisition price. A conservative appraisal of their economic value is about $100 million. Other expenses include an $8 million loss from litigation allowed as a tax deduction.

b. Reconcile the firm's tax payment. Assume the normal tax rate for the Company is 35%.

c. Compute these ratios for Hampstead:

(1) Gross profit margin

(2) EBIT ratio

(3) Net profit margin

(4) Total asset turnover

(5) Earning-power-of-total-investment ratio

(6) Net-profits-to-common-equity ratio

(7) Total debt-to-equity ratio.

Problem 7.3. As an investment analyst, you are reviewing the financial statements for the Boheme Corporation, a manufacturer of casual attire.

Boheme Corporation Statement of Income for the Year 2002 (thousands)		
Sales		$200,000
Less:		
Cost of goods sold	$100,000	
Selling, general & admin. expenses	65,000	
Depreciation	10,000	175,000
Earnings before interest and taxes		$25,000
Less Interest		5,000
Earnings before taxes		$20,000
Less Taxes		4,000
Net income		$16,000
Preferred dividends		1,000
Net income to common shareholders		$15,000
Common dividends		5,000
Earnings retained		$10,000

Boheme Corporation Balance Sheet as of December 31, 2002 (thousands)	
Current Assets:	
Cash	$5,000
Receivables	5,000
Inventories	10,000
Total current	$20,000
Fixed assets:	
Gross plant	$100,000
Less:	
Accumulated depreciation	30,000
	$70,000
Other assets	10,000
Total assets	$100,000
Current liabilities	
Accounts Payable	$10,000
Wages Payable	6,400
Total Current	$16,400
Bonds (10's due 2011)	50,000
Preferred stock (10%, $10 par)	10,000
Common stock ($1 par)	10,000
Retained earnings	13,600
	$100,000

For tax purposes, the firm uses accelerated depreciation. Charges amounting to $22,000 were deducted for 2002. You have determined the following additional data from the 2000 and 2001 annual reports of the company:

	2000	2001
Sales (thousands)	$50,000	$100,000
Gross margin	40.0%	45.0%
EBIT margin	14.0%	10.0%
Net margin	3.2%	4.0%
Asset turnover	.625×	1.1×
Return on investment		
(EBIT/Total assets)	8.75%	11.0%
Total debt to total capital	.75	.74
Interest coverage	1.4×	2.0×

a. Make any necessary adjustments to the 2002 statements.

b. Reconcile the firm's tax payment. Assume a normal tax rate of 35 percent.

c. Compute the above seven ratios for 2002.

Problem 7.4. You are considering the purchase of Boheme common stock (see 7.3 above). In order to value the security, you have attempted to project future sales and earnings for 2003. You are convinced that sales will increase according to the past pattern (i.e., up by 100%) and the EBIT margin will be about the same as it was in 2002. You assume debt levels and the tax rate will remain constant.

a. Determine sales, earnings before interest and taxes, and net income to common shareholders for 2003.

b. Earnings per share of common were $.16 in 2000 and $.40 in 2001. Determine EPS for 2002 and project EPS for 2003.

c. The firm paid no dividend in 2000 or 2001 but did pay out one-third of 2002 earnings. Cash dividends per share are expected to double in 2003.

(1) How much in dividends were paid in 2002?

(2) What projected pay-out rate is expected for 2003?

d. The P/E multiple for Boheme averaged 20× in 2000 and 2001. What was the average price of the stock for each year?

e. As of January 2, 2003, Boheme was selling at $15 per share. What is the current P/E multiple based on 2002 earnings?

f. What may account for this change in the multiple?

g. Based on the current price of Boheme and your expectations about the future of the company, would you buy it?

References

[1] Batra, R. (1987). *The Great Depression of 1990.* New York: Simon and Schuster.

[2] Bogle, J. C. (1999). *Common Sense and Mutual Funds: New Imperatives for the Intelligent Investor.* New York: John Wiley & Sons.

[3] Box, G. E. P. and G. M. Jenkins (1976). *Time Series Analysis: Forecasting and Control,* San Francisco: Holden-Day.

[4] Fama, E. F. and French, K. (1992). "The cross-section of expected stock returns," *Journal of Finance,* June, 427–465.

[5] Graham, B., Dodd, D. and Cottle, S. (1962). *Security Analysis,* 4th ed. New York: McGraw-Hill Book Company.

[6] Hicks, J. R. (1946). *Value and Capital*, 2nd ed., London: Oxford University Press.

[7] Keynes, J. M. (1936). *The General Theory of Employment, Interest, and Money*. New York: Harcourt, Brace and World.

[8] Lowenstein, R. (1996). *Buffett: the Making of an American Capitalist*. New York: Random House.

[9] Press, E. (1999). *Analyzing Financial Statements: 25 Keys to Understanding the Numbers*. New York: Lebhar-Friedman Books.

[10] Taylor, W. M. and Williams, E. E. (1991). "Market microstructure and Post Keynesian theory," *Journal of Post Keynesian Economics*, Winter, 233–247.

[11] Thompson, J. R. (1989). *Empirical Model Building*. New York: John Wiley & Sons.

[12] Tukey, J. W. (1977). *Exploratory Data Analysis*. Reading, MA: Addison-Wesley.

[13] Williams, E. E. and Findlay, M. C. (1974). *Investment Analysis*. Englewood Cliffs, NJ: Prentice-Hall.

[14] Williams, E. E. and Thompson, J. R.(1996). *The Economics of Production and Productivity: A Modeling Approach*. Austin, TX: Capital Book Company.

Chapter 8

Empirical Financial Forecasting

8.1 Introduction

The subject of forecasting is as old as recorded history. We recall how the Patriarch Joseph achieved his important position as Minister of Planning for Pharaoh as a result of predicting that Egypt, and the adjacent territories, would experience seven years of plenty followed by seven years of famine. It is interesting to note that Joseph achieved his promotion, not as a result of the accuracy of his forecast, but rather on the spot, largely as a result of the forecast itself. This is a recurring aspect of forecasting, namely, that a forecast is only valuable prospectively. Should Pharaoh have tarried for the full fourteen years to determine the accuracy of the forecast, then it would have been worth relatively little. Happily, Pharaoh accepted the forecast straightaway, to the benefit of Pharaoh, the forecaster and, most importantly, to the people of the region.

Joseph had the benefit of divine revelation, an insight which few forecasters can credibly claim. In many other ancient cultures, the mapping of future events was regarded not so much as a special revelation, but as a holistic science. The future was regarded as more or less fixed. If one could only understand the movements of the celestial bodies and their correlations with events on earth, then, since there was an absolutely predictable regularity to the movements of the planets, there was the possibility of reliable predictability of the events which affect the course of human lives. For example, we have abundant evidence that the Mayans spent a large fraction of the product of human industry on the maintenance of a hierarchy of priestly astronomers whose task was simply to predict the inevitable. The Mayans were not alone in this fixation, and we note that even today, billions of dollars are expended annually in some cultures to predict the favorable dates for weddings, new business ventures, etc. So numerous have been the astrologers in some of these societies that the name of astrologer is as common as that of Smith (Schmidt, Kowalski, etc.). For example, "Joshi" (astrologer) is one of the more common Mahratta surnames (and in Bengali "Ghosh," also meaning astrologer). We occasionally see obvious cultural transferals of this tendency into Western societies. The recent cyclical theories of the tenured SMU Professor of Economics, Ravi Batra in his best selling book *The Great Depression of 1990* [1] is one such example. Another is that of businessman H. E. Figgie [4] who forecasted in 1992 that the entire financial structure of the United States would collapse by 1995 if the U.S. federal government did not balance its budget and reduce the international trade deficit. Even absent such cultural transferals, some American investors appear to place mystical faith in this or that financial guru.[1]

A characteristic of the Hellenistic society (which is usually referred to as Western Civilization) is the belief that individuals are not the pawns of some fatalistic kismet but that they can, by human action, make a difference in the flow of events. Such a belief in free will should cause a healthy skepticism regarding the predictability of most complex phenomena which involve human beings. It is true that we can predict the position of the moon 2,000 years hence with phenomenal accuracy. However, any prediction as to human exploitation of the resources of the moon is a matter of very unreliable conjecture. Does this mean, then, that no "best guessing" should be attempted? By no means. It simply means that we should carefully assess every attempt at forecasting made by ourselves or others as to how much credence we can reasonably place in the forecast. Our beliefs in the forecast will depend upon many factors. One major factor is the planning horizon of the forecast. For example, we are fairly confident in predicting that the profits of tourism to Mars by the year 2010 will be zero. We should have no great confidence in anyone's prediction as to what this profit will be by the year 2500.

Another factor to be considered in the reliability of a forecast is the process considered. Let us, for example, consider the prediction as to the nominal value of a U.S. Treasury bond paying 5% per year, starting with one unit of capital at time zero. If compounding is carried out annually, then we can easily compute the total principal at the t years as simply

$$P(t) = P(0)[1 + .05]^t. \tag{8.1}$$

---

[1]Stock market prognosticators are perhaps the worst of the gurus. At almost the peak of the last bull market, two rather widely quoted "analysts" published an article in the *Wall Street Journal* entitled, "Stock Prices Are Still Far Too Low" [5]. They predicted a much higher DJIA, suggesting a figure of 36000 (and *tomorrow* not 10 or 20 years from now) was a "perfectly reasonable level for the Dow." Of course, the Dow was near its peak at this time, and the bear market of 2000−2002 was just around the corner! Needless to say, these gentlemen remain respected as forecasters and analysts and still write regularly for the *Wall Street Journal* and other publications.

Should the interest be compounded k times per year, then the principal is given by

$$P(t) = P(0) \left[1 + \frac{.05}{k} \right]^{kt} . \tag{8.2}$$

Should the interest be continuously compounded, we find in the limit as $k \to \infty$,

$$P(t) = P(0)e^{.05t} . \tag{8.3}$$

If we wish to use an arbitrary interest rate of α, then the principal is given by

$$P(t) = P(0)e^{\alpha t} . \tag{8.4}$$

Forecasts based on compound interest would appear to be completely reliable. Of course, the country might be overthrown or the government could default on its bonds, and it is this fact that still causes so many citizens of other countries to invest a substantial portion of their resources in gold. But, inasmuch as the United States has not defaulted on any of its bonds for well over a hundred years, we might do well to make the assumption that (8.4) will serve our needs. From the standpoint of such well-defined systems as compound interest, it might seem that nontrivial forecasting is not required.

This would indeed be the case if there were no inflation. In the Middle Ages, a ruler who denigrated his currency in any way was deemed to have lost his legitimacy. As medieval philosophers reasoned, if currency could arbitrarily lose its value, then money would not be a reliable means of exchange and the country could be driven essentially to a barter system. Furthermore, individuals who had saved money for their families against the eventuality that they would, for reasons of death, sickness or age, lose their ability to earn for them, would have labored in vain. Thus, denigration of currency was tantamount to stealing from widows and orphans, one of the crimes "crying to heaven for vengeance."

Unfortunately, this taboo against inflating the currency is not followed by most modern states. Consequently, the value of the principal in constant dollars after a period of time t , assuming an interest rate of α and an inflation rate of i is given by

$$P(t) = P(0)e^{(\alpha - i)t} . \tag{8.5}$$

Now, we see immediately a problem, for although α may be fixed by contract, it is likely that i will vary in time. Consequently, we might more properly write

$$P(t) = P(0)e^{\int_0^t (\alpha - i(\tau))d\tau} . \tag{8.6}$$

But we have simply symbolically assumed away the problem, for the reality is that, in most situations, we will not know $i(t)$. Part of our task here will be to forecast things like a firm's revenues so that we will be able, realistically, to predict such things as future earnings and dividend payments.

Let us now consider situations in which the system to be forecast is not well understood but has the property of increasing or decreasing in time. In those cases where we have such a series, the use of the human eye can be invaluable in assessing a pattern not understood *a priori*. The best known exponent of the use of the eye to identify patterns is also the founder of modern time series analysis, John W. Tukey. The technique, named by Tukey, *Exploratory Data Analysis*, is actually a collection of techniques, only a few of which we shall have time to cover in this chapter. (For a more extensive introduction to EDA, see [10, pp. 133–148] or [11].)

Let us attempt to carry out an analysis of the tabulated data below in Table 8.1.

Table 8.1. Increasing Data Set.	
t	$Y(t)$
0	3.0000
1	3.6642
2	5.4664
4	6.6766
5	8.1548
6	9.9604
7	12.1656
8	14.8591
9	18.1489
10	22.1672
11	27.0751
12	33.0695

Let us consider the following plot of the data in Table 8.1:

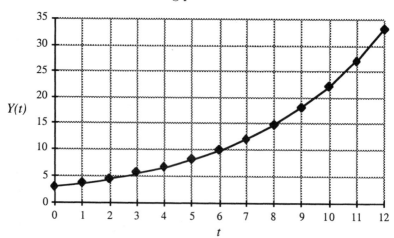

Figure 8.1 Untransformed Data.

We note that there seems to be practically no noise in the system. But the relationship between time t and $Y(t)$ is not immediately apparent. The fact is that the only functional relationship which can be perceived by the human eye is the straight line. Let us address a means of utilizing the ability of the human eye to recognize a straight line. We might suppose that since linear relationships are not all that ubiquitous, the fact that we can recognize straight lines is not particularly useful. Happily, one can frequently reduce monotone relationships to straight lines through transformations. We note that $Y(t)$ appears to be growing very fast. Let us see what happens when we plot $\ln(Y(t))$ versus t as in Figure 8.2.

We note here that we have been fortunate in that our first attempt to transform the data to linearity worked. Since $\ln(Y(t))$ is linear in t, the following must hold

$$\ln(Y(t)) = a + bt . \tag{8.7}$$

Exponentiating both sides, we have

$$Y(t) = e^{a+bt} = e^{a}e^{bt} = ke^{bt} . \tag{8.8}$$

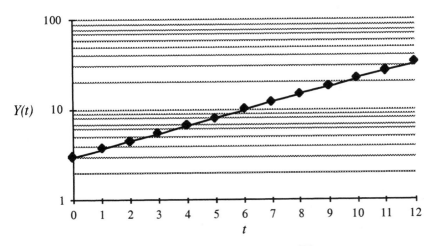

Figure 8.2. Semilogarithmic Plot.

From Table 8.1, it is now a fairly easy matter to solve for k and b. Since the $\ln(Y(t))$ line reveals little in the way of "jiggle" we can use virtually any two points to give the values of these parameters. Generally speaking, when fitting a straight line in this fashion, it is best to pick points near the extremes of the data set. So, using $t = 0$ and $t=12$, we have the following two equations in two unknowns:

$$\ln(3) = a = \ln(k) \tag{8.9}$$

$$\ln(33.07) = \ln(k) + 12b. \tag{8.10}$$

Solving, we obtain immediately the relationship

$$Y = 3e^{.2t} .$$ (8.11)

How comfortable should we feel in using (8.11) for forecasting purposes? The answer, as with answers to most important questions, is ambiguous. We should feel very much encouraged that the empirical model is so faithful to the data base. We have used a two parameter model to fit a curve through 13 points, and the fitted curve passes virtually through the points. If the fit had been very close with a 10 parameter model, our confidence should be lessened somewhat, when we remember that we can always find an $n+1$ parameter curve which will pass precisely through n points.

In general, we should not feel very confident about any model unless we can understand it theoretically as well as empirically. For example, should we, upon further investigation, discover that the mechanism in question is simply an annuity starting with \$3 million and paying interest at the rate of 20% per year, then that would give us a different level of confidence than would Table 8.1. Furthermore, any merely empirical relationship is likely to be believed with more confidence for shorter forecast lead times than with longer ones. We would feel pretty good about a revenue stream of a company which had the track record exhibited in Table 8.1 for a period, say, of one year. Unless we understood the company better, our confidence would be much less for a lead, say, of twenty years. Unfortunately, we shall of necessity, more often than we would wish, deal with very empirically perceived models. Now we have transformed the relationship between t and $Y(t)$ to a linear one. By recalling how we transformed the data, we can complete our task of identifying the functional relationship between t and $Y(t)$. So, then, we recall that we started with an unknown functional relationship

$$Y(t) = f(t).$$ (8.12)

We will not usually get our transformation to linearity after trying simply a semilog plot. Consider another possible functional relationship between Y and t:

$$Y(t) = 3t^{.4} .$$ (8.13)

Here, simply taking the logarithm of $Y(t)$ will not give a linear plot, since

$$\ln(Y) = \ln(3) + .4\ln(t)$$ (8.14)

is not linear in t. But, as we see immediately from (8.14), we would get a straight line if we plotted $\ln(Y)$, not versus t, but versus $\ln(t)$. And, again, as soon as the transformation to linearity has been achieved, we can immediately infer the functional relationship between t and Y and compute the parameters from the linear relationship between $\ln(Y)$ and $\ln(t)$.

From the above we see that simply using semilog and log-log plots will enable us to infer functional relationships of the forms

$$Y = ae^{bt} \tag{8.15}$$

and

$$Y = at^b , \tag{8.16}$$

respectively.

The transformation to essential linearity has been in use for over a century in the empirical modeling of complex systems in mechanics and thermodynamics. We shall consider a more general family of transformations than simply the logarithmic. Let us first consider the curves in Figure 8.3.

We note that curve A is growing faster than linearly for t large. Thus, if we wish to investigate transformations which will bring its rate of growth to that of a straight line, we need to use transformations which will reduce its rate of growth.

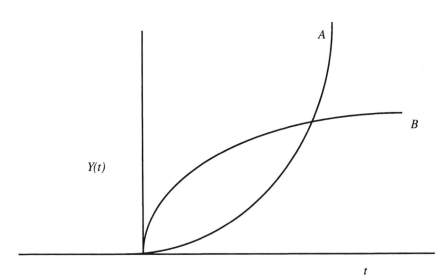

Figure 8.3. Concavities Up and Down.

For example, let us consider the following transformations in increasing order of tendency to reduce the apparent rate of growth:

- $Y^{\frac{1}{2}}$

- $Y^{\frac{1}{4}}$

- $\ln(Y)$

- $\ln(\ln(Y))$.

Considering the curve B which is growing more slowly than linearly, we could try, in decreasing order of severity:

- $\exp(e^Y)$

- $\exp(Y)$

- Y^4

- Y^2 .

The two groups of transformations, pooled together, give us Tukey's [11] *transformational ladder*

- $\exp(e^Y)$

- $\exp(Y)$

- Y^4

- Y^2

- Y

- $Y^{\frac{1}{2}}$

- $Y^{\frac{1}{4}}$

- $\ln(Y)$

- $\ln(\ln(Y))$.

The shape of the original Y curve points us up or down the transformational ladder. Using the transformational ladder to find more complicated functional relationships between Y and t becomes much more difficult. For example, it would require a fair amount of trial and error to infer a relationship such as

$$Y = 10 + 2.13t^{1.5} + 17.3t^2 . \tag{8.17}$$

Moreover, in practice, our data will be contaminated by noise. Nevertheless, for a great many situations, particularly in the exploratory stages, the use of the transformational ladder will bring us quickly to insights into what is going on.

For more complicated problems, such as, say, the output of (9.12) contaminated by noise, we can still be try to use least squares to go through a complex hierarchy of possible models, fitting the parameters as we go.

$$S(\text{Model(in } t)) = \sum (Y(t) - \text{Model})^2 . \tag{8.18}$$

Such an approach has not, to date, proved very effective. Generally speaking, if the transformational ladder approach works, it does so based on a simple graphing via semi-log and log-log plotting.

We now turn to another of the perception based notions of John Tukey [11]: namely, the fact that the eye expects continuity, that adjacent points should be similar. This notion was used by NASA starting with the early Mariner pictures of the surface of the moon. For example, let us suppose we have a noisy monochromatic two dimensional photograph with light intensities measured on a Cartesian grid. The Mariner voyages photographs were smoothed with the algorithm as shown in (8.19) and Figure 8.4.

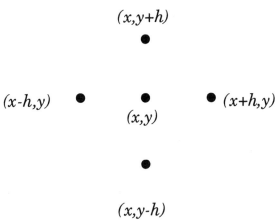

Figure 8.4. Digital Filter.

One procedure, which is both intuitive and simple, is to replace the intensity of reflective light at a point by an average of that intensity with an average of intensities about the point of interest:

$$I(x,y) \leftarrow \frac{4I(x,y) + I(x-h,y) + I(x,y-h) + I(x,y+h) + I(x+h,y)}{8},$$

$$(8.19)$$

where $I(x,y)$ is the light intensity at grid point (x,y).

For a one-dimensional data set, we have that used by Tukey many years before the Mariner voyages in connection with time series data, namely, the Hanning rule

$$I(x) \leftarrow \frac{2I(x) + I(x-h) + I(x+h)}{4}. \qquad (8.20)$$

This rule is frustrated whenever one of the points is miswritten or seriously contaminated by "noise". Let us consider below a data set where all of the values really should be "10", but one of the values has been miswritten as "1,000" We show below the data set followed by successive Hanning smooths:

Table 8.2. Repeated Hanning Smooths.		
Data	H	HH
10	10	10
10	10	10
10	10	71.88
10	257.50	257.75
1,000	505.00	381.25
10	257.50	257.75
10	10	71.88
10	10	10
10	10	10

We see that the *outlier* value of 1,000 has effectively contaminated the entire data set. This problem plagued analysts in many areas from space photography to stock market forecasting for many years. Tukey has proposed a smooth based on medians of groups of three down the data set, that is, we use the rule

$$I(x) \leftarrow Med(I(x - h), I(x), I(x + h)) \ . \tag{8.21}$$

In our discussion, the endpoints will simply be left unsmoothed. However, somewhat better rules are readily devised. In the data set above, the smoothing by threes approach gives us what one would presumably wish, namely, a column of ones.

Happily, Tukey's median filter is readily used by the computer. Unlike the older smoothing devices, such as Hanning, it can be "turned loose" on the data without fear that it will oversmooth. It is a very localized filter, so that typically if we apply it until no further changes occur (this is called the 3R smoother), we will not spread values of points throughout the data set. We note that repeated applications of the Hanning filter will continue to change the values throughout the set until a straight line results. On the other hand, the 3R smoother does tend to move from plateau to plateau, giving artificial flat spots in the smooth. Consequently, it is frequently appropriate to use the 3R filter followed by one application of the Hanning filter (H). The combined use of the 3RH filter generally gets rid of the wild points (3R), and the unnatural plateaus of the 3R are smoothed by the H. Still more elaborate algorithms are, of course, possible. We could, if we believed that two wild points could occur in the same block of three points, simply use a 5R (median of five values) filter. Or, we could use a 7R smooth, etc.

Table 8.3. Various Smooths.				
Quarter	Production	3	3R	3RH
1	50.00	50.00	50.00	50.00
2	52.00	52.00	52.00	52.25
3	56.00	55.00	55.00	54.50
4	55.00	56.00	56.00	56.25
5	58.00	58.00	58.00	58.00
6	60.00	60.00	60.00	61.50
7	65.00	63.00	63.00	62.75
8	63.00	65.00	65.00	64.75
9	66.00	66.00	66.00	66.25
10	68.00	68.00	68.00	68.50
11	73.00	72.00	72.00	70.75
12	72.00	73.00	73.00	74.00
13	78.00	78.00	78.00	76.75
14	81.00	78.00	78.00	78.00
15	76.00	80.00	78.00	77.50
16	80.00	76.00	76.00	76.25
17	75.00	75.00	75.00	75.25
18	73.00	75.00	75.00	74.50
19	75.00	73.00	73.00	73.50
20	73.00	73.00	73.00	72.25
21	70.00	70.00	70.00	71.50

In Table 8.3 we perform a 3RH smooth on a data set of number of quarterly unit sales of an aircraft manufacturer. A graph quickly shows how the 3RH smooth approximates closely what we would do if we smoothed the raw data by eye. At this point, we should mention that all the smooths of EDA are curve fits (and hence purely empirical), not derived models. We clearly find the 3RH smooth a more visually appealing graph than the raw data. But the data was measured precisely; the fluctuations really were there. So, in a sense, we have distorted reality by applying the 3RH smooth. Why have we applied it nevertheless? The human visual system tends to view and store in memory such a record holistically. Whether we smoothed the data or not, our eye would attempt to carry out more or less equivalent operations to those of 3RH. The human eye expects continuity and we do not readily perceive data digitally. The smooth gives us a benchmark (the forest) around which we can attempt to place the trees.

Another advantage of an empirical smooth like the 3RH is that, in a situation where there is a fair amount of noise in a time series, it is better to forecast along a smooth than along the wiggly data. In the aircraft data set examined in Figure 8.5, it would probably not be unreasonable simply to draw the arrowed forecast line freehand through the last epoch of the 3RH smooth as shown in Figure 8.6 below.

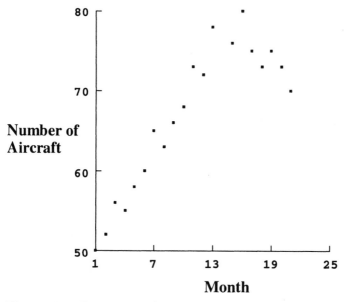

Figure 8.5. Raw Data (number of aircraft produced per month).

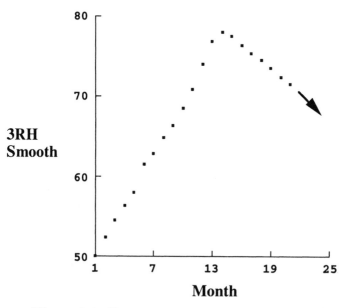

Figure 8.6. Freehand Forecast Based on a 3RH Smooth.

To put much credence in such a forecast past a few days would be ill-advised. Note that we have chosen to make our forecast along a straight line. (In most empirical analyses, it will be preferred to carry out extrapolation forecasts by the use of a straight line.) Now we have admittedly engaged in a very simple tool for obtaining forecasts. We shall shortly examine much more complex techniques for forecasting. However, the reader should be warned that, absent insights into the generation of the process at hand, there is no advantage to using a complex methodology rather than a simple one for the purposes of forecasting. Many an investor has lost his entire investment because he assumed that a bit of forecasting software was reliable simply because it was complex.

8.2 Forecasting as Regression

The subject of linear regression has been one of the most popular techniques for dealing with empirical models. In a forecasting context, the argument can be advanced as follows. Suppose that we believe that Y is a function which is dependent upon time. Then we may write this fact down as shown in equation (8.22) below.

$$Y(t) = f(t). \qquad (8.22)$$

Now, assuming that $f(t)$ has derivatives of all orders, we may expand $f(t)$ as a Taylor's series expansion about $t = t_0$ where t_0 is any time origin we select:

$$Y(t) = f(t_0) + (t - t_0)f'(t_0) + \frac{(t - t_0)^2}{2!}f''(t_0) + \dots . \qquad (8.23)$$

Now, if we knew $f(t)$ explicitly in functional form, we could carry out the evaluation of each of the derivatives at t_0 . Then, we would have

$$Y(t) = a_0 + a_1 t + a_2 t^2 + \dots . \qquad (8.24)$$

This form is called "linear" because it is linear in the coefficients. Now we note the effect of the factorials in the denominators of the terms in (8.23) to give diminishing effects to the higher-order terms in $(t - t_0)$ for time series which are not growing at a very fast rate. So we might be tempted to disregard some of the higher order terms in t. Consequently, we might try something like

$$Y(t) = a_0 + a_1 t + \epsilon, \qquad (8.25)$$

where ϵ represents the error due to truncation of the series.

Now since we typically will not know $f(t)$ or its derivatives, we will have to estimate them from the data. There is no unique method for carrying this out. However, one of the simplest, developed by Karl Friedrich Gauss is that of least squares. According to this procedure, if we have a set of n data points $\{t_i, f(t_i)\}_{i=1}^{n}$, we select a_0 and a_1 so as to minimize the sum of squares of deviations of $a_0 + a_1 t$ from $Y(t)$. That is, we find the parameters which minimize the sum of squares

$$S(a_0, a_1) = \sum_{i=1}^{n} [Y(t_i) - (a_0 + a_1 t)]^2 . \qquad (8.26)$$

To find estimates of the parameters can be accomplished by differentiating S with respect to each and setting the derivatives equal to zero.

$$\frac{\partial S}{\partial a_0} = \sum_{i=1}^{n} [Y(t_i) - \hat{a}_0 - \hat{a}_1 t_i] = 0 \qquad (8.27)$$

$$\frac{\partial S}{\partial a_1} = \sum_{i=1}^{n} [t_i (Y(t_i) - \hat{a}_0 - \hat{a}_1 t_i)] = 0 \ . \qquad (8.28)$$

This gives the estimates

$$\hat{a}_1 = \frac{\sum_{i=1}^{n} t_i Y(t_i) - \frac{\sum_{i=1}^{n} t_i \sum_{i=1}^{n} Y(t_i)}{n}}{\sum_{i=1}^{n} t_i^2 - \frac{(\sum_{i=1}^{n} t_i)^2}{n}} \qquad (8.29)$$

$$\hat{a}_0 = \frac{\sum_{i=1}^{n} Y(t_i)}{n} - \frac{\hat{a}_1 \sum_{i=1}^{n} t_i}{n} \ . \qquad (8.30)$$

Here, as elsewhere in our discussion, a "hat" over a parameter indicates an estimate rather than the true value of the parameter. Another popular linear regression model is one which looks upon a stock price, say, as having a linear trend on top of a sinusoidal cycle as in

$$Y(t) = a + bt + c \cos \left[\frac{2\pi t}{p} \right] , \qquad (8.31)$$

where the period p is known.

The coefficients here are computed as above. If p is unknown and must be estimated as part of the least squares algorithm, however, then no simple closed form for a,b,c, and p is available, and we must resort to a nonlinear optimization technique, such the Nelder-Mead algorithm given in Appendix A.18.

It would be nice if the linear regression approach worked well, for there is no difficulty in finding, in the polynomial model, say, the least squares estimators of the coefficients for all powers of t, up to the $n + 1$st power (there is no unique solution if we use more coefficients than we have data points). Unfortunately, this is not the case. Linear regression seldom works, although the less scientists understand about the system in question, the more likely they seem to try linear regression. In the case of the forecasting problem, let us examine some possible inadequacies of linear regression. Fitting a straight line through the data of Figure 8.2 would give something like the situation shown in Figure 8.7 below

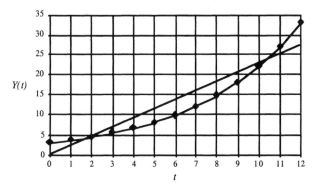

Figure 8.7. Linear and Quartic Fits.

The straight line fit here, in a smoothly monotonic situation, is not good at all. It would be dangerous to extrapolate along the indicated line. The quartic fit is much better. Naturally, in this case, we have been dealing with the growth of an instrument according to the laws of compound interest. If we plotted $\ln(Y(t))$ rather than $Y(t)$ versus t we would obtain an excellent fit, as in Figure 8.2. In econometric situations, growth and decay are more generally expected, at least over short epochs, to be exponential rather than linear in time. So if we insist on using linear regression, we should try it, not only on $Y(t)$, but on $\ln(Y(t))$ as well.

Would we feel very comfortable about forecasting along the continuation of either the linear or the quartic fits? Possibly for very short time horizons, certainly not for long horizons. Well, we could go further and fit a polynomial of the 20th degree which would pass precisely through the data points. Would this give us increased confidence about forecasting into the future? By no means. It is to be remembered that almost every forecasting analysis we pursue should be buttressed by some considerations of stability. At the most basic level, we can investigate stability by perturbing a few of the data points. If the resulting forecast is widely divergent from the forecast using the unperturbed points, then generally speaking, the forecasting strategy is unstable. For high-order polynomial strategies in the above example, the results are very unstable. Is there, then, no philosopher's stone which allows us to use mathematical power to give precise predictions of the future? Alas, there is not. But we can do much better than linear regression. A major problem with forecasting is that, properly viewed, the coefficients in equation (8.24) are themselves varying in time, so perhaps we should rewrite (8.24) as

$$Y(t) = a_0(t) + a_1(t)t + a_2(t)t^2 + \dots . \qquad (8.32)$$

Note that in the case of compound interest, the important aspect is that the generating mechanism is not changing in time. Were it not for the time varying aspect of an empirical process, we could indeed try something like polynomial fitting. Once we had figured out what was going on formally, even if we did not understand the process, we could use our polynomial fit to predict the future. As we pointed out early on in this chapter, such *stationarity* in time does not appear to be in the nature of most processes dealt with in business.

In such a world of nondeterminism, what should we do? One very practical attitude might be that we should place more weight on data in the near past than on data in the remote past. Barring catastrophes, i.e., abrupt rather than gradual changes in the generating mechanism, we should expect the market value of a stock tomorrow to be very near to the value today, less so to the value last month, still less so to the value three months ago. It is this notion of smooth rather than abrupt decay in the quality of a time series model which we shall exploit in the next chapter to give us procedures preferred to approaches based on linear regression

8.3   Data Analysis of Ameritape, Inc.

At this point, we return to the data we gathered for Ameritape, Inc. Sales, operating income, the ratio of operating income to sales, EPS, and DPS for the years 1991 through 2002 are replicated in Figure 8.8. What are we to make of these data?

Year	Sales (S)	Oper. Inc. (OI)	OI/S	EPS	DPS
1991	$3,087,493	$611,323	.20	$.56	$.20
1992	3,272,743	641,457	.20	.60	.20
1993	3,436,379	673,531	.20	.63	.24
1994	3,642,563	713,942	.20	.67	.24
1995	3,533,286	642,548	.20	.61	.24
1996	3,781,402	813,001	.20	.77	.24
1997	4,137,536	861,781	.20	.88	.24
1998	4,355,301	1,061,823	.19	.99	.36
1999	4,042,539	779,656	.19	.74	.36
2000	4,525,059	651,472	.14	.70	.36
2001	4,912,094	956,403	.19	1.02	.36
2002	5,438,195	1,172,792	.22	1.24	.40

Figure 8.8. Sales, Operating Income, EPS, and DPS 1991-2002 (in Thousands Except Per Share Numbers).

It is natural to suppose that both S and OI in Figure 8.8 grow as noisy compound interest. To see whether this may be so, in Figure 8.9 we plot $\log(S)$ versus (Year-1991).

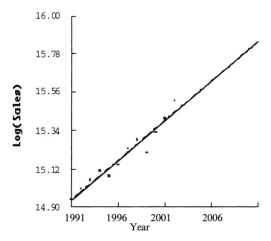

Figure 8.9. Semilog Plot of S versus Time.

The fit seems reasonably good when we fit the least squares regression to it (see Section 8.2). The equation of this line is:

$$\log(S) = 14.94091 + 0.04562 \times (\text{Year - 1991}). \tag{8.33}$$

Exponentiating both sides give us

$$S = \exp[14.94091 + 0.04562 \times (\text{Year - 1991})] \tag{8.34}$$

Now, for log(Operating Income) versus (Year - 1991), the picture is much noisier

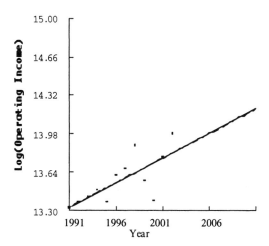

Figure 8.10. Semilog Plot of OI versus Time.

The least squares fit is given by

$$\log(OI) = 13.32705 + 0.04383 \times (\text{Year - 1991}), \qquad (8.35)$$

or

$$OI = \exp[13.32705 + 0.04383 \times (\text{Year - 1991})]. \qquad (8.36)$$

The situation for EPS is rather noisy to use for forecasting purposes. Nevertheless, the least squares line is given by

$$\log(EPS) = -0.58671 + .05713 \times (\text{Year - 1991}), \qquad (8.37)$$

or

$$EPS = \exp[-0.58671 + .05713 \times (\text{Year - 1991})]. \qquad (8.38)$$

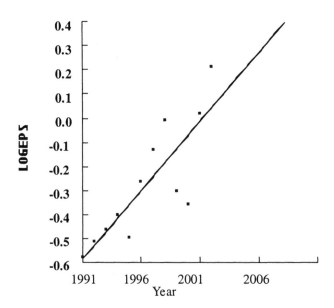

Figure 8.11. Semilog Plot of *EPS* versus Time.

Finally, let us consider the situation with regards to *DPS*. We note that

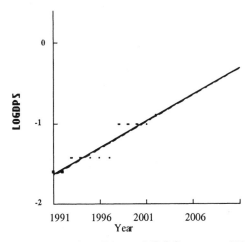

Figure 8.12. Semilog Plot of DPS versus Time.

the dividend payout is lumpy. That is, increases are apparently made only when it appears to the Board of Directors of Ameritape that sustained earnings increases justify raising the dividend. With the recent increase in 2001 to $.40 per share, we might note that the dividend has doubled since 1991.

Using the least squares fits from (8.34) and (8.36), we have the projections:

Year	Sales	Operating Income
2003	$5,327,270	$1,038,210
2004	$5,575,930	$1,084,730
2005	$5,836,200	$1,133,330
2006	$6,108,610	$1,184,110
2007	$6,393,740	$1,237,160

Figure 8.13. Sales and Operating Income Projections 2003–2007 (in Thousands of Dollars).

One last set of data must be located to complete the picture, and that is the past record of stock prices. These are provided in Figure 8.14.

Year	EPS	High	Low
1998	$.99	$17.82	$14.85
1999	$.74	$11.84	$11.10
2000	$.70	$11.20	$10.50
2001	$1.02	$18.36	$16.32
2002	$1.24	$22.32	$19.84

Figure 8.14. EPS and Stock Prices 1998–2002.

8.4   Some Conclusions and a Philosophy

At this point, we are ready to consider whether we wish to include Ameritape, Inc. as part of our portfolio. In order to do this, we might consider some of the methods of perhaps the greatest investor of all time—Warren Buffett. As some readers may know, his partners turned a $10,000 investment in his fund in 1955 into $250 million today (assuming the partnership interest was converted into Berkshire Hathaway stock in 1969). In this book the authors can only summarize some of the Buffett story and his investment strategies. For those eager to learn more about the canny Warren Buffett, an excellent book to read is Roger Lowenstein's *Buffett: The Making of an American Capitalist* [7].

Buffett begins by trying to understand a company, its business and products (services), the nature of its competition, and what risks it faces. He tries to determine whether the economics of the business are likely to be as good, better or worse over the next few years as they currently are. His intent is to be a long-term investor, and he only wants to buy into situations where he is comfortable with the people who manage the business. The Buffett strategy is to find stocks that he will never have to sell. An ideal stock is one where he may well want to buy even more shares in the future. He employs "filters" to screen companies. Many of the ratios discussed earlier in Chapter 7 are employed in these screens which allow him to discard most of the ideas he sees quickly. He tries to stay alert to anomalies because he believes that most investors are conditioned to reject them, and where they do exist is possibly where an opportunity might reside.

Buffett never buys stock in companies that burn cash. Instead, he hopes to find businesses that drown the investor in cash. The ideal business is one where there is little change, particularly in technology. Obviously, this makes longer-run forecasting easier and more believable. Buffett has avoided high-tech companies because he does not feel comfortable with them, does not understand them, and could not feel confident with projections made for a decade in the future, even if he liked the people involved. He admires and has friendships with people like Bill Gates and Michael Dell, but would not invest in their companies early on because of the uncertainties. He prefers to invest in companies like Coca-Cola, Gillette, and Disney because these companies offer products and services where longer-term demand is fairly predictable.

Now, given the wisdom of Warren Buffett, we can return to Ameritape. Is this really one of those companies that fits the Buffett criteria? At fist glance, it may appear to do so. In the first place, it is definitely not high tech. It is a rather prosaic business that has not really changed very much over the past several decades, yet it has still managed to grow and throw off a lot of cash (some of it paid out in dividends to stockholders and some of it reinvested in the business). We really cannot answer the question, however, until we complete the Ameritape story in Chapter 10.

Problems

Problem 8.1. A simplified income statement and balance sheet for Dynamic Sinculator is given below:

Income Statement		Balance Sheet	
Revenue	$100,000,000	Assets	$50,000,000
Expenses	59,000,000		
EBIT	41,000,000	Bonds	$20,000,000
Interest	1,000,000	Net worth	30,000,000
EBT	40,000,000		$50,000,000
Taxes	20,000,000		
N.I.	20,000,000		
Dividends	4,000,000		
R.E.	$16,000,000		

Other data:

  Market price per common share $80
  Number of shares outstanding   5 million

a. Compute the following relationships:
   (1) EBIT/Total assets
   (2) Debt/Equity
   (3) Retention rate
   (4) Average tax rate
   (5) EPS
   (6) DPS
   (7) P/E multiple
   (8) Dividend yield
   (9) Coupon rate on bonds

b. Assuming that all the relationships are constant in the future, determine the dividend growth rate for Dynamic Sinculator.

c. What would the earnings growth rate be? The rate of growth in the share price?

d. What would the annual growth in assets be?

e. What would the growth in sales be, given a constant total-asset turnover?

Problem 8.2. International Transportation and Telecommunication has 10 million shares outstanding and earns $2 per share. As a "growth" conglomerate, the firm's stock sells at 50 times current earnings. The Sinking Offshore Shipping Co. has 5 million shares outstanding and sells at $10 per share based on current EPS of $1. International plans to merger with S.O.S. by offering shares in the ratio of current market prices.

a. What is the ratio of exchange of International shares for S.O.S. shares?

b. How many International shares would be outstanding after the merger?

c. What will be the new International EPS?

d. What apparent "growth" has taken place?

e. Given the fact that International is heavily engaged in transportation, what diversification effects might the merger make possible?

f. Suppose the merged company were riskier than the premerger International. What price and P/E multiple might an efficient market pay for the merged company's stock?

g. Suppose the market retained the old P/E multiple that existed for premerger International. What would the new market price of the merged company be?

h. Trace the "growth effect" if International could find yet another Sinking Offshore Shipping Co.

Problem 8.3. Consider the population figures plotted for the United States in Figure 8.15.

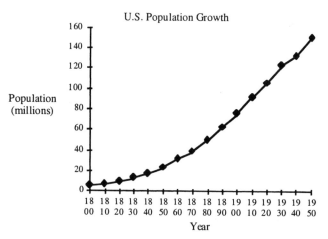

Figure 8.15. USA Population Growth

Use the data from the figure to predict the population of the United States in 1990. (It might be better if the data were given us in tabular form, but frequently we have to read rough figures from a graph.) Actually the true value was around 249,000,000. Analyze the reasons for your prediction error.

Problem 8.4. Experience indicates that we need to be flexible about smoothing.

(a) Show a hypothetical data plot for which a hanning smooth carried out until the smooth looks satisfactory will be disastrous, but a 3R smooth will work well.

(b) By means of a sketch and an argument show an hypothetical data plot for which a 3R smooth is unsatisfactory but a hanning smooth is satisfactory.

(c) In recording a series of observations–all having magnitude between 1 and 10–a value of 20 is inadvertently and mistakenly added to each of two adjacent values. Neither hanning nor 3R works here. Suggest something which does.

(d) After completing–say–a 3RH3 smooth of stock prices versus time, a market analyst obtains a satisfactorily smoothed graph. However, before presenting the graph to his boss, he should carry out an additional step to help guard against a false sense of security. Describe this step.

References

[1] Batra, R. (1987). *The Great Depression of 1990*. New York: Simon and Schuster.

[2] Bogle, J. C. (1999). *Common Sense and Mutual Funds: New Imperatives for the Intelligent Investor*. New York: John Wiley & Sons, Inc.

[3] Box, G. E. P. and Jenkins, G. M. (1976). *Time Series Analysis: Forecasting and Control*. San Francisco: Holden-Day.

[4] Figgie, H. E. (1992). *Bankruptcy 1995: The Coming Collapse of America and How to Stop It*. Boston: Little, Brown and Company.

[5] Glassman, J. K. and Hassett, K. A. (1999). "Stock Prices Are Still Far Too Low," *Wall Street Journal*, March 17 (op. ed.).

[6] Keynes, J. M. (1936). *The General Theory of Employment, Interest, and Money*. New York: Harcourt, Brace and World.

[7] Lowenstein, R. (1996). *Buffett: The Making of an American Capitalist*. New York: Random House.

[8] Press, E. (1999). *Analyzing Financial Statements: 25 Keys to Understanding the Numbers*. New York: Lebhar-Friedman Books.

[9] Taylor, W. M. and Williams, E. E. (1991). "Market microstructure and Post Keynesian theory," *Journal of Post Keynesian Economics*, Winter, 233–247.

[10] Thompson, J. R. (1989). *Empirical Model Building*. New York: John Wiley & Sons.

[11] Tukey, J. W. (1977). *Exploratory Data Analysis*. Reading, MA:

Addison- Wesley.

[12] Williams, E. E. and Findlay, M. C. (1974). *Investment Analysis.* Englewood Cliffs, NJ: Prentice-Hall.

[13] Williams, E. E. and Thompson, J. R. (1996). *The Economics of Production and Productivity: A Modeling Approach.* Austin, TX: Capital Book Company.

Chapter 9

Stock Price Growth as Noisy Compound Interest

9.1 Introduction

There are two major ways of minimizing risk for a portfolio of securities by "diversification." One way is that of diversification among a number of securities. If we have ten stocks, each with the same growth rate and each with the same volatility, dividing our investment among the ten stocks rather than putting all our investment in any one of them is almost a "free lunch" (assuming their returns are not perfectly correlated with each other). Of course, the lunch is not entirely free. Such diversification should save us from losing everything in an Enron but it might kill our hopes of becoming a Microsoft millionaire (as occurred to many Microsoft clerical personnel who had retirement plans invested heavily in the stock of their employer, which is the other side of the coin from the experience of Enron employees). Diversification of this sort has been used for a long time (in the nineteenth century many farmers planted corn as well as wheat in the event that hail storms zapped the more profitable wheat).

Unfortunately, a bear market will tend to cause most stocks in the portfolio to drop. Just as an extended drought will zap both corn and wheat, a bear market will hurt stocks generally. (An old politically incorrect adage of Wall Street is "When the paddy wagon comes, good girls are arrested as well as the bad.") What other variable can we use for "diversification"? The answer is **time**. As we shall demonstrate in this chapter, investors over longer periods of time, have the advantage of the fact that in roughly 70% of the years, the index of large cap U.S. stocks rises rather than falls. And there is the further encouraging news that in over 40% of the years, the index rises by over 20%. In 30% of the years, the market rises by over 25%. And in 25% of the years, the index has risen by over 30%. Over the

225

roughly 75-year period such records have been kept, the United States has lived through the Great Depression, the Second World War, the Cold War, Korea, Vietnam, assorted massive sociological changes, shifts toward and away from free markets, and assorted epidemics. These can all be viewed as the political/economic/sociological analogs of major "droughts." It is true that we have yet to experience Martian invasion, attacks by genetically engineered viruses or suitcase nuclear devices, or the costs of mounting the Sixth Crusade. We hope such events do not occur, but events of comparable angst have occurred to other countries of the West over the past 75 years. Poland was occupied by Russia and Germany in September of 1939, and the Russian occupation only ended (sort of) in June of 1989. It is hard to imagine a course of action (other than attempting to take one's self and one's money out of Poland and moving to, say, the United States) which would have saved an investor in the Warsaw Stock Exchange. And it is hard today to imagine a safe harbor for one's self or one's property in the event that the United States falls.[1] Past performance is not an infallible guide for looking over the risk profile and we do not claim it to be. Moreover, readers have surely by this point learned that we have not produced any schemes for getting rich quick. We hope, however, to give clues by which an investor can hope to become wealthy at a moderate rate of speed. Let us consider a federally insured certificate of deposit with interest rate r (interest reinvested) with a time horizon of, say, five years. Calling the value of the CD at time t, $X(t)$, we can compute the change in the value of the CD over an increment of time Δt via

$$\Delta(X(t)) = rX(t)\Delta(t). \qquad (9.1)$$

We recognize, in the limit as Δt goes to zero, one of the most venerable of simple differential equations:

$$\frac{dX(t)}{X(t)} = rdt \qquad (9.2)$$

with solution

$$X(t) = X(0)\exp(rt). \qquad (9.3)$$

This is the equation of compound interest of which Einstein is reputed to have said, "There is no magic in special relativity. Compound interest, now that is magic."

---

[1] This discussion represents a rather extreme example of the distinction between risk and Knight-Keynes uncertainty which we covered in Chapter 5. Although we will not proceed with a discussion of the investment merits of a moneybelt filled with gold coins or diamonds (augmented by a Uzzi), it should be noted that a number of stock exchanges, from Berlin to Cairo to Buenos Aires, have flourished and then disappeared for various reasons over the last 150 years. The American experience over the last 75 years may be subject to survivor bias.

9.2   Stock Progression Using Binomial Trees

If we wish to consider the growth of a stock with value (at time t) of $S(t)$, then it is tempting to model it using something like noisy compound interest. One such model is

$$\frac{\Delta S}{S} = \mu \Delta t + \sigma \epsilon \sqrt{\Delta t}. \tag{9.4}$$

Here μ is the growth rate of the stock (i.e., the noisy interest rate), σ (the *volatility*) is a measure of the variability of the process as time increases, ϵ is a variable which takes the value 1 with probability 1/2 and the value -1 with probability 1/2. There is a small problem with this model. Ideally, we would like to have the price move in such a way that, on the average, going from time 0 to time $\Delta(t)$, the value of the stock will be the same as that obtained by compound interest, namely, $S(0)\exp(\mu\Delta(t))$. But,[2]

$$
\begin{aligned}
E[S(\Delta(t))] &= S(0)\exp[\mu\Delta(t)] \left[\frac{1}{2} e^{\sigma\sqrt{\Delta(t)}} + \frac{1}{2} e^{-\sigma\sqrt{\Delta(t)}} \right] \\
&= S(0)\exp[\mu\Delta(t)] \left[\frac{1}{2} \left(1 + \sigma\sqrt{\Delta(t)} + \frac{1}{2}(\sigma\sqrt{\Delta(t)})^2 + o(\Delta(t)) \right) \right. \\
&\qquad \left. + \frac{1}{2} \left[1 - \sigma\sqrt{\Delta(t)} + \frac{1}{2}(\sigma\sqrt{\Delta(t)})^2 + o(\Delta(t)) \right] \right. \\
&= S(0)\exp[\mu\Delta(t)] \left[1 + \frac{1}{2}\sigma^2\Delta(t) \right] \\
&= S(0)\exp[\mu\Delta(t)] \exp\left(\frac{1}{2}\sigma^2\Delta(t) \right).
\end{aligned}
$$

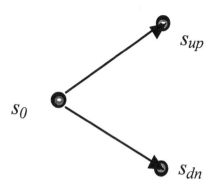

Figure 9.1. Simple Binomial Tree.

[2]We recall the series expansion $e^z = 1 + z + (1/2)z^2 + o(z)$, where $o(z)$ goes to zero so quickly as z goes to zero that even when divided by z, it still goes to zero.

We can correct for this small anomaly by reducing the payoffs slightly. Accordingly, we shall be looking at

$$
\begin{aligned}
S(t + \Delta(t)) &= S(t) \exp \left[(\mu - \frac{1}{2}\sigma^2)\Delta(t) \right] \exp[\sigma \sqrt{\Delta(t)}] \text{ with prob } 1/2 \\
&= S(t) \exp \left[(\mu - \frac{1}{2}\sigma^2)\Delta(t) \right] \exp[-\sigma \sqrt{\Delta(t)}] \text{with prob } 1/2.
\end{aligned}
$$
(9.5)

We will, standing at time 0, where the value of the stock is $S(0)$, look forward to time t and attempt to determine what will be the probability distribution of $S(t)$. In order to carry out this task, we will make the Δt values very small and go, step by step, from time 0 to time t. Now, let us suppose that we divide the interval from 0 to t into n subintervals of equal length, so that $\Delta t = t/n$. As we go from time 0 to time t, some of the steps in S will be positive, some negative. Let us consider the random variable X_n = total number of the n jumps which were positive. Then, X_n is a binomial random variable with probability $p = 1/2$. Thus, the expected value and variance of X_n are given by

$$
\begin{aligned}
E(X_n) &= np = \frac{n}{2} \\
Var(X_n) &= np(1-p) = \frac{n}{4},
\end{aligned}
$$

respectively.

We note that by the Central Limit Theorem, as n goes to infinity,

$$
\begin{aligned}
Z &= \lim_{n \to \infty} \frac{X_n - n/2}{\sqrt{n/4}} \\
&= \lim_{n \to \infty} \frac{2X_n - n}{\sqrt{n}} \\
&= \mathcal{N}(0, 1),
\end{aligned}
$$

that is, is distributed as a Gaussian random variable with mean 0 and variance 1. So then, when we look at the distribution of

$$
S(t) = S(0) \exp \left[\left(\mu - \frac{1}{2}\sigma^2 \right) t + \sigma \sqrt{t} \frac{2X_n - n}{\sqrt{n}} \right],
$$
(9.6)

in the limit as Δt goes to zero (equivalently, n goes to infinity),we have

$$
\begin{aligned}
S(t) &= S(0) \exp \left[\left(\mu - \frac{1}{2}\sigma^2 \right) t + \sigma \sqrt{t} Z \right] \qquad\qquad (9.7)\\
&= S(0) \exp(\mu t) \times \exp \left(\sigma \sqrt{t} Z - \left(\frac{1}{2}\sigma^2 t \right) \right) \qquad (9.8)\\
&= S(0) \exp(W(t)), \qquad\qquad\qquad\qquad\qquad\qquad (9.9)
\end{aligned}
$$

where Z is Gaussian with mean 0 and variance 1, and $W(t)$ is a Gaussian random variable with mean $(\mu - \sigma^2/2)t$ and standard deviation $\sigma\sqrt{t}$. Equation (9.8) is interesting because it has a deterministic part (the same precisely as one gets with compound interest) multiplied by a stochastic part, hence the noisy compound interest notion which defines this chapter.

When a stock price $S(t)$ obeys the law given in three equivalent forms (9.7)–(9.9), we say that $S(t)$ is *lognormal* with *growth rate* μ and *volatility* σ. We note that the expected value of $S(t)$ is given by

$$
\begin{aligned}
E(S(t)) &= S(0)\int_{-\infty}^{\infty} \exp\left(\left(\mu - \frac{1}{2}\sigma^2\right)t + \sigma\sqrt{t}Z\right)\frac{1}{\sqrt{2\pi}}e^{-Z^2/2}dZ\\
&= S(0)\exp\left(\left(\mu - \frac{1}{2}\sigma^2\right)t\right)\frac{1}{\sqrt{2\pi}}\int_{-\infty}^{\infty}e^{\sigma\sqrt{t}Z}e^{-Z^2/2}dZ\\
&= S(0)\exp\left(\left(\mu - \frac{1}{2}\sigma^2\right)t\right)\int_{-\infty}^{\infty}\exp\left[-\frac{1}{2}(Z-\sigma\sqrt{t})^2\right]dZ e^{\frac{1}{2}\sigma^2 t}\\
&= S(0)\exp\left[\left(\mu - \frac{1}{2}\sigma^2\right)t + \left(\frac{1}{2}\sigma^2\right)t\right]\\
&= S(0)e^{\mu t}. \quad (9.10)
\end{aligned}
$$

It is a straightforward matter to show that

$$
\begin{aligned}
Var(S(t)) &= E[S(t) - E(S(t))]^2\\
&= S(0)^2\exp[2\mu t][e^{\sigma^2 t} - 1].
\end{aligned}
$$

9.3 Estimating μ and σ

From (9.7), we have, for all t and Δt

$$
r(t+\Delta t, t) = \frac{S(t+\Delta t)}{S(t)} = \exp\left[\left(\mu - \frac{\sigma^2}{2}\right)\Delta t + Z\sigma\sqrt{\Delta t}\right]. \quad (9.11)
$$

Defining $R(t+\Delta t, t) = \log(r(t+\Delta t, t))$, we have

$$
R(t+\Delta t, t) = \left(\mu - \frac{\sigma^2}{2}\right)\Delta t + \epsilon\sigma\sqrt{\Delta t}.
$$

Then

$$
E[R(t+\Delta t, t)] = \left(\mu - \frac{\sigma^2}{2}\right)\Delta t. \quad (9.12)
$$

Suppose we have a stock that stands at 100 at week zero. In 26 subsequent weeks we note the performance of the stock as shown in Table 9.1. Here $\Delta t = 1/52$. Let

$$
\bar{R} = \frac{1}{26}\sum_{i=1}^{26}R(i) = .002931.
$$

Table 9.1. 26 Weeks of Stock Performance			
Week=i	Stock(i)	$r(i)$=Stock(i)/Stock($i-1$)	$R(i)$=log($r(i)$)
1	99.83942	0.99839	-0.00161
2	97.66142	0.97818	-0.02206
3	97.54407	0.99880	-0.00120
4	96.24717	0.98670	-0.01338
5	98.65675	1.02503	0.02473
6	102.30830	1.03701	0.03634
7	103.82212	1.01480	0.01469
8	103.91875	1.00093	0.00093
9	105.11467	1.01151	0.01144
10	104.95000	0.99843	-0.00157
11	105.56152	1.00583	0.00581
12	105.44247	0.99887	-0.00113
13	104.21446	0.98835	-0.01171
14	103.58197	0.99393	-0.00609
15	102.70383	0.99152	-0.00851
16	102.94174	1.00232	0.00231
17	105.32943	1.02320	0.02293
18	105.90627	1.00548	0.00546
19	103.63793	0.97858	-0.02165
20	102.96025	0.99346	-0.00656
21	103.39027	1.00418	0.00417
22	107.18351	1.03669	0.03603
23	106.02782	0.98922	-0.01084
24	106.63995	1.00577	0.00576
25	105.13506	0.98589	-0.01421
26	107.92604	1.02655	0.02620

By the strong law of large numbers, the sample mean $\bar{R}$ converges almost surely to its expectation $(\mu - \sigma^2/2)\Delta t$. Next, we note that

$$[R(t + \Delta t, t) - E(R(t + \Delta t, t))]^2 = \epsilon^2 \sigma^2 \Delta t, \qquad (9.13)$$

so

$$\text{Var}[R(t + \Delta t, t)] = E[R(t + \Delta t, t) - \left(\mu - \frac{\sigma^2}{2}\right)\Delta t]^2 = \sigma^2 \Delta t. \qquad (9.14)$$

For a large number of weeks, this variance is closely approximated by the sample variance

$$s_R^2 = \frac{1}{26 - 1} \sum_{i=1}^{26} (R(i) - \bar{R})^2 = .000258.$$

For a large number of weeks, this variance is closely approximated by the sample variance

$$s_R^2 = \frac{1}{26-1} \sum_{i=1}^{N-1} (R(i) - \bar{R})^2.$$

Then $\hat{\sigma}^2 = .000258/\Delta t = .000258 \times 52 = .013416$, giving as our volatility estimate $\hat{\sigma} = .1158$. Finally, our estimate for the growth rate is given by

$$\hat{\mu} = \bar{R} \times 52 + \frac{\hat{\sigma}^2}{2} = .1524 + .0067 = .1591.$$

9.4 Time Indexed Risk Profiling

9.4.1 Investing in a Sure Thing

Much of classical investment analysis is based on the assumption that all we need to know as an investor are μ and σ and r (the interest rate on a U.S. Treasury bill). Let us consider the case where an investor puts $100,000 into a an IRA (Individual Retirement Account). The $100,000 is then invested in a U.S. Treasury bond with a 20-year maturity. Although these bonds typically pay interest semi-annually (and the interest coupons must then be re-invested at prevailing market rates), we shall simplify and assume the bond pays 6% interest which is continuously compounded. We can easily write down the value of the bond at time t, via

$$S(t) = 100,000 \exp(.05t). \tag{9.15}$$

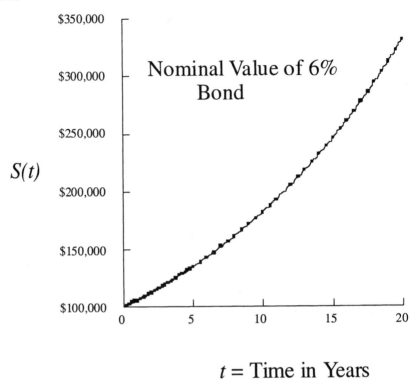

t = Time in Years

Figure 9.2. Nominal Value of Zero Volatility Security.

For such an investment, it would appear that only the growth rate μ is required. There is no volatility in the future nominal value of the bond (which we plot in Figure 9.2). Of course, we do not know what the rate of inflation might be in future years. Hence the real value of the bond is very much in doubt. Since the bond is tax protected until the bond and accrued interest are withdrawn from the IRA, the investor will not be subjected to paying taxes until the money is taken from the account. But what will that rate be when the investor decides to take the money? Will the country will have moved to a "flat tax" in the 18% range? Or perhaps the country will have moved into the rank of "social democracies" with top marginal tax rates approaching 90%? These are big questions.

Let us suppose that the inflation rate ρ is 0%, 2%, 4%, 6%. The real value at time t then will be given by

$$S_R(t) = 100,000 \exp\{(.06 - \rho)t\}. \tag{9.16}$$

We show these results in Figure 9.3.

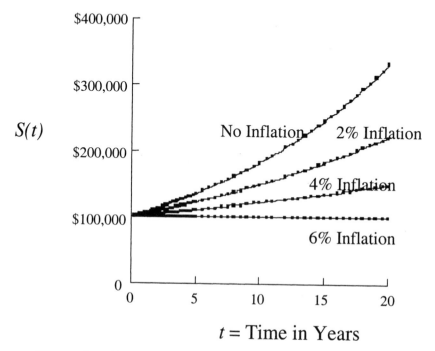

t = Time in Years

Figure 9.3. Real Value of Zero Volatility Security.

Without inflation, the $100,000 investment grows, in 20 years, to $332,012 in terms of Year Zero dollars. If the inflation rate is 6%, then, in Year Zero dollars, the investment is only worth the original $100,000.

As yet, we have not looked at the tax situation. Consider the situation of 6% inflation. Our investor will be obliged to pay taxes on the nominal increase in the value of the bond. Of course, we have no idea what the tax rate will be in 20 years. Let us suppose it turns out to be 35%. We show for a cash-in 20 years out, in Year Zero dollars, the value of the investment assuming varying rates of inflation. For no inflation, the value is $250,808. For 2% inflation, it is $168,122. For 4% inflation it is $112,695. For 6% inflation, it is $75,542. We recall that in Year Zero, our investor had $100,000. For 6% inflation, he has lost, with his 20 year bond, one-fourth of his stake. With 4% inflation, he has made around 1/2 percent per year. Even with the rather optimistic 2% inflation, he has little to show for his 20 years of conservative investing.

Things could be worse. Suppose our investor had been investing his $100,000 outside the tax shielding of an IRA. That means, he must pay tax yearly. Then the real after-tax value of his investment, year by year, is given (computing taxes once a year) by

$$S_{AT}(t) = 100000 \times (1 + .06(.65))^t \times \exp((.06 - i)t). \qquad (9.17)$$

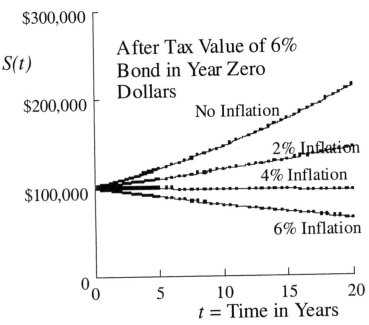

Figure 9.4. Real After Tax Value of Zero Volatility Security.

Note that the real value after 20 years of the 6% bond with 0% inflation and 35% tax rate is $214,937. For 2% inflation, this has shrunk to $144,646. For 4% inflation, the investor has for his 20 years of patience $98,094, less than his original stake. For 6% inflation, he has $67,018, around 2/3 his original stake.

Thus, we can see the outcome of investing in a "sure thing" like U.S. Treasury bonds. Because of taxes and inflation, the investor may actually be robbed by the system if he decides to invest in "guaranteed" securities. The "sure thing" in such investing is that it will most probably not protect the investor against the ravages of taxes and inflation.

9.4.2   Investing in an Index Fund

Let us consider a different strategy of investing. The investor invests his $100,000 (with equal dollar weightings) into a randomly chosen 25 of the larger cap stocks from the S&P 500. The stocks will be held until the time of liquidation. Mergers and splits will be handled as additions to the portfolio, as will acquisitions by purchase of stock in the acquiring firm. The purchase cost of the portfolio should run around $250 with a discount broker. The annual return in dividends will be of the order of 2%; and on these, the investor must pay taxes at the ordinary income tax rate (say, 35%). In this instance, we shall assume the investor is presently paying taxes rather than simply deferring them (as in the previous example). The residual (after paying taxes) is invested into one of the stocks in the portfolio using random selection. Let us suppose the stocks in the fund enhance their value (exclusive of dividends) on the order of 8% per year. Then, the nominal value of this portfolio in Year Zero dollars before capital gains taxes will be given by

$$S(t) = S(0) * (1 + .08 + .65 \times .02)^t. \tag{9.18}$$

After 20 years, this becomes $592,111. Now, at the time the portfolio is converted to cash, the investor will pay a 20% capital gains tax on the value in excess of the pretax value of the portfolio.[3] Here, things will be complicated a bit by the fact that we have been reinvesting dividends into stocks which remain within the portfolio. This is really not a problem, for the reinvested dividends after the 35% tax is taken out, will simply be the portfolio value of the year multiplied by $.65 \times .02$. The invested dividends at the end of year one will be $100,000 \times .013 = \$1,300$. The total amount of invested dividends after 20 years will be given by

$$100,000(.013) + 100,000(1.093)(.013) + \ldots$$

$$+100,000(.013)(1.093)^{18} = 1300 \times \frac{(1.093)^{18} - 1}{1.093 - 1} = 55,304 \qquad .$$

Assuming 8% growth of the stocks in the portfolio, the value in nominal dollars invested in the portfolio will then be $100,000 + $55,304, or $155,304. Therefore, at the end of 20 years, the investor will have to pay taxes on $592,111 - \$155,304 = \$436,807$. The tax rate on capital gains held one year or more is assumed to be 20%, hence: $(.20)(\$436,807)=\$87,361$, and $\$592,111-\$87,361= \$504,750$. So, in nominal dollars, our investor will have, at the end of 20 years, an after tax portfolio value of $504,750. At an inflation rate of 2%, this will have value in Year Zero dollars of $339,682. At an inflation rate of 4%, this portfolio will have after tax value in Year Zero dollars of $230,361. At an inflation rate of 6%, the value is reduced to $157,383. Using the 2% rate as an optimistic benchmark, we see that for the Treasury bond, after 20 years, the investor would have, in Year Zero dollars, $144,646. With the portfolio, the investor would have $339,682. For the mildly pessimistic 4% scenario, with the Treasury bond, the investor would have $98,094 in Year Zero dollars. With the portfolio, this

---

[3]The U.S. federal income tax on capital gains is complicated. Rates may be as low as 10% or as high as 20% + (including Alternate Minimum Taxes). States with an income tax (the vast majority of the 50), tax these gains at various rates. In California, the state and federal combined rate approaches 30%.

number would be $230,361. At 6%, the numbers compare at $67,018 versus $157,383.

It may be appropriate to consider the government as being responsible for a significant part of the inflation due to increases in the money supply (which the government uses to pay for its programs). Hence, in the case of the 4% inflation scenario for the stock portfolio, the government will have received over the 20 years, taxes plus "inflation tax" in excess of the profit of the taxpayer in Year Zero dollars. Note that the investor will have increased his $100,000 investment by only $130,361 in Year Zero dollars. The investor has profited much less than the state!

Computations such as those shown above tend to depress many persons.[4] Such is reality, unfortunately, for the citizens who work hard to be productive and expect a reasonable return on their investment. It is not depressing for the tens of millions of persons on the receiving end of the plethora of government sponsored *transfer payments*. And it is downright pleasant for the few million of persons who enjoy their positions in the inner circles of government, banking, the law, etc., who get their cut of the nation's productivity as surely as any Soviet nomenklatura member in the bad old days.

Unfortunately, like the soldier who insisted on playing in a crooked poker game because it was the "only game in town," it is our function in this book to help the reader do the best he or she can under admittedly suboptimal circumstances. Even though the government has made rules which do not compare favorably in their rates to those of Mafia protection rings, we do not call for the Libertarian Revolution in this book.

Now, we have seen that the "risk free" interest scheme is simply dreadful. The stock portfolio scheme is much better. Why then, does not everyone draw out his or her savings account and invest in stocks? The short answer is: **Risk**. History shows that the portfolio strategy we have advocated in this section would have worked well for all 20 year periods since 1927 in the United States (but not so well on, say, the Warsaw Stock Exchange over the same period).[5] But in the short run, portfolios can decline. Moreover

---

[4]It should be noted that nominal interest rates (and stock returns) may well not be invariant to expected inflation rates. It also may be true that even taxation rates may influence nominal yields on all investments. Economists disagree on these issues. Some say market rates reflect everything, inflation, taxes, etc. perfectly. Others say the process is much less perfect. It is clear that an investor who purchased, say, 5% U.S. Treasury bonds in the early 1970s got killed after-tax and after-inflation, by the 1980s (and this ignores the market value depreciation of the bonds due to higher interest rates, no doubt caused by inflation and high tax rates). It is equally clear that stocks, over the long-haul, have allowed investors positive, after-tax real (adjusted for inflation) returns. If one is unfortunate enough not to be able to hold stocks (due to the inability to bear risk) or if the equity markets do not replicate past performance in the future, what is the alternative? Consume all and die?

[5]Fama and French [3] have recently argued that the returns on U.S. equities in excess of the riskless rate (the so-called "equity premium") since 1950 were high relative to historical experience, were unexpected, and are unlikely to be repeated in the future.

it is possible they can decline over the next 20 year period as well. We now want to show how we can evaluate risk in ways that are usable by the institutional as well as the individual investor.

9.4.3   Time Based Risk Profiling

Let us suppose we are considering $100,000 invested into a portfolio (within an IRA and hence untaxed until liquidation) which grows according to a geometric Brownian motion with $\mu = \sigma = .1$. Now we are no longer in the realm of certitude. For any time in the future, the value of the portfolio will be

$$P(t) \;\; = \;\; P(0) \times \exp\left[\left(\mu - \frac{1}{2}\sigma^2\right)t + \sigma\sqrt{t}Z\right] \qquad (9.19)$$

$$= \;\; 100,000 \times \exp\left[\left(.1 - \frac{(.1)^2}{2}\right)t + .1\sqrt{t}Z\right], \qquad (9.20)$$

where Z is a normal (Gaussian) random variable with mean zero and variance one.

We can no longer talk about a certain value of the portfolio. We can, however, give a profile of value, given the assumptions of the investment. We recall that for Z a *standard normal variable* with mean zero and variance one, we have

$Z_{critical}$	$P(Z > Z_{critical}) = P(Z < -Z_{critical})$
2.3263	.01
1.6449	.05
1.2816	.10
1.0364	.15
.8416	.20
.6745	.25
.5244	.30
.2533	.40
0	.50

It is standard procedure to describe a security simply by giving its μ and σ. Even for the simple geometric Brownian model considered here, that will not be sufficient information for most investors. One thing an investor would like to know is the expected value of the portfolio at varying time horizons. That is easy here, since

$$E(P(t)) = 100,000\exp(.10t).$$

The investor might also like to know what will be the median trace (tracking) of the portfolio's value. In other words, for each year we would like to

---

[5](continued) It might also be noted that the time span 1927−2002 contains fewer than four *non-overlapping* twenty year periods.

find the value where 50% of the time the investment value will be less than that value, and 50% of the time it will be greater than that value. From (9.20), we see that the answer is

$$P_{.50}(t) = 100,000 \exp(.095t).$$

Of course, this is less than the expected value. Most people are surprised by this fact. They suppose the 50 percentile (median) and mean are the same or nearly so. By the growth assumptions of geometric Brownian motion, we arrive at a log normal distribution (see Appendix A at the end of this book). That distribution has a long right hand tail. Typically, the mean will be greater than the median except for the case where $\sigma = 0$. The reader might care to look ahead where a nonparametric analysis via resampling from the Ibbotson Index reveals a curve which is, in fact, log normal. (For a discussion of the resampling approach used, the reader is referred to Appendix A.) Beyond this, it would be reasonable for an investor to ask other questions. For example, suppose we look at a value of the portfolio which will be exceeded 90% of the time? Again, from (9.20), we quickly find

$$P_{.10}(t) = 100,000 \exp(.095t - .1\sqrt{t} \times 1.285).$$

At twenty years, this is a respectable \$668,589 . This compares with \$332,012 for a 6% Treasury bond. In Table 9.2 and Figure 9.5, we show the time indexed value of the portfolio for various percentiles.[6]

---

[6]Note that almost 90% of the time, the stock portfolio will outperform the Treasury bond even before taking taxes into account.

Table 9.2. Twenty Years of Portfolio Performance with $\mu = .10$ and $\sigma = .10$					
Time	10%	30%	Median	Mean	70%	90%
0	100000	100000	100000	100000	100000	100000
1	93286	104352	109966	110517	115882	129628
2	95825	112288	120925	122140	130226	152598
3	100008	121439	132976	134986	145610	176813
4	105232	131679	146228	149182	162385	203196
5	111312	143022	160801	164872	180791	232294
6	118182	155527	176827	182212	201044	264571
7	125832	169276	194449	201375	223365	300485
8	134276	184373	213828	222554	247988	340511
9	143548	200933	235137	245960	275165	385164
10	153696	219087	258571	271828	305171	435008
11	164776	238980	284340	300417	338310	490662
12	176854	260773	312677	332012	374911	552810
13	190006	284644	343838	366930	415341	622214
14	204316	310787	378104	405520	460003	699715
15	219876	339416	415786	448169	509339	786251
16	236789	370766	457223	495303	563840	882863
17	255167	405094	502789	547395	624044	990709
18	275133	442684	552896	604965	690547	111108
19	296822	483845	607997	668589	764007	1245400
20	320379	528916	668589	738906	845147	1395260

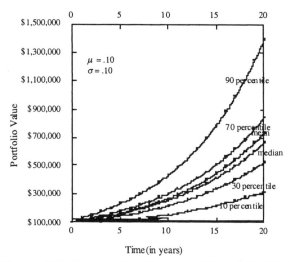

Figure 9.5. Percentile Traces of Portfolio Value.

We can concentrate on the *cumulative distribution* of the portfolio values at the end of twenty years as shown in Figure 9.6. When this distribution (or its first derivative) is estimated using historical data or is based on some scenario, we call the result a *simugram*. It is an attempt to estimate the probabilities of the various things which might happen.

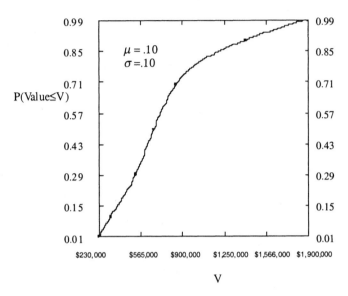

Figure 9.6. Distribution Function of Portfolio at 20 Years.

The fact that for each value of t, we can produce a distribution function of portfolio values gives us a means of *risk profiling* the investment. For the example given here (a portfolio with $\mu = .10$ and $\sigma = .10$), the situation is rather clear to the investor. If the model assumptions are correct, then the investor knows that he or she will have, at the end of 20 years, 90% of the time, portfolio equity at least as great as \$320,379. It is information of this sort, rather than the standard deviation *volatility* of the portfolio which conforms more closely to notions of *risk*. The investor also knows that with probability 30% his or her portfolio will have grown to \$845,147 or more. All in all, the profile shown in Figure 9.6 will be attractive to most investors when compared to investing in a Treasury bond.

And yet, people do invest in U.S. Treasury securities. Our portfolio model may have been overly optimistic. U.S. stock prices on the whole have grown by 10.6% compounded annually for the last 75 years. On the other hand, our volatility estimate has been on the low side. From 1982-1997, volatility of the stock market was .132. On the other hand, using the period 1926-1997, the volatility was .204 [1]. Suppose we redo our analysis with a volatility of .25 and a growth rate of .10. Then (9.19) becomes

$$P(t) = 100,000 \exp\left[\left(.10 - \frac{.25^2}{2}\right)t + .25\sqrt{t}Z\right]. \qquad (9.21)$$

Table 9.3. Twenty Years of Portfolio Performance with $\mu = .10$ and $\sigma = .25$				
Time	10%	30%	Median	70%	90%
0	100000	100000	100000	100000	100000
1	77743	93964	107117	122110	147587
2	72923	95336	114740	138094	180535
3	70548	97956	122906	154211	214121
4	69350	101308	131653	171087	249927
5	68872	105214	141023	189019	288755
6	68896	109594	151059	208211	331203
7	69300	114414	161810	228838	377807
8	70011	119659	173325	251061	429096
9	70981	125326	185661	275041	485617
10	72179	131422	198874	300945	547948
11	73585	137957	213027	328947	616710
12	75183	144948	228188	359232	692574
13	76964	152413	244428	391994	776269
14	78922	160375	261823	427446	868590
15	81055	168858	280457	465811	970403
16	83360	177891	300417	507334	1082650
17	85838	187503	321797	552276	1206380
18	88491	197726	344699	600917	1342700
19	91322	208597	369230	653561	1492860
20	94334	220152	395508	710536	1658200

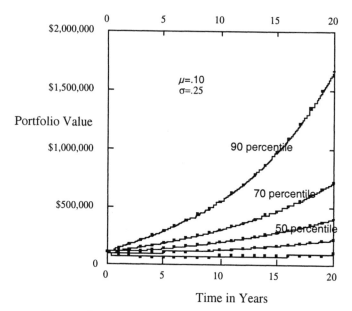

Figure 9.7. Percentile Traces of Portfolio Value.

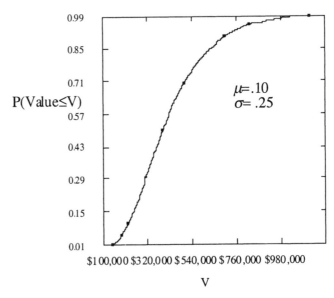

Figure 9.8. Distribution Function of Portfolio at 20 Years.

Although the mean value after 20 years remains a healthy $738,906, the median value has dropped to $395,508. The lower 10 percentile value has dropped to $222,628, less than the $332,012 guaranteed by a 6% Treasury bond. Now, suppose, instead of showing the figures and tables above, we had simply stated that in both cases $\mu = .10$ with, in the first case, $\sigma = .10.$, in the second $\sigma = .25$. Except for those who have computed these tables and figures numerous times, the dramatic change in the portfolio prospectus would not have been reflected by simply giving the mean and volatility. The investor would be much better off looking at the time indexed risk profile, as we have done here. And note, this is for the simplest stochastic investment case, that of a simple geometric Brownian walk. In the more complex situations, some of which we shall shortly show, the time indexed risk profile becomes a necessity.

Now, in the real world, bear markets happen. Generally, they happen swiftly and modeling them as an unpatched part of a geometric Brownian process is not a good idea. Moreover, they usually happen across the market so that diversification of a stock portfolio provides only limited protection. We shall choose to retain the Brownian process "patched" in the following fashion. There are many ways to approximate these jumps. For example, Dorfman [2] claims that a drop of 10% typically happens about once a year, a drop of between 15% and 20% has happened seven times in the last 50 years, and a correction of 20% or more has happened eight times in the last 50 years. Our procedure will be [7,8]: on the average of once every five years, the market will drop 20% and on the average of once every year, the market will drop 10%. This is slightly more optimistic than the Dorfman scenario but a rather close approximation. The combination of these two types of "bear jumps" will cause roughly a .15 diminution of growth. So, if we are looking for a long term nominal growth of .10, we will have to enhance the α to .25. In other words, the growth μ is equal to α minus the Poissonian bear jumps. The volatility for the Brownian walk part of the process is .10. We see such a situation in Figure 9.9. The median value of the portfolio is around $670,615, a growth rate of around 9.5%. The lower thirty percentile is $461,544, a growth rate of around 7.64%. The lower twenty percentile is $350,788, reflecting a return of around 6.4%. The lower ten percentile is around $260,225, reflecting an annual growth of 4.8%. The simulated expected value of the portfolio at the end of twenty years a rosy $799,000. but the risk profile indicates a more gloomy scenario. The simulated coefficient of variation is 1.74.

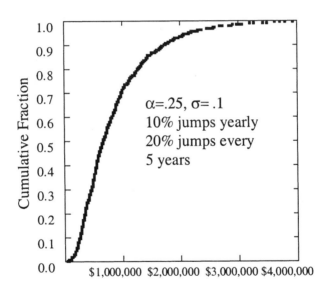

Value of $100,000 Portfolio after 20 years

Figure 9.9. Distribution Function of Portfolio after 20 Years.

Let us move to a standard deviation of .25 for the Brownian part of the process (retaining $\alpha = .25$ and the bear jump pattern of 20% down once every five years and 10% down annually). The lower ten percentile is $78,306. The lower twenty percentile is $148,828, an annual return of only 2%. The lower thirty percentile is $221,435, an annual return of 4%. The lower forty percentile is $312,275, an annual return of 5.7%. The median is $421,729, an annual return of 7.2%. The simulated coefficient of variation is 1.92.

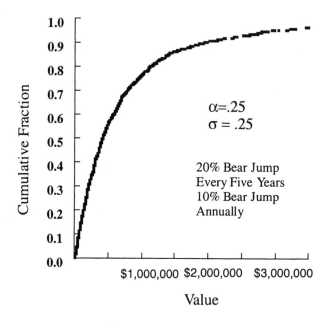

Figure 9.10. Distribution Function of Portfolio after 20 Years.

9.5   A Resampling Approach

So far, we have used some mathematical assumptions in our modeling of stock prices. Our basic model has been the usual Brownian geometric walk with the addition of bear jump Poissonian "patches." Now we shall look at historical records of the large company index (Ibbotson Associates: *Stocks, Bonds, Bills and Inflation 2001 Yearbook*). We show, in Table 9.4 the year to year ratio increases in the Ibbotson large company index (including dividends plus increases in stock prices from 1926 through 2000).

Using sampling of twenty of these ratios, with replacement, 5,000 times, we show in Figure 9.11 the cumulative distribution function of the value of an indexed based investment of $100,000 twenty years later. (For a discussion of resampling procedures, we refer the reader to Appendix A). The average of these increases is 1.13364. The median is 1.16810. To transform these ratios into continuous growth (at per year rates), we simply take the natural logarithm. Using the method of moments techniques in Section 9.3, we have the volatilities shown in Table 9.7.

Table 9.6. Large Stock Index Year Over Year Increases (Including Dividends)			
Year	Increase	Year	Increase
1926	1.11620	1964	1.16480
1927	1.37490	1965	1.12450
1928	1.43610	1966	0.89940
1929	0.91580	1967	1.23980
1930	0.75100	1968	1.11060
1931	0.56660	1969	0.91500
1932	0.91810	1970	1.04010
1933	1.53990	1971	1.43100
1934	0.98560	1972	1.18980
1935	1.47670	1973	0.85340
1936	1.33920	1974	0.73530
1937	0.64970	1975	1.37200
1938	1.31120	1976	1.23840
1939	0.99590	1977	0.92820
1940	0.90220	1978	1.06560
1941	0.88410	1979	1.18440
1942	1.20340	1980	1.32420
1943	1.25900	1981	0.95090
1944	1.19750	1982	1.21410
1945	1.36440	1983	1.22510
1946	0.91930	1984	1.06270
1947	1.05710	1985	1.32160
1948	1.05500	1986	1.18470
1949	1.18790	1987	1.05230
1950	1.31710	1988	1.16810
1951	1.24020	1989	1.31490
1952	1.18370	1990	0.96830
1953	0.99010	1991	1.30550
1954	1.52620	1992	1.07670
1955	1.31560	1993	1.09990
1956	1.06560	1994	1.01310
1957	0.89220	1995	1.37430
1958	1.43360	1996	1.23070
1959	1.11960	1997	1.33360
1960	1.00470	1998	1.28580
1961	1.26890	1999	1.21040
1962	0.91270	2000	0.90890
1963	1.22800		

Year	μ	σ	Year	μ	σ
	**Table 9.7. Ibbotson Large Stock Index**				

Table 9.7. Ibbotson Large Stock Index
μ (Including Dividends) and σ

Year	μ	σ	Year	μ	σ
1926	0.10993	0.11798	1964	0.15255	0.03985
1927	0.31838	0.13038	1965	0.11734	0.08560
1928	0.36193	0.16555	1966	−0.10603	0.11051
1929	−0.08796	0.32487	1967	0.21495	0.11965
1930	−0.28635	0.27484	1968	0.10490	0.12837
1931	−0.56810	0.47468	1969	−0.08883	0.13045
1932	−0.08545	0.63357	1970	0.03932	0.20436
1933	0.43172	0.51663	1971	0.35837	0.13481
1934	−0.01450	0.22225	1972	0.17379	0.06531
1935	0.38981	0.15963	1973	−0.15853	0.14316
1936	0.29207	0.14341	1974	−0.30748	0.23396
1937	−0.43124	0.23919	1975	0.31627	0.17478
1938	0.27094	0.42202	1976	0.21382	0.13136
1939	−0.00411	0.29154	1977	−0.07451	0.09550
1940	−0.10292	0.29437	1978	0.06354	0.16708
1941	−0.12319	0.14373	1979	0.16924	0.13337
1942	0.18515	0.14602	1980	0.28081	0.18309
1943	0.23032	0.15564	1981	−0.05035	0.12923
1944	0.18024	0.07712	1982	0.19400	0.18413
1945	0.31071	0.12841	1983	0.20302	0.09729
1946	−0.08414	0.18971	1984	0.06081	0.13610
1947	0.05553	0.09503	1985	0.27884	0.11859
1948	0.05354	0.19931	1986	0.16949	0.17930
1949	0.17219	0.10062	1987	0.05098	0.32354
1950	0.27543	0.10740	1988	0.15538	0.09988
1951	0.21527	0.11992	1989	0.27376	0.12009
1952	0.16865	0.11214	1990	−0.03221	0.18407
1953	−0.00995	0.09333	1991	0.26659	0.15397
1954	0.42278	0.12566	1992	0.07390	0.07315
1955	0.27429	0.12017	1993	0.09522	0.06076
1956	0.06354	0.14693	1994	0.01301	0.10559
1957	−0.11406	0.12720	1995	0.31794	0.05080
1958	0.36019	0.06137	1996	0.20758	0.08288
1959	0.11297	0.08002	1997	0.28788	0.15119
1960	0.00469	0.13557	1998	0.25138	0.21275
1961	0.23815	0.08793	1999	0.19095	0.12391
1962	−0.09135	0.20038	2000	−0.09552	0.16284
1963	0.20539	0.09662			

In the plot of μ versus σ we note some interesting years outside the apparent cluster. The two years with both high volatility and high growth are 1933 (the "Happy Days Are Here Again" optimism which characterized the start of the Roosevelt era) and 1938 (the year after the bottom of the Great Depression). Nine of the eleven outliers are depression years. Both 1974 and 1987 had significant bear epochs. With these eleven years removed, the correlation between μ and σ is $-.142$. With all 75 years left in the data base, the correlation is $+.184$. [7]

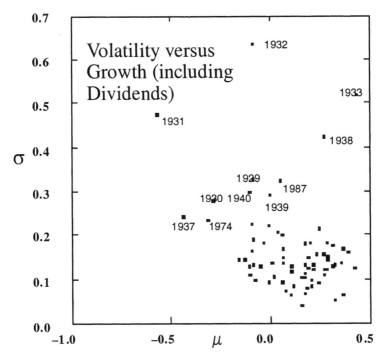

Figure 9.11. 75 Years of Ibbotson Index Growth and Volatility.

Let us consider a portfolio with initial value of $100,000. We pick five of the 75 index annual growths at random (with replacement), say, $\mu_1, \mu_2, \mu_3, \mu_4, \mu_5$. Then a simulated portfolio value after the five years is given by

$$V = \$100,000 \exp(\mu_1 + \mu_2 + \mu_3 + \mu_4 + \mu_5).$$

In Figure 9.12, we show the picture obtained by sorting a thousand such simulations according to percentiles.

---

[7]Interpreting these correlations in terms of the ability of differential volatility (σ) to explain differential stock market returns (μ) over time, the implied R-squared is less than 4% (and the relation goes negative when the outliers are excluded). Recall from Chapter 5 that even the EMH advocates have ceased to use standard risk measures (e.g., beta) to explain the cross-section of individual stock returns. The difference in return between stocks and short term Treasuries (i.e., the equity premium) has been massive over the last 75 years, and this is commonly ascribed to risk. A risk-based explanation for differences in return among stocks only, either intertemporally or cross-sectionally, remains elusive. This is causing some of us to wonder if there is more to the equity premium than risk. See also Fama-French, cited above.

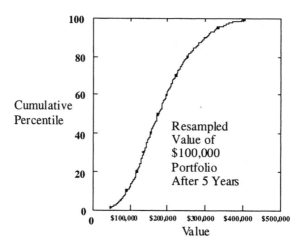

Figure 9.12. Distribution Function of Portfolio after 5 Years Using Resampling (Resampling Cumulative Simugram).

The mean value of a $100,000 portfolio after five years is $192,676. The median value is $175,530 (growth rate of .1125). However, the lower ten percentile is $92,747 (growth rate of −.015).

Next, we consider the same scenario except looking 20 years into the future. The results are quite optimistic. The median value is $873,100, an annual increase of 10.8%. Even the lower ten percentile value of $285,590 represents a growth rate of 5.2%.

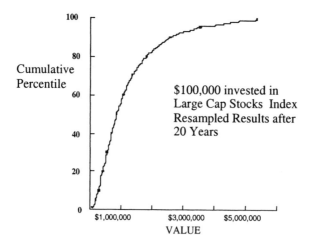

Figure 9.13. Distribution Function of Portfolio after 20 Years Using Resampling.

9.5.1   A Partially Privatized Social Security Plan

President George W. Bush and others have suggested a partial privatization of Social Security. Under the present plan, workers and their employers together "contribute" roughly 15.3% of a worker's salary (up to a given ceiling) to the FICA fund. The worker's portion of the contribution is not tax exempt. After roughly 45 years of employment, the worker may start to draw a Social Security stipend until the time of his or her death. Part of the stipend may be subject to income tax, even though no tax exemption was given the worker while paying his or her FICA tax.

The Bush plan would permit a worker to use a portion of his or her FICA setasides to invest in the stock market, and it is actually quite similar to the recently adopted Swedish plan where a worker gets back from the state 2% of his or her salary for investing in bond or stock funds. Here, we will be looking at a more aggressive approach where something less than half of a worker's FICA payments are privatized. Typically, it is assumed that some restrictions leaning toward fiscal conservatism should be applied. So, we will use, as an example, a contribution of $2,000 per year for each worker over a period of 45 years invested in something like the Ibbotson Index [5]. Let us note in Figure 9.12 what the resampled (strictly speaking, this is not a bootstrap) results look like when we take 5,000 resampled concatenations of the index, assuming that $2,000/year will be added to the fund.

The results are quite promising. The mean value of such a fund is $4.84 million dollars. The median value is $2.724 million. The lower twenty percentile is $1.105 million. Even the lower ten percentile is $695 thousand. The lower five percentile is $464 thousand. The lower one percentile is $225 thousand. Of course, there is the problem of inflation. Even so, we realize that the "contributions" would be indexed for inflation. We are using an annual contribution half the full FICA contribution. If we assume a 4% inflation rate and index the annual contributions by 4%, the median value after 45 years is $4.258 million ($728 thousand in current dollars). The lower ten percentile is $1.2043 million ($206 thousand in current dollars). The mean value is $7.0389 million ($1.2050 million in current dollars). Half the current Social Security maximum payout for a person reaching 65 is around $12,000. A lifetime annuity at age 65 may be purchased for approximately 12.5 times the annual payout. Hence, the mean translates into a current annual payout of $96,404, roughly eight times the current annual half maximum Social Security payment. The fifty percentile translates into a current annual payout of $58,240 in present value dollars, nearly five times the current annual half Social Security payment. The lower ten percentile translates into a current annual payout of $16,480 in present value dollars. Even the pessimistic ten percentile gives a better present value payout than the current Social Security half maximum payout. The median payout in current dollars bests the current Social Security half maximum payout by a factor of nearly five.

It should be noted, by the way, that 4% is unrealistically high for inflation, particularly so in recent years. The lower the inflation rate, the better the privatized plan works relative to Social Security. Another obvious benefit of privatization is giving citizens ownership in real bonds and/or stocks rather than a promise of something potentially quite arbitrary put not so safely in a highly porous lockbox. Over long stretches of time, historically, investing in things like the S&P 500 index or in the broadly based Wilshire 5000 index, is very much like buying a certificate of deposit paying over 10%. The time averaged investment of pension funds in highly diversified index funds gives the investor the benefit of the only two "free lunches" in security investment of which we are aware: diversification to protect against Enron catastrophes and time to smooth out losses from bear years.

One benefit of the introduction of some measure of privatization in a time of a very down market would be a stimulation of the market. It would also have a desirable effect for the younger workers who would be buying stocks at a bargain and for those already retired or nearing retirement in raising the price of equities. We observe, moreover, that in privatizing part of Social Security, we will have moved toward a situation where workers have obtained partial ownership of real companies, rather than slips of paper kept vaguely in a highly porous lock box.

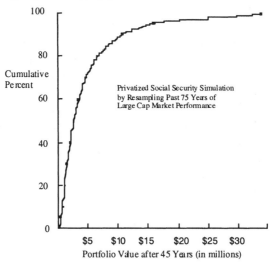

Figure 9.14. Resampling Based Distribution Function of Privatized Social Security Index Fund After 45 Years of Work.

It is unlikely that worries about the volatility of the market will be the reason for not going ahead with privatization. We have seen that over time long term investing gets around this problem. The actual reason will be the loss of control of money by the political elites. Who implicitly owns assets is not nearly as important as who controls them. The attempt to take over the health care system in the United States by the federal government during the Clinton Administration was unsuccessful, in part, when it was understood by the citizenry that over 10% of the American economy would have been transferred from the private sector to that of the state. History has shown an extreme reluctance of politicians to give up control of assets once they have achieved such control. And the Social Security program in the United States is nearly 70 years old. On the other hand, it is possible that, like the Swedish socialists, the American bureaucracy will recognize that they have little choice but to privatize an increasingly expensive and inefficient program. In any event, even if the partially privatized FICA plan is never introduced, the study in this section could be of use to a person thinking of making regular investments into a tax deferred index fund.

9.5.2 Index Funds as High Interest Paying Money Market Accounts

Figure 9.13 and Figure 9.14 give some support for investing in an index fund broadly composed of large cap corporations. Note that we used an approach which has few modeling assumptions in both these cases. We have assumed that the future increases in such an index will be similar to those in the past.

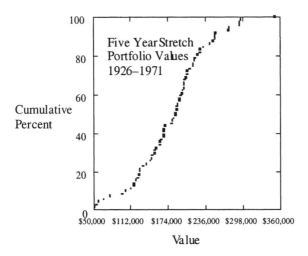

Figure 9.15. Resampling Based Distribution Function of $100,000 invested for Five Contiguous Years.

Some may object to taking single year rates from the Ibbotson history [5]. What happens when we have long patches of decline? Might not inclusion of these in an appropriate fashion introduce more pessimism into both long term investments in index funds and the hypothetical privatized

Social Security plan. In Figure 9.15 we show the cumulative percentiles one obtains when examining a $100,000 investment for five contiguous years starting in 1926 and going through 1996 (we are limited by 1996, since we are looking at 1996 and the four following years).

Now the lower ten percentile represents almost no gain at all. On the other hand, the median and mean both correspond to an annual gain (again, including dividends) of around 12%. Still, there is the troubling lower ten percentile.

Next, we carry out 10,000 resamplings of two five year stretches from the Ibbotson Index with an initial $100,000 investment in the index. This time we show a histogram of the results in Figure 9.16. The lower five percentile is slightly better than break even. The lower ten percentile now pays over 2.7%. The lower twenty percentile pays 4.7%. The median pays 11.3%. The mean pays 11.7%. Perhaps we can say that at ten years we really have reached the point where we can talk meaningfully about "the long term." An investor in the index fund for ten years would appear likely to be pleased with his or her end results and has little chance of awful results. In other words, risk would appear to have been reduced to bearable levels. We note the shape of the histogram (which has been based on no distributional assumptions) has the characteristic shape of the log normal density function.

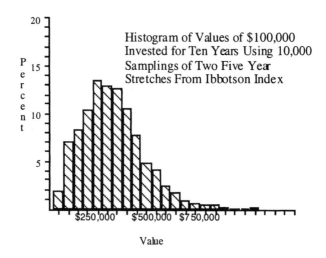

Figure 9.16. Resampling Based Density Histogram (Simugram) of $100,000 Invested for 2 Five Contiguous Year Stretches.

Next, let us look at an index fund starting with $100,000 randomly selecting four five year stretches from the Ibbotson index of large cap stocks. We show these results in Figure 9.17.

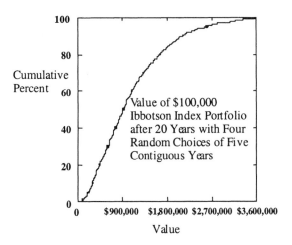

Figure 9.17. Resampling Based Distribution Function of Initial $100,000 Invested in Ibbotson Index for 20 Years.

The riskiness is reduced still further. The lower ten percentile of performance is a growth of 5.3%. The median growth shows over 11% annual growth and the mean over 12%.

Finally, we return to the notion of resampling a 45 year investment from an annual $2,000 per year invested in the index fund. Here we pick randomly nine contiguous five year stretches. For each year, we add $2,000 and we suppose the annual increase in the fund is the average of the five year stretch in which the year lies. We note here that the lower ten percentile is $1.32 million, the mean $4.4653 million and the median $2.5380 million.

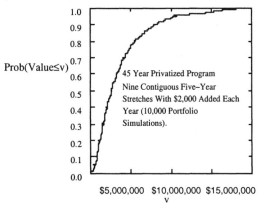

Figure 9.18. Resampling Based Distribution Function of Privatized Social Security Index Fund After 45 Years of Work by Selecting Randomly 9 Five–Year Contiguous Stretches from the Ibbotson Index.

In any event, it would appear that if history of growth is the best guide to the future, large cap index funds appear to be very attractive.[8] Our results show that, over the long haul, they appear to act like money market accounts paying high rates of interest (over 10%) with relatively small chance that the investor will be disappointed.[9]

9.5.3   Gaussian Model versus Resampling

Let us now ask how closely the resampling procedure (nonparametric) tracks what might have been obtained had one estimated the growth and volatility from the Ibbotson Index and used the lognormal distribution to give a distribution of outputs of $100,000 invested in the Ibbotson Index. Using the methodology of Section 9.3, and the figures in Table 9.7, we have as estimates for μ and σ, .12672 and .19468 respectively.

We note in Figure 9.19, that for one year out, the resampling results give significantly different results from those of the Gaussian model. The lower percentiles, particularly, have smaller values than those obtained from the Gaussian model.

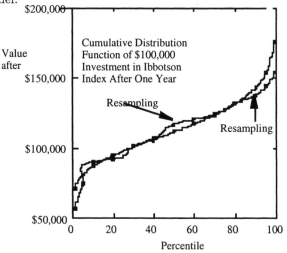

Figure 9.19. Resampling and Gaussian Model Distributions One Year in the Future.

---

[8]Like everybody else, we are extrapolating the past here (because that is about the only feasible approach). Recall, however, that the theory says we should be discounting expectations of the future (which, in principle, could be very different from the past or simply wrong). Only a general, long-run equilibrium would assure that the two approaches would converge on today's stock prices. Indeed, even to keep the neoclassical fantasy alive, Fama-French (cited above) are forced to inform us that we are "expecting" a substantially lower equity premium going forward than we have experienced over the last 50 years.

[9]In the financial economics literature, this result is called the equity premium "puzzle."

However, when we look three years out in Figure 9.20, the differences are much less.

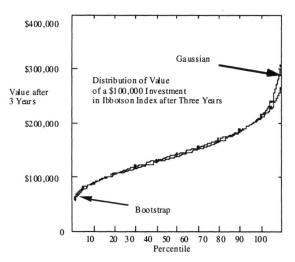

Figure 9.20. Resampling and Gaussian Model Distributions Three Years in the Future.

And by ten years out, as we see in Figure 9.21, the two curves almost fall on top of each other.

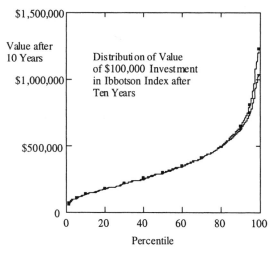

Figure 9.21. Resampling and Gaussian Model Distributions Ten Years in the Future.

9.6   Stock Progression Using Differential Equations

9.6.1   Ito's Lemma

Although the binomial tree model can suffice for modeling stocks and derivatives, persons comfortable with partial differential equations will find it easier and more concise to deal with models based on Ito's Lemma.

Following Hull [4], let us suppose we have a continuously differentiable function of two variables $G(x, t)$. Then, taking a Taylor's expansion through terms of the second order, we have

$$\begin{aligned}\Delta G \quad \approx \quad & \frac{\partial G}{\partial x}\Delta x + \frac{\partial G}{\partial t}\Delta t \\ & + \frac{1}{2}\frac{\partial^2 G}{\partial x^2}(\Delta x)^2 + \frac{1}{2}\frac{\partial^2 G}{\partial t^2}(\Delta t)^2 + \frac{\partial^2 G}{\partial x \partial t}\Delta x \Delta t. \quad (9.22)\end{aligned}$$

Next let us consider the *general Ito process*

$$dx = a(x, t)dt + b(x, t)dz \quad (9.23)$$

with discrete version

$$\Delta x = a(x, t)\Delta t + b(x, t)\epsilon\sqrt{\Delta t}, \quad (9.24)$$

where dz denotes a Wiener process, and a and b are deterministic functions of x and t. We note that

$$(\Delta x)^2 = b^2\epsilon^2\Delta t + \text{ terms of higher order in } \Delta t. \quad (9.25)$$

Now

$$\text{Var}(\epsilon) = E(\epsilon^2) - [E(\epsilon)]^2 = 1.$$

So, since by assumption $E(\epsilon) = 0$,

$$E(\epsilon^2) = 1.$$

Furthermore, since ϵ is $\mathcal{N}(0, 1)$, after a little algebra, we have that $\text{Var}(\epsilon^2) = 2$, and $\text{Var}(\Delta t \epsilon^2) = 2(\Delta t)^2$. Thus, if Δt is very small, through terms of order $(\Delta t)^2$, we have that it is equal to its expected value, namely,

$$(\Delta x)^2 = b^2\Delta t. \quad (9.26)$$

Substituting (9.24) and (9.26) into (9.22), we have

Lemma 9.2 (Ito)

$$\Delta G = \left(\frac{\partial G}{\partial x}a(x, t) + \frac{\partial G}{\partial t} + \frac{1}{2}\frac{\partial^2 G}{\partial x^2}b^2\right)\Delta t + \frac{\partial G}{\partial x}b\epsilon\sqrt{\Delta t} \quad (9.27)$$

or

$$dG = \left(\frac{\partial G}{\partial x}a(x, t) + \frac{\partial G}{\partial t} + \frac{1}{2}\frac{\partial^2 G}{\partial x^2}b^2\right)dt + \frac{\partial G}{\partial x}b\,dz. \quad (9.28)$$

9.6.2   A Geometric Brownian Model for Stocks

Let us return to (9.4)

$$\frac{\Delta S}{S} = \mu \Delta t + \sigma \epsilon \sqrt{\Delta t}. \tag{9.29}$$

Again, σ (the *volatility*) is a measure of the variability of the process as time increases. Here we will formally take ϵ to be a normal variate with mean zero and variance 1. In the limit, as Δt goes to zero, such a process is uniquely defined and is commonly referred to as a geometric Brownian process.

$$d(\ln S) = \mu dt + \sigma dz. \tag{9.30}$$

Alternatively, we have

$$dS = \mu S dt + \sigma S dz. \tag{9.31}$$

Now, in Ito's lemma we define $G = \ln S$. Then we have

$$\frac{\partial G}{\partial S} = \frac{1}{S}; \frac{\partial^2 G}{\partial S^2} = -\frac{1}{S^2} \; ; \frac{\partial G}{\partial t} = 0.$$

Thus G follows a Wiener process:

$$dG = \left(\mu - \frac{\sigma^2}{2} \right) dt + \sigma dz. \tag{9.32}$$

This tells us simply that if the price of the stock at present is given by $S(0)$, then the value t units in the future will be given by

$$\begin{aligned} S(t) &= S(0) \exp \left[(\mu - \frac{\sigma^2}{2})t + \epsilon \sigma \sqrt{t} \right] \\ &= S(0) \exp \left[\mathcal{N} \left((\mu - \frac{\sigma^2}{2})t, t\sigma^2 \right) \right] \\ &= \exp \left[\mathcal{N} \left(\log(S(0)) + (\mu - \frac{\sigma^2}{2})t, t\sigma^2 \right) \right], \end{aligned} \tag{9.33}$$

where $\mathcal{N}(\log(S(0)) + (\mu - \sigma^2/2)t, t\sigma^2)$ is a normal random variable with mean $\log(S(0)) + (\mu - \sigma^2/2)t$ and variance $t\sigma^2$. Thus, $S(t)$ is a normal variable exponentiated (i.e., it follows the *lognormal* distribution). The expectation of $S(t)$ is given by $S(0) \exp[\mu t]$. In the current context, the assumption of an underlying geometric Brownian process (and hence that $S(t)$ follow a lognormal distribution) is somewhat natural. Let us suppose we consider the prices of a stock at times t_1, $t_1 + t_2$, and $t_1 + t_2 + t_3$. Then if we assume $S(t_1 + t_2)/S(t_1)$ to be independent of starting time t_1, and if we assume $S(t_1 + t_2)/S(t_1)$ to be independent of $S(t_1 + t_2 + t_3)/S(t_1 + t_2)$, and if we assume the variance of the stock price is finite for finite time, and if we assume that the price of the stock cannot drop to zero, then, it can be shown that $S(t)$ must follow geometric Brownian motion and have the lognormal distribution indicated.

9.7   Conclusions

Treasury bills, insured bank accounts and Treasury notes and bonds pay, historically, at a mere fraction of the long term appreciation plus dividends available in the stock market. The willingness of investors to accept less return for less risk is, of course, completely reasonable in many circumstances.

For the investor in stocks, there are two basic weapons for reducing volatility while maintaining the average ten plus percent annual gains historically associated with the United States stock market. One of these is diversification of investment across a portfolio of stocks. This will protect the investor against the disaster experienced by persons who had everything invested in Enron.

On the other hand, diversification will not protect investors very well from across the board bear markets, such as we have seen in 2000−2002. This brings us to the second weapon: time. An investor who is willing to invest in, say, an S&P 500 index fund for 20 years has (using history as our guide), a relatively small chance of being disappointed.

In the evaluation of an investment strategy, most investors will not find it meaningful simply to balance growth versus volatility. Volatility is not a very intuitive surrogate for risk. A much better way, in our view, is to examine the *risk profile* of the strategy. That means, using historical data, distribution functions of the value are graphed versus time. Thus, the investor will be able to look, say, five years out and estimate the probability that his investment will have grown by at least, say, seven percent annually. And the investor will be able to examine, say, the lower ten percentile of the value of his portfolio at the time of one's retirement.

In modeling the risk profiles of investment strategies, we advocate, for longer time horizons, say five years or more, the use of the simple lognormal model associated with geometric Brownian motion. We were pleasantly surprised to discover that one obtains almost identical results in the time sliced distribution functions for the large cap Ibbotson index when we use a resampling of historical annual return rates. Thus, the highly parametric Gaussian model gives the same results as that obtained by a highly nonparametric resampling strategy. For shorter time horizons, certainly for times of less than two years, the risk profiling becomes more difficult for the parametric model. It may be appropriate, having historically estimated growth and volatility, to add on a Poisson jump process to take account of the possibility of bear jumps.

Problems[10]

Problem 9.1. Let X be a random variable with the property that $Y = \ln X$ has a normal distribution with mean μ and variance σ^2. Prove that:
(a) The mean of X is $\exp(\mu + \sigma^2/2)$.
(b) The variance of X is $[\exp(\sigma^2) - 1] \exp(2\mu + \sigma^2)$.

Problem 9.2. Using equation (9.5),write a simulation program for a single stock with specified growth and volatility. With all time units in years, use $\mu = .1$ and $\sigma = .1$, and $S(0) = 100$ to obtain a thousand values of $S(1)$. Carry this out for a binomial step every tenth of a year and also for every hundredth of a year. Give histogram presentations of the distribution of the $S(1)$ outputs.

Problem 9.3. Using equation (9.9), obtain 1,000 one step simulations of $S(1)$ for the conditions given in Problem 9.2.

Problem 9.4. Using the weekly stock prices below, estimate μ and σ.

Table 9.1. 26 Weeks of Stock Performance		
Week=i	Stock(i)	$r(i)$=Stock(i)/Stock($i-1$)
1	100.00000	
2	101.29874	1.01299
3	100.54802	0.99259
4	102.08259	1.01526
5	104.42557	1.02295
6	105.12951	1.00674
7	102.69395	0.97683
8	101.87883	0.99206
9	103.08261	1.01182
10	103.36549	1.00274
11	104.47395	1.01072
12	104.19463	0.99733
13	105.96339	1.01698
14	105.70573	0.99757
15	105.29111	0.99608
16	106.64880	1.01289
17	104.50578	0.97991
18	105.32778	1.00787
19	106.51591	1.01128
20	105.99684	0.99513
21	106.90942	1.00861
22	106.67928	0.99785
23	109.19090	1.02354
24	110.04130	1.00779
25	109.48404	0.99494
26	107.96238	0.98610

---

[10]For those uncomfortable with simulations, the reader will find assistance in Appendix A at the end of this book.

References

[1] Bogle, J. C. (1999). *Common Sense and Mutual Funds: New Imperatives for the Intelligent Investor.* New York: John Wiley & Sons.

[2] Dorfman, J. R., (1997). *The Wall Street Journal,* Monday Apr 7, Sec: C p: 1.

[3] Fama, E. and French, K. (2002). "The equity premium," *Journal of Finance,* April.

[4] Hull, J. C. (1993). *Options, Futures, and Other Derivative Securities,* Englewood Cliffs, NJ: Prentice-Hall.

[5] Ibbotson Associates (2001). *Stocks, Bonds, Bills and Inflation 2001 Yearbook.* New York: Ibbotson.

[6] Smith, C. W. (1976). "Option pricing: a review," *Journal of Financial Economics,* Vol. 3, 3–51.

[7] Thompson, J. R. (1999). *Simulation: A Modeler's Approach.* New York: John Wiley & Sons, 115–142.

[8] Williams, E. E. and Thompson, J. R. (1998). *Entrepreneurship and Productivity.* New York: University Press of America.

Chapter 10

Investing in Real World Markets: Returns and Risk Profiles

10.1 Introduction

The basic equations which we outlined in Chapter 6 and later developed in Chapter 7 highlight the importance of being able to identify future cash flow streams as the basic determinants of the value of an investment in common stock. Let us reiterate equation (7.1):

$$P(0) \;=\; \frac{D(0)}{(1+r)^0} + \frac{D(1)}{(1+r)^1} + \frac{D(2)}{(1+r)^2} + \dots \qquad (10.1)$$

$$\;=\; \sum_{t=0}^{\infty} \frac{D(t)}{(1+r)^t},$$

where

$$
\begin{aligned}
P(0) &= \quad \text{Price (Value) today} \\
D(t) &= \quad \text{Dividend } t \text{ years in the future} \\
r &= \quad \text{Market's (investor's) required rate of return.}
\end{aligned}
$$

Now, of course, the big issue is reward, and just how much risk one must assume to achieve reward. Part of the task of trying to figure out reward involves forecasting, and our initial attempt at evaluating the hypothetical company Ameritape, Inc. was directed at determining just what sort of reward we might achieve were we to own shares in that company (and, conversely, what kinds of losses we might suffer).

10.2   Ameritape Revisited

At this point, we shall consider our forecasts of sales and operating income from Chapter 8 (Figure 8.13). Notice that the projection shows sales growing to about $6.4 billion in 2007. Similarly, operating income is projected to grow to about $1.3 billion in 2007. The ratio of forecasted operating income to sales is given in Figure 10.1.

Year	Sales (S)	Operating Income (OI)	OI/S
2003	$5,327,270	$1,038,210	.19
2004	$5,575,930	$1,084,730	.19
2005	$5,836,200	$1,133,330	.19
2006	$6,108,610	$1,184,110	.19
2007	$6,393,740	$1,237,160	.19

Figure 10.1. Sales and Operating Income Projections and Their Ratio 2003–2007.

Of course, the average ratio of .19 is close to the historical ratio (see Figure 7.17).

Now, our objective will be to forecast EPS and DPS given the sales and operating income forecasts made in Figure 8.13 and repeated in Figure 10.1. From the income statements for 1998–2002 (Figure 7.11), we may observe that there are several other items we must consider in order to do this. First, we need a projection of "other income (expense)." Next, we need an income tax calculation. And, finally, we need a projection of the number of shares outstanding. The first item is the most difficult to project. We do not have much information about the sources of "other income" and "other expenses" but the most likely explanation is that these are "interest income" (earned on the firm's temporary cash balances) and "interest expense" (based on the indebtedness of the company). In Ameritape's case, it would require a complete forecast of balance sheets for 2003–2007 to get an accurate measure of these data. An analysis of balance sheet ratios (discussed in detail in Chapter 7) would lead to the conclusion that the sales and operating income levels we project should provide sufficient *free cash flow* (after-tax income plus depreciation and amortization) to finance extra net working capital and capital expenditure requirements for the Company. This most likely means that the firm will not have any major financing required (hence, interest expense should remain fairly constant). The capital structure ratios (see Chapter 7) should also be easily maintained with the growth of retained earnings so that additional equity (sale of new shares) will not be required.

Given the above, it might be a reasonable assumption to let "other income (expense)" remain at 2001–2002 levels. A calculation of the ratio of "income tax" to "income before tax" in Figure 7.11 suggests that ratio has been a fairly constant .34 over 1998–2002. Since this is close to the statutory federal income tax rate for a Subchapter "C" corporation (essentially all publicly held corporations), it is a fair rate to apply to projected "income before tax." Finally, since the number of shares outstanding should not change, the 2002 figure of 640 million shares should be applicable to the years 2003–2007. Thus, in Figure 10.2 we provide the calculations to arrive at income before taxes for 2003–2007.

Year	Operating Income(1)	Other Income (2)	Income Before Tax (1) +(2)
2003	$1,038,210	$28,000	$1,066,210
2004	$1,084,730	$28,000	$1,112,730
2005	$1,133,330	$28,000	$1,161,333
2006	$1,184,110	$28,000	$1,212,110
2007	$1,237,160	$28,000	$1,265,160

Figure 10.2. Income Before Tax Projections 2003–2007 (in Thousands of Dollars).

Figure 10.3 uses $(1 - t)$, or $(1 - .34) = .66$ as the after tax income ratio constant, while this number is divided by 640 million to arrive at EPS.

Year	Income Before Tax	Net Income	EPS
2003	$1,066,210	$703,699	$1.10
2004	$1,112,730	$734,402	$1.15
2005	$1,161,330	$766,478	$1.20
2006	$1,212,110	$799,993	$1.25
2007	$1,265,160	$835,006	$1.30

Figure 10.3. Income Before Tax, Net Income, and EPS Projection 2003–2007 (in Thousands of Dollars Except Per Share Amounts.)

At this point, it may be worth reflecting on the historical EPS data for 1998–2002 (Figure 7.11) together with the projections made in Figure 10.3. These are provided in Figure 10.4

Year	EPS (P=Projection)
1998	$.99
1999	$.74
2000	$.70
2001	$1.02
2002	$1.24
2003	$1.10 (P)
2004	$1.15 (P)
2005	$1.20 (P)
2006	$1.25 (P)
2007	$1.30 (P)

Figure 10.4. EPS Historical and Projections 1998–2007.

Now we shall note that our projections are *trend-line* projections based on historical relationships. This is not a specific forecast for 2003 (which might well be above $1.24 in fact). Nevertheless, these data reflect the fact that past sales and income patterns may produce negative (as well as positive) growth years. This is often ignored by analysts who often begin with the base year (2002 in this case) and forecast forward. Let us see what this kind of analysis might produce (Figures 10.5–10.7).

Year	EPS	DPS
2003	$1.31	$.40
2004	$1.39	$.48
2005	$1.48	$.48
2006	$1.57	$.54
2007	$1.66	$.54

Figure 10.5. EPS and DPS Projections 2003–2007.

Year	EPS	High	Low
1998	$.99	$17.82	$14.85
1999	$.74	$11.84	$11.10
2000	$.70	$11.20	$10.50
2001	$1.02	$18.36	$16.32
2002	$1.24	$22.32	$19.84

Figure 10.6. EPS and Stock Prices 1998–2002.

Price/earnings multiples are then frequently computed by dividing the high and low stock prices by the EPS for each year. The results of these calculations are provided in Figure 10.7.

Year	High Multiple	Low Multiple
1998	18×	15×
1999	16×	15×
2000	16×	15×
2001	18×	16×
2002	18×	16×

Figure 10.7. High and Low Multiples 1998–2002.

Now, a number of calculations are often made to demonstrate returns an investor might expect to earn in the future if, say, a 6% EPS/DPS growth rate applied (the rate assumed in Figure 10.5) and assuming prices at the high and low ends of the multiple range for terminal year 2007. For example, suppose we are making our investment at the beginning of year 2003, expect dividends to be paid in 2003–2007 as per Figure 10.5, and expect EPS and P/E multiples as per Figures 10.5 and 10.7. Suppose further that the market price today is $21.08 per share (17× EPS for 2002). Two immediate calculations could be made using equation (10.1) assuming for sake of

calculation ease, dividends being paid one, two, three, four, and five years
hence with the stock being sold at either the high or low multiple in five
years:

$$\$21.08 = \frac{\$.40}{(1+r)^1} + \frac{\$.48}{(1+r)^2} + \frac{\$.48}{(1+r)^3} + \frac{\$.54}{(1+r)^4} + \frac{\$.54}{(1+r)^5} + \frac{P_5}{(1+r)^5},$$

where $P_5 = (15)(\$1.66) = \$24.90$ or $(18)(\$1.66) = \$29.88$.

In the low multiple case ($P_5 = \$24.90$), the solution for r is about .055, or
a 5.5% return. In the high multiple case ($P_5 = \$29.88$), r is about .103,
or a 10.3% return. Notice that simply changing one variable (the terminal
price earnings ratio), we vary our return by almost double. Not only does
this give us some feel for our expected returns from investing in Ameritape,
assuming a 6% earnings growth rate, it also gives us some understanding
of the risk to which we might be exposed.

Let us change our focus a bit in order to get a better handle on risk.
Our real exposure with owning Ameritape is that we might lose some (or
perhaps all) of our money! Unfortunately, it is very difficult to forecast
major events which might completely undermine the historical record, and
that is all we have to go on in most cases. Even predicting the Enron
bankruptcy would have been next to impossible until the event was almost
happening. Nevertheless, we can construct, using simulation, a distribution
of possible outcomes where we have some reasonable expectations about the
future.

In Ameritape's case, we have some confidence in the future dividends
that will be paid. This would provide us with an average return of about
2.3% (say $\$.49/\$21.08$) on our investment. This, of course, would not set
a minimum on the return we might earn since all sorts of things (even if
our earnings and dividend forecasts were perfectly correct) can cause stock
prices to decline. Suppose that we could determine, over a period of many
years, returns (dividends plus stock price gains) for Ameritape. Suppose
further that we could compute the μ and σ of this distribution. These
statistics could be used to simulate future returns. Although the past is
no guarantee of the future, past history tempered with reasonable analysis
and expectations about the future can certainly be a guide. For example,
if one had purchased Ameritape in 1998 at, say, its average price of $16.34
and had collected the dividends indicated in Figure 7.11, one would have
earned:

$$\$16.34 = \frac{\$.36}{(1+r)^1} + \frac{\$.36}{(1+r)^2} + \frac{\$.36}{(1+r)^3} + \frac{\$.36}{(1+r)^4} + \frac{\$.40}{(1+r)^5} + \frac{\$21.08}{(1+r)^5}$$

assuming today's price of $21.08 per share. *Ex post facto*, that is a return
of about 7.2%. If one had become nervous with the EPS (and stock price)
declines in 1999 and 2000 and had sold, the results would have been much
worse. Suppose you had bought at $16.34 but had sold at the low in the year

2000 of $10.50. Assuming you received the 1998, 1999, and 2000 dividends, the return would have been a rather dismal:

$$\$16.34 = \frac{\$.36}{(1+r)^1} + \frac{\$.36}{(1+r)^2} + \frac{\$.36}{(1+r)^3} + \frac{\$10.50}{(1+r)^3},$$

solving for $r = -.111$, or -11.1%.

But suppose you had bought at the low in 2000, sold at the high in 2002, and had collected all your dividends. Your return would have been an impressive $r = .285$, or 28.5% (by solving)

$$\$11.20 = \frac{\$.36}{(1+r)^1} + \frac{\$.36}{(1+r)^2} + \frac{\$.40}{(1+r)^3} + \frac{\$22.32}{(1+r)^3}.$$

10.3 Forecasting: A Reality Check

Interestingly, the projected returns assuming an earnings base at the year 2002, and a 6% growth rate, suggest a return from holding the stock in a range of from 5.5% to 10.3%, depending on the price/earnings multiples in the terminal holding year (say five years out). We may be able to narrow that range a bit by revisiting our admittedly more conservative projections in Figure 10.3. Let us adopt these EPS projections, assume the $.40 dividend for 2002 is not reduced, and then try to project what the Board of Directors might do with the dividend in the future. In order to do this, let us consider the dividend payout ratio over 1991−2002. A review of EPS and DPS from Figure 7.17 will allow us to make the following computations:

Year	DPS	EPS	DPS/EPS
1991	$.20	$.56	.36
1992	$.20	$.60	.33
1993	$.24	$.63	.38
1994	$.24	$.67	.36
1995	$.24	$.61	.39
1996	$.24	$.77	.31
1997	$.24	$.88	.27
1998	$.36	$.99	.36
1999	$.36	$.74	.49
2000	$.36	$.70	.51
2001	$.36	$1.02	.35
2002	$.40	$1.24	.32

Figure 10.8. Dividend Payout 1991−2002.

The dividend payout ratio has ranged from a low of .27 in 1997 to a high of .51 in 2000. The average (arithmetic mean) is .37. One might expect that a payout of this magnitude should apply in the future. Total EPS over 2003 to 2007 is expected to be $6.00. Applying the .37 payout ratio, we would probably be safe in assuming an average dividend of: (.37)($6.00)/5=$.44. Given the lumpiness of the payout, a projection similar to that in Figure 10.9 might be appropriate:

Year	Dividend
2003	$.40
2004	$.40
2005	$.44
2006	$.48
2007	$.48

Figure 10.9. Dividend Payout Projection 2003–2007.

The average price/earning multiple over 1998–2002 (see Figure 10.7) was 16.3×. Applying that to the $1.30 EPS projection for 2007 produces: (16.3)($1.30)=$21.19. If the price of the stock today (beginning of year 2003) was $21.08, our expected yield from owning the stock would be:

$$\$21.08 = \frac{\$.40}{(1+r)^1} + \frac{\$.40}{(1+r)^2} + \frac{\$.44}{(1+r)^3} + \frac{\$.48}{(1+r)^4} + \frac{\$.48}{(1+r)^5} + \frac{\$21.19}{(1+r)^5},$$

or about 2%. This rather unimpressive yield comes almost exclusively from the dividend, since we are forecasting almost no change in the price of the stock over the period. Our longer-run earnings forecast of $1.30 per share is only marginally above the $1.24 earned in 2002, and we expect the price/earnings multiple to decline from 17× to the 16.3×. Given these prospects, it appears that we should not include Ameritape in our portfolio, unless, of course, we can find other reasons to expect sales and profit patterns in the future to be better than those experienced in the past.

Given that Ameritape is a mature, dividend paying company, the analysis from Chapter 6 can highlight certain of the issues raised thus far. To begin with, observe that we have two growth rates: the five year historical rate computed in Chapter 7 of about 6% and the regression projections in Chapter 8 and above which show a decline and recovery over the next 5 years for essentially no growth. We are also assuming a payout ratio of 37% and a P/E multiple around 17X. Now we recall (6.7):

$$m = \frac{1-b}{r-\mu}, \tag{10.2}$$

Where

$$m = \text{P/E multiple } (= 17\text{X here})$$
$$(1-b) = \text{payout ratio } (= 0.37 \text{ here})$$
$$r = \text{return}$$
$$\mu = \text{growth rate} = Xb \text{ where } X \text{ is the average return on retentions.}$$

Solving for r, we obtain

$$r = \frac{1-b}{m} + \mu$$

The first term, which is $(D/E)/(P/E) = D/P$, is the dividend yield in disguise. With our numbers, it is $0.37/17 =$ about 2.2%. The second term is growth, for which our numbers are 6% or 0%. This is another way to look at why we are getting values for r of 8% or 2%.

Now consider that the assumed retention ratio b is 0.63 and the growth μ equals the earnings rate on retentions (X) times b. Clearly, if $\mu = 0$, $X = 0$, which is quite unacceptable. Even if $\mu = 6\%$, $X = (0.06/0.63 = 9.5\%)$, which is a rather low average return on reinvestment for the company's asset investment.

The book value per share at the end of 2002 was \$3.27 (\$2.1 billion divided by 640 million shares). This does not appear to be a company with substantial intangible or other hidden assets which are not reflected on the balance sheet. A current market price of \$21.08 would be 6.4 times book value. On its face, this would appear to be an absurd premium to pay for a low-tech firm in an industry with few barriers to entry.

Growth stock proponents often advocate the PEG (price earnings to growth) ratio, which is the ratio of the P/E multiple to the 5 year earnings growth rate (expressed as an integer). They say a growth stock is a bargain if it can be had for less than 1 (or 1.5 or some other number depending upon how overpriced the market is at the time). Even though Ameritape is a slow growth, cyclical company, at the most optimistic growth estimate of 6%, its PEG is almost 3 (i.e., 17/6). This would indicate that, even viewed as a growth company, it is overpriced.[1]

10.4   Other Scenarios

Many other outcomes could be computed for future projected values that might provide more optimistic scenarios. Suppose that after a thorough review of past data (and future projections), you are convinced that the return distribution for Ameritape is $\mu = .08$ (including reinvested dividends) and $\sigma = .08$.[2] Suppose further that you could buy the stock at today's price of \$21.08 and you had a five-year time horizon in mind. A risk profile can be obtained (see Figure 10.10) without simulation assuming that the security price follows the standard lognormal distribution developed in

---

[1] Ameritape's over valuation reflects a phenomenon evidenced at the peak of a bull market (such as that at the end of 1999 and the beginning of the year 2000). Prices for these over priced stocks plummeted during the 2000–2002 bear market. Most likely, if Ameritape were a real company, its market price would have likely declined rather dramatically by the end of 2002.

[2] The past data we have analyzed suggest $\mu = .02$ and $\sigma = .02$, not a very risky portrait, yet riskier than Treasury bills with similar returns.

(8.7) and (8.8):

$$S(t) \;=\; S(0) \exp\left[\left(\mu - \frac{1}{2}\sigma^2\right)t + \sigma\sqrt{t}Z\right] \tag{10.3}$$

$$\;=\; S(0)\exp(\mu t) \times \exp\left(\sigma\sqrt{t}Z - \left(\frac{1}{2}\sigma^2 t\right)\right). \tag{10.4}$$

| Table 10.1. Tables of the Cumulative Normal Distribution ||
$Z_{critical}$	$P(Z > Z_{critical}) = P(Z < -Z_{critical})$
2.326	.01
1.645	.05
1.282	.10
1.037	.15
.842	.20
.575	.25
.524	.30
.253	.40
0	.50

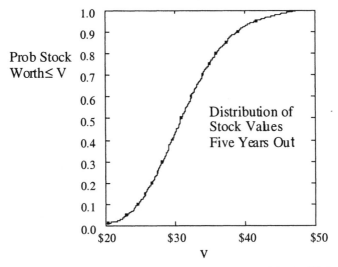

Figure 10.10. Percentile Traces of Ameritape Share Value.

Next, let us suppose that we wish to superimpose on the simple Brownian (Gaussian) model for Ameritape a "bear jump process" which takes the market down 20% on the average once every five years and down 10% on the average once every year. It is possible that our estimates of growth and volatility have been measured during a "bear free period" (e.g., 1993–1999). But we know that bear jumps are endemic to the market process, so we might well include a "patch" which adds the jump process onto the Brownian model. This gives a very dismal picture indeed as we see in Figure

10.11. It might be more realistic to make some enhancement of the growth rate in the Brownian part of the model. The bear jump process described has roughly the growth effect of a negative annual growth rate of .14. Accordingly, we see, in Figure 10.11, what happens if we add .14 to the growth rate of .08 to obtain .22. Note the effect is that the lower ten percentile of the stock value is $24.61 for the pure Brownian growth model, but $18.07 for the Brownian growth model (growth enhanced to .22) with bear jumps. The lower twenty percentiles are $26.62 and $22.59, respectively. Because some of the simulations will experience few if any bear jumps, the upper ninety percentiles become $38.93 versus $45.65. The means are nearly the same for both the unpatched and patched models. But the model patched with bear jumps is much more risky.[3] Unlike the simple Brownian model, we require simulation to obtain the simugrams of the patched Brownian models shown in Figure 10.11.

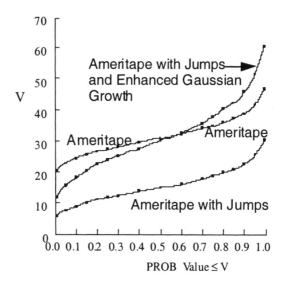

Figure 10.11. Percentile Traces of Brownian Models of Ameritape and Brownian Models Enhanced with a Bear Jump Process.

---

[3] If one would argue that for longer time periods, one might simply make do with the unpatched Brownian model albeit with inflated volatility, we would not disagree. However, for short periods this is not a good idea. It seems fairly clear, if one looks back over many years, the unpatched Brownian model does not correspond to reality as well as does the Brownian model patched with bear jumps.

10.5 The Time Indexed Distribution of Portfolio Value

Let us construct, using simulation, the distribution of 1000 possible outcomes of an investment of $10,000 in a stock with $\mu = .10$ and $\sigma = .10$ after ten years. We show the results in Figure 10.12. This is not, of course, a *histogram* in the usual sense of the term. A histogram is a relative count register of the number of historical observations which fall into the intervals of value observed in the past. Here, we have taken parameter values and used them in a model to obtain simulations. So, each simulation gives an observation of simulated value. So we will use the expression *simugram*. (Of course, for this simple case, as we have seen in Chapter 8, one can obtain the density as a Gaussian probability integral. Very quickly, we will be moving to a situation where such a "closed form solution" is no longer practical.)

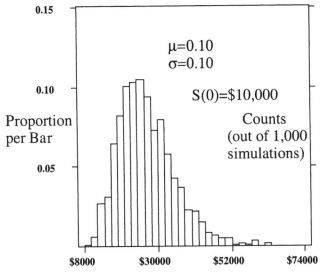

Figure 10.12. Simugram of Investment Value at Ten Years.

Next, if we can find 20 stocks, each with $\mu = .10$ and $\sigma = .10$, then *assuming they are stochastically independent of each other*, we might take the $10,000 and invest $500 in each of the stocks. The distribution of value at ten years (using 200 possible outcomes) is shown in Figure 10.13. The sample means for both the one-stock investment and the diversified 20-stock mutual fund are $27,088 and $26,715, respectively. But the standard deviation for the one-stock investment ($8,343) is roughly $\sqrt{20}$ times that for the mutual fund investment ($1,875). A portfolio that has such such stochastic independence would be a truly *diversified* one. Generally speaking, one should expect some dependency between the tracks of the stocks.

Now let us recall the general equation for geometric Brownian motion:

$$\frac{\Delta S_i}{S_i} = \mu \Delta t + \sigma \epsilon \sqrt{\Delta t}. \tag{10.5}$$

Let us modify (10.4) to allow for a mechanism for dependence:

$$\frac{\Delta S_i}{S_i} = \mu \Delta t + \sigma \epsilon_i \sqrt{\Delta t}. \tag{10.6}$$

We shall take η_0 to be a Gaussian random variable with mean zero and variance 1. Similarly, the 20 η_i will also be independent Gaussian with mean zero and variance 1. Then we shall let

$$\epsilon_i = c(a\eta_0 + (1-a)\eta_i). \tag{10.7}$$

We wish to select c and a so that a is between zero and 1 and so that $\mathrm{Var}(\epsilon_i) = 1$ and any two ϵ_i and ϵ_j have positive correlation r. After a little algebra, we see that this is achieved when

$$a = \frac{\rho - \sqrt{\rho(1-\rho)}}{2\rho - 1} \tag{10.8}$$

and

$$c^2 = \frac{1}{a^2 + (1-a)^2}. \tag{10.9}$$

At the singular value of $\rho = .5$, we use $a = .5$.

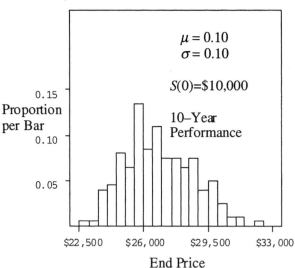

Figure 10.13. Simugram of Idealized Mutual Fund Value at Ten Years.

Let us examine the situation with an initial stake of $500 per stock with $\mu = \sigma = .10$ and $\rho = .8$ as shown in Figure 10.14. We employ 500 simulations. We note that the standard deviation of the portfolio has grown to 7,747. This roughly follows the rule that the standard deviation of a portfolio where stocks have the same variance and have correlation ρ, should be $\sqrt{1 + (n - 1)\rho}$ times that of an uncorrelated portfolio.

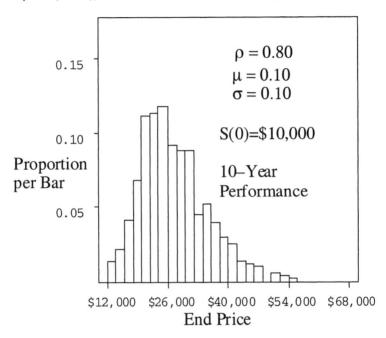

Figure 10.14. Simugram of Mutual Fund with Correlated Stock Prices.

10.6   Negatively Correlated Portfolios

Is there anything more likely to reduce the variance of a mutual fund portfolio than the assumption that the stocks move in a stochastically independent fashion? We recall that if we have two random variables X_1 and X_2, each with unit variance and the same unknown mean μ, the variance of the sample mean is given by

$$\text{Var}((X_1 + X_2)/2) = \frac{1}{4}[2 + 2\rho]. \qquad (10.10)$$

Here the variance can be reduced to zero if $\rho = -1$. Let us consider a situation where we have two stocks each of value $5000 at time zero which grow according to

$$\frac{\Delta S_1}{S_1} = \mu \Delta t + \sigma \epsilon \sqrt{\Delta t} \qquad (10.11)$$

$$\frac{\Delta S_2}{S_2} = \mu \Delta t - \sigma \epsilon \sqrt{\Delta t},$$

where ϵ is a Gaussian variate with mean zero and unit variance. Then the resulting portfolio (based on 500 simulations) is exhibited in Figure 10.15.

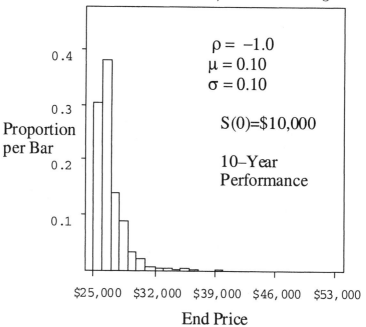

Figure 10.15. Simugram of Two-Stock Portfolio with $\rho = -1$.

We note that the standard deviation of this two-stock portfolio is 1701, even less than that observed for the 20-stock portfolio with the assumption of independence of stocks. Now, the assumption that we can actually find stocks with negative correlation to the tune of -1 is unrealistic. Probably, we can find two stocks with rather large negative correlation, however. This is easily simulated via

$$\frac{\Delta S_1}{S_1} = \mu \Delta t + (a\epsilon_0 + (1-a)\epsilon_1)c\sqrt{\Delta t}\sigma \qquad (10.12)$$

$$\frac{\Delta S_i}{S_2} = \mu \Delta t - (a\epsilon_0 + (1-a)\epsilon_2)c\sqrt{\Delta t}\sigma,$$

where

$$a = \frac{\rho + \sqrt{-\rho(1-\rho)}}{2\rho + 1},$$

$c = 1/\sqrt{a^2 + (1-a)^2}$ and ϵ_0, ϵ_1, and ϵ_2 are normally and independently distributed with mean zero and variance 1. Let us consider, in Figure 10.10, the situation where $\rho = -.5$.

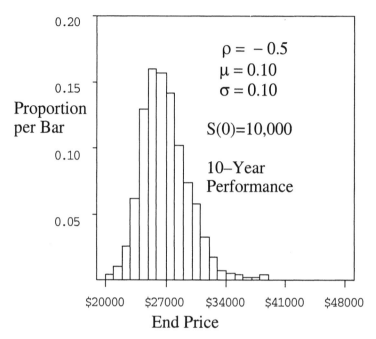

Figure 10.16. Simugram of Two-Stock Portfolio with $\rho = -.5$.

We note that the standard deviation here has grown to $2,719. When it comes to utilizing negative correlation as a device for the reduction of the variance of a portfolio, a number of strategies can be considered. We know, for example, that if one wishes to minimize the variance of a sample mean, we can pose the problem as a constrained optimization problem to find the optimal correlation matrix, where we impose the constraint that the covariance matrix be positive definite. Our problem here is rather different, of course.

We could try something simple, namely take our two-stock negatively correlated portfolio and repeat it 10 times, (i.e., see to it that the stocks in each of the ten subportfolio have zero correlation with the stocks in the other portfolios). Here, each of the 20 stocks has an initial investment of $500. In Figure 10.17, we show the profile of 500 simulations of such a portfolio. The standard deviation of the pooled fund is only $1,487.

How realistic is it to find uncorrelated subportfolios with stocks in each portfolio negatively correlated? Not very. We note that we can increase the sizes of the subportfolios if we wish, only remembering that we cannot pick an arbitrary correlation matrix—it must be positive definite. If we have a subportfolio of k stocks, then if the stocks are all to be equally negatively correlated, the maximum absolute value of the correlation is given by $1/(k-1)$.

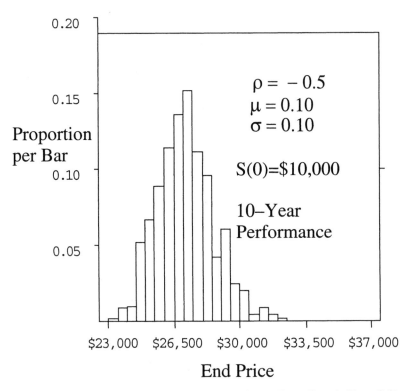

Figure 10.17. Simugram of 10 Independent Two-Stock Portfolios with $\rho = -.5$.

Let us consider another type of randomness in the stock market. Superimposed over Brownian geometric motion of stocks there are periodic bear market downturns in the market overall. It is unusual for bull markets to exhibit sharp sudden rises. But 10% corrections (i.e., rapid declines) are quite common, historically averaging a monthly probability of as much as .08. Really major downturns, say 20%, happen rather less frequently, say with a monthly probability of .015.

10.7 Bear Jumps

In Figure 10.18 we see the simugram of 500 simulations with the jumps modeled as above, $\sigma=.10$, $\rho = 0$, and $\mu = .235$. The mean here is 27,080, very similar to that of the situation with independent stocks, with $\mu=.10$ and $\sigma=.1$. However, we note that the standard deviation is a hefty 11,269.

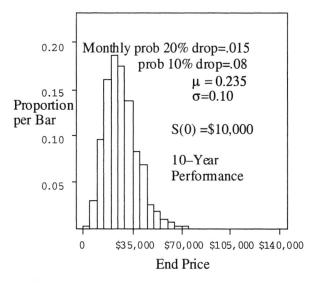

Figure 10.18. Simugram of Portfolio of 20 Independent Stocks
with Bear Jumps.

We note that these general (across the market) downward jumps take away
some of the motivation for finding stocks which have local negative correlation in their movements. (For example, had our portfolio had a .8 correlation between the stochastic increments, the standard deviation would
only have increased from 11,269 to 13,522.) It is rather easy to advise an
investor not to put all of one's assets in Apple Computers. It is also clear
that the investor ought not invest in a portfolio consisting of Apple, Dell,
Compaq, SUN and HP, for these stocks tend to move together. However,
when a big bear market hits, our investor is going to be damaged even if he
or she includes department stores, utilities, drug companies and restaurant
chains. Still, a diversified investor, in the bear market which started in the
last half of 2000, generally fared better than the investor who only invested
in a few high tech stocks. Now, we have arrived at a situation where nearly
25% of the time, our portfolio performs worse than a riskless security invested at a 6% return. If we increase the volatility σ to .5, then nearly 40%
of the simulated portfolios do worse that the fixed 6% security.

Let us return to looking at the situation where a $100 million university
endowment, consisting of 20 stocks ($5 million invested in each stock) with
stochastically independent geometric Brownian steps, with μ=.235, σ=.1
and with monthly probabilities .08 of a 10% drop in all the stocks and a
.015 probability of a 20% drop in all the stocks. We shall "spend" the
portfolio at the rate of 5% per year. Let us see in Figure 10.18 what the
situation might be after 10 years. With probability .08, after 10 years, the
endowment will have shrunk to less than $50 million. With probability .22,
it will have shrunk to less than $75 million. With probability .34, it will
have shrunk to less than the original $100 million. Given that universities

tend to spend up to their current cash flow, it is easy to see how some of those which tried such a strategy in the 1960s went through very hard times in the 1970s and 1980s.

We have discussed how a portfolio of stocks has, historically, been a superior investment over the long run for retirement purposes. The endowment problem just discussed applies to those who have followed such a policy as they near retirement (especially if the timing is mandatory). If the date falls soon after a bear jump, the options are ugly: cash out into a fixed annuity and consume much less than one had anticipated or consume from the equity portfolio and hope for a big, quick recovery before the portfolio is literally "eaten up." Retirement flexibility can moderate this risk (and is why a lot of people who hoped to retire in 2000–2002 did not). Failing this, at least a partial shift out of equities as one approaches mandatory retirement is indicated.[4]

It is very likely the case that broad sector downward jumps ought to be included as part of a realistic model of the stock market. The geometric Brownian part may well account for the bull part of the market. But downward movements may very well require a jump component. By simply noting variations on geometric Brownian drift in stock prices, analysts may be missing an essential part of the reality of the stock market, namely, large broad market declines which occur very suddenly.

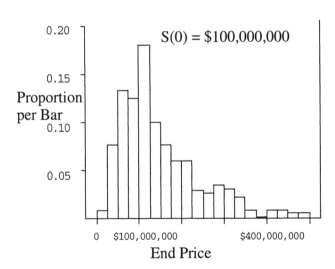

Figure 10.19. Simugram of $100 Million Endowment with 5% Payout.

---

[4]A friend of one of the authors who recently retired said his 401k Plan was now a "201k" plan.

10.8   Conclusions

Persons who indulge in market purchases based on fine notions of a Brownian regularity and rapidly attained equilibria unmoderated by the modifications which are almost always necessary if we are to look at real stocks from the varying standpoints of real people and real institutions, should not be surprised when they lose their (or, more commonly, other people's) shirts.

The whole area of modeling markets is an exciting one. Anyone investing or anyone advising an investor may make ruinous mistakes if they do not ask questions such as "what happens if ...". Such questions, typically, are not easy to answer if we demand closed-form solutions. Simulation enables us to ask the questions we need to ask rather than restricting ourselves to looking at possibly irrelevant questions for which we have formulas at hand.

Problems

Problem 10.1. Let us consider two investors.

(a) Mr. A is considering an "index fund." This will be essentially a mutual fund of stocks randomly selected from a broad category. For example, we could form an index fund of computer-related stocks. Or, we could form an index fund of stocks selected from the American Stock Exchange. The stocks might be purchased in arbitrary amounts or weighted according to the total market value of each particular firm in relation to that of the total value of all stocks on the exchange. A person who chooses to buy stocks in this way is, in effect, assuming that all other purchasers of stocks are behaving rationally (i.e., they are willing to pay more for promising stocks than for stocks seemingly on the downward track). Accordingly, the index fund manager should be perfectly happy to buy in random fashion; he needs no market analysis: the rational behavior of the other players in the market implicitly guides him. He does not worry about any idea of playing a zero-sum game. The technological improvements, present in whatever group that is being indexed, tend to move the value of the index higher.

(b) Ms. B thinks A's strategy is rather dim. She picks the same category as A—say, computer-related stocks. Then she looks at the fundamentals of her choices and invests her capital in a portfolio of stocks where the growth rates and other fundamentals appear to be most favorable.

Which strategy do you favor? Give your reasons.

Problem 10.2. Write a simulation program for a single stock with specified growth and volatility. Note that for many traverses of a loop, older languages such as Fortran and C tend to perform much more quickly than more modern user-friendly programs.

Problem 10.3. Write a simulation program for a portfolio of k stocks, correlated with each other with correlation equal to ρ but with arbitrary growth rates and volatilities for each stock.

Problem 10.4. Take the program in Problem 10.3 and add on the possibility of bear jumps across the market.

Problem 10.5. Stocks are tied to real companies, and these companies can get better or worse. Let us consider a mutual fund which is solely managed by a computer. At any given time, there will be 20 stocks in the fund. Let us suppose that these stocks are selected randomly from stocks which, over the last six months, have exhibited a growth rate μ of .25 and a volatility of .15. We will assume, as in the past, that there is an across-the-board bear jump mechanism whereby a sudden drop of 10% happens, on the average, of once a year and a sudden drop of 20% happens on the average once every five years. Let us suppose that there is a mechanism which happens, on the average once every three years, whereby the growth rate of a stock can drop to a level which is uniformly distributed between $-.05$ and .2 with a corresponding increase in the volatility to a level uniformly distributed between .2 and .4. Since the computer does not read newspapers, it needs to have an automatic mechanism for removing, at the earliest possible time, from the portfolio stocks whose fundamentals have failed in this fashion. When such a stock is removed, it will be replaced by another with growth rate .25 and volatility .15.

(a) Assuming no transaction costs, devise a good automatic scheme for the computer to follow. Show simulation studies to substantiate the wonderfulness of your paradigm.

(b) Assuming transaction costs equal to 1.5% of the value of the stock traded on both the sell and the buy, devise a good strategy for the computer to follow and show its effectiveness by a simulation study.

Problem 10.6. Consider a portfolio of n stocks, each with starting value $S(0)$ and the same μ and σ. Suppose for each stock

$$\frac{\Delta S_i}{S_i} = \mu \Delta t + \sigma \epsilon_i \sqrt{\Delta t}$$

$$\epsilon_i = c(a\eta_0 + (1-a)\eta_i)$$

$$a = \frac{\rho - \sqrt{\rho(1-\rho)}}{2\rho - 1}$$

$$c^2 = \frac{1}{a^2 + (1-a)^2}.$$

Show that the variance of the portfolio's value at time t is given by

$$\text{Var} = nS(0)^2 e^{(2\mu t)}[\exp(\rho\sigma^2 t) - 1] + nS(0)^2 e^{(2\mu t)}[\exp(\sigma^2 t) - \exp(\rho\sigma^2 t)].$$

References

[1] Graham, B., Dodd, D., and Cottle, S. (1962). *Security Analysis*. New York: McGraw-Hill.

[2] Markowitz, H. (1956). "The optimization of a quadratic loss function subject to linear constraints," *Naval Research Logistics Quarterly*, March–June, 111–133.

[3] Markowitz, H. (1959). *Portfolio Selection*. New York: John Wiley & Sons.

[4] Shleifer, A. (2000). "Arbitrage is inherently risky," *The Wall Street Journal*, December 28, A10.

[5] Siconolfi, M., Raghavan, A., and Pacelle, M. (1998). "All bets are off: How the salesmanship and brainpower failed at Long-Term Capital," *The Wall Street Journal*, November 16, A1, A18–A19.

[6] Thompson, J. R. (1999). *Simulation: A Modeler's Approach*. New York: John Wiley & Sons.

[7] Tukey, J. W. (1962). "The future of data analysis," *Annals of Mathematical Statistics*, Vol. 33,13.

[8] Williams, E. E. and Thompson, J. R. (1998). *Entrepreneurship and Productivity*. New York: University Press of America.

Chapter 11

Common Stock Options

11.1 Introduction

An *option* is a contract between two parties wherein one party grants the other the right to buy (sell) a specified asset, at a specified price, within a specified period of time (American) or at the end of the period (European). A *call* option gives the purchaser the right to buy a certain common stock (or other security) at a stated price for a given period of time. A *put* option allows the holder to sell a stock at a stated price for a given period of time. A *straddle* is the purchase of a put and call on the same security for the same period of time at the same price. A *spread* is similar to a straddle except the put price is set below the call price. If the option is written at the current market price of the stock, which is usually the case, the spread put would be below the market price, and the call would be above it. Combinations of options are also possible. A *strip* is a straddle plus a put. A *strap* is a straddle plus a call.

Buyers pay positive prices for call (put) options because they believe the price of the stock optioned will rise (fall) sufficiently at some time during the option period to recover the cost of the option and provide an acceptable rate of return. Call (put) options may be written at prices below (above) the current market price of the optioned shares. In this case, the option has an intrinsic arithmetic value, i.e., the difference between the existing stock price and the specified exercise, or strike, price of the option. All options with any time to expiration will, of course, trade at prices above this value, the difference being called its time value. The time value, when added to the intrinsic value, gives the market price or option premium.

Options are written at various strike prices with varying expiration dates. For example, on April 30, 2002, suppose there are five contracts to purchase (sell) Cisco stock. The stock closed the previous day at \$14.13, and the following were available (closing prices for calls and puts):

Strike Price	Expiration	Call	Put
$12.50	July '02	$2.45	$0.85
$15.00	May '02	$0.50	$1.45
$15.00	June '02	$0.90	$1.75
$17.50	June '02	$0.20	$3.70
$17.50	July '02	$0.40	$3.90

The price of an option will depend upon supply and demand factors in the market for the option. If speculators believe an upward movement in a common stock's price is imminent, they will bid up the price of options on that stock. Because the probability of an upward movement increases with time, longer options usually sell for higher prices than short-period ones. We shall discuss methods for pricing options (including the famous "Black-Scholes" method) below [2].

Unfortunately, many people tend to view the option market strictly in terms of a small investor buying a call on a volatile stock in the hope of achieving a leveraged gain. Although such things do occur, they present a very limited view of the role of the market. Paper gains can be locked in by the purchase of a put if the investor has doubts about the short-term price movement of the stock. Furthermore, a portfolio that must hold a stock for tax or other reasons but has little enthusiasm for its short-run price performance can earn "premium income" by selling calls against the security. If, on the other hand, the portfolio manager is enthusiastic about a security, he can earn premium income writing puts. Although this is hardly an area for the novice, it should be iterated that options, properly employed, can increase the return or reduce the risk of the portfolio (but not both, simultaneously, as some enthusiasts preach; there really is no such thing as a free lunch). Consider the following examples:

1. A stock sells for $40 per share. A 90-day call option (with strike price of $40) can be purchased for $200 on 100 shares. What profit would be made if the stock sold for $45 within 90 days and the option were exercised (or, more likely, sold)? For $42? $41? $38?

at $45: $4500 −($4000 + $200) = $300; $300/$200 = 150% gain

at $42: $4200 −($4000 + $200) = $0; $0/$200 = 0% (breakeven)

at $41: $4100 −($4000 + $200) = −$100; −$100/$200 = 50% loss

at $38: $3800 − ($4000 + $200) = −$400; −$200/$200 = 100% loss (option not exercised)

2. Suppose a stock sells for $40 and a put option could be purchased at a price of $400 per 100 shares for one year at a strike price of $40 per share. What returns will be generated if the stock sells at $30 during the year, assuming the option is exercised? At $36? At $40?

at $30: ($4000 − $3000) − $400 = $600; $600/$400 = 150% gain

at $36: ($4000 −$3600) − $400 = $0; $0/$400 = 0% (break even) at $40: ($4000 − $4000) − $400; −$400/$400 = 100% loss (no exercise)

11.2 Black–Scholes and the Search for Risk Neutrality

11.2.1 Option Writers and Bookies

Let us consider the game indicated in Figure 11.1

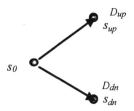

Figure 11.1. Game Unlinked to Probabilities.

There are two prizes D_{up} and D_{dn}. What is the fair price D_0 to pay the casino to play this game? As we have seen in Chapter 2, if the prizes are reasonably small relative to the wealth of the player, the fair value might be taken to be the expected value:

$$D_0 = p_{up}D_{up} + p_{dn}D_{dn}. \tag{11.1}$$

We can be sure, in the real world, that the casino operator will charge a bit more than D_0, say $D_1 > D_0$ to play this game, for a zero rate of return on his or her investment (casino, employees, utilities, security, etc.) would soon put him or her out of business. It is not reasonable, one might argue, for the player to become involved with the wager, since in paying D_1 to play, he or she will, over the long run, lose money. Yet casinos do a thriving business. In general, the casino organization will be happy, in the case of many of its games, to operate on a reasonably small margin $(D_1 − D_0)/D_1$ of profit, for, over time, it is almost certain to return, on the game, a profit margin per wager equal to the profit margin times the total bets wagered on the game.

In casino games of chance, we know precisely the probabilities associated with each game. But what if one is engaging in a game where one does not know the probabilities associated with winning and losing? Suppose one is taking bets, for example, on a horse race. There are horse racing experts who will set preliminary odds of the sort, Purple Martian has a ten percent chance of winning the race. So, preliminary odds might be that Purple Martian, for a one dollar ticket, will pay ten dollars if it wins the race. But as the bets come in, the track may note that too many people are betting on Purple Martian at the bet as stated. That means, if Purple

Martian should win, the track would actually lose money. So, a woman who bet on Purple Martian would know that the ticket she bought might pay less than the rate stated at the time of the purchase. The track has the right to readjust payoffs right up until the time of the start of the race to make sure it has locked in a profit, irrespective of which horse actually wins the race. Similarly, if the track finds that too few people are betting on Purple Martian, it will typically improve the payoff. The whole idea of the track, then, is to readjust payoffs for all the horses, continuously in time, to guarantee a profit for the track. Now, in the world of bookmakers, competition from several "agencies" will tend to drive the payoffs from all to be similar, for why buy a ticket which pays $10 if another bookmaker is paying $15? If communications are as rapid as they have been for some decades, there will be, essentially, a national market for betting on Purple Martian to win. The only thing which can restrict (partially) the creation of a nearly efficient market in the victory of Purple Martian is interference by some governmental or quasi-governmental agency (e.g., organized crime) to restrict trade in Purple Martian tickets. Notice, then, that the world of horse racing bookmakers provides us with an example of a market in outcomes in which the probabilities are not very well known, and that the bookmakers actually set the rates based, not to their retrospective guesses as to the probabilities, but rather on volumes of purchases of tickets for specific pay-off rates.

Notice that the pay-off rates are computed by the bookmakers with little reliance on probability estimates. From the standpoint of an individual buying Purple Martian tickets, however, the decision is generally based on the probability of a Purple Martian win as intuited by the individual buying tickets. Here, the bookmaker has a very different strategy from that of the woman making the wager. The bookmaker is attempting to hedge his bets so that he makes a return based, essentially, on commissions.[1]

It is clear how one might argue that if we want to determine a "fair" price to promise to deliver a stock T units in the future, when that price is to be paid on the day of delivery, it can be determined by buying the stock at today's known price $S(0)$ and then borrowing the money for that purchase at the going interest rate r. That would mean that the *futures price at time* T is $S(0) \exp(rT)$. That means that, assuming the broker could borrow the money $S(0) \exp(rT)$ for rate r, then he could have a perfect hedge and make a profit exactly equal to his commission for the transaction. He would be in somewhat the same position of a bookie at the races.

From the standpoint of the buyer, however, he or she probably has reason to believe that at time T the stock will be worth more than $S(0) \exp(rT)$.

---

[1]The aforementioned description of horse race bookmaking follows closely the way investment bankers price new stock issues (IPOs). The initial price increases or decreases depending on the "book" (even the terminology is similar). Substantial initial interest (buy orders) will raise the IPO price. If there is weak interest, the IPO price will be set lower. The size of the issue may also be adjusted depending on interest in the deal.

(There are other reasons, of course, for the buyer to purchase the future instead of simply buying the stock today. For example, it might well be that other investments will bring into the futures purchaser's hands a sum of money at time T at least as great as $S(0)\exp(rT)$, a sum which the purchaser does not have in hand today. This rather obvious hedging price whereby the vendor can sell a futures contract goes back so far in history that it can probably be called a "folk theorem."

11.2.2   Alternative Pricing Models

A major distinction between the futures contract discussed above and an option is that the former transaction must occur (and with sufficient margin can be made riskless to the writer of the contract) while the latter will only occur at the option of the holder (who will not exercise unless it is to his or her advantage). Thus, the writer of a call option could not obtain a riskless profit by simply buying and holding the optimal shares, because he or she would be stuck holding these shares if their price did not rise sufficiently to make exercise profitable.

Before deriving the Black-Scholes pricing formula for the price of a *European call option*, let us look at two older pricing formulas C_A and C_B giving the arguments for each. What should be the fair value for an option to purchase a stock at an exercise price X starting with today's stock price $S(0)$ and an expiration time of T? If the rate of growth of the stock is μ, and the volatility is σ, then, assuming the lognormal distribution for $S(t)$, the stock value at the time T should be

$$S(0)\exp\left(\mathcal{N}\left((\mu - \frac{\sigma^2}{2})T, T\sigma^2\right)\right),$$

where $\mathcal{N}(a, b)$ is a Gaussian random variable with expectation a and variance b. If we borrow money at a fixed riskless rate r to purchase the option, then the value of the option could be argued[2] to be equal to C_A in

Method A $C_A = \exp(-rT)E[\mathrm{Max}(0, S(T) - X)]$,

where E denotes expectation.

On the other hand, it could also be argued that the person buying the option out of his or her assets is incurring an opportunity cost by using money to buy the option which might as well have been used for purchasing the stock so that the value of the option should be given by

Method B $C_B = \exp(-\mu T)\ E[\mathrm{Max}(0, S(T) - X)]$.

---

[2]The use of the expectation criterion as the measure of value is questionable. We recall, for example, the St. Petersburg Paradox discussed in Chapter 2 shows how expectation of gain can give ridiculous results.The question of appropriate criteria is extremely important, and we have addressed it extensively throughout this book. For the moment, we stick with expectation, for that is, unfortunately, the standard view.

Smith [6] has shown:

Lemma 11.1. If S is lognormal with growth rate μ and volatility σ and if

$$
\begin{aligned}
Q &= \lambda S - \gamma X \text{ if } S - \psi X \geq 0 \\
&= 0 \qquad \text{if } S - \psi X < 0,
\end{aligned}
$$

then

$$
\begin{aligned}
E(Q) &= \int_{\psi X}^{\infty} (\lambda S - \gamma X) f(S) dS \\
&= e^{\mu T} \lambda S(0) \Phi \left(\frac{\log(S(0)/X) - \log(\psi) + [\mu + (\sigma^2/2)]T}{\sigma \sqrt{T}} \right) \\
&\quad - \gamma X \Phi \left(\frac{\log(S(0)/X) - \log(\psi) + [\mu - (\sigma^2/2)]T}{\sigma \sqrt{T}} \right), \quad (11.2)
\end{aligned}
$$

where λ, γ, and ψ are arbitrary parameters and Φ is the standard Gaussian cumulative distribution function.[3]

Then we have for **Method A**, taking $\psi = 1$ and $\lambda = \gamma = e^{-rT}$,

$$
\begin{aligned}
C_A &= e^{-rT} \{ e^{\mu T} S(0) \Phi \left(\frac{\log(S(0)/X) + [\mu + (\sigma^2/2)]T}{\sigma \sqrt{T}} \right) \\
&\quad - X \Phi \left(\frac{\log(S(0)/X) + [\mu - (\sigma^2/2)]T}{\sigma \sqrt{T}} \right) \}. \quad (11.3)
\end{aligned}
$$

For **Method B**, taking $\psi = 1$ and $\lambda = \gamma = e^{-\mu T}$,

$$
\begin{aligned}
C_B &= e^{-\mu T} \{ e^{\mu T} S(0) \Phi \left(\frac{\log(S(0)/X) + [\mu + (\sigma^2/2)]T}{\sigma \sqrt{T}} \right) \\
&\quad - X \Phi \left(\frac{\log(S(0)/X) + [\mu - (\sigma^2/2)]T}{\sigma \sqrt{T}} \right) \}. \quad (11.4)
\end{aligned}
$$

Black and Scholes [2] assumed that, by dynamic hedging, the option could be made riskless to the writer. Although the derivations follow, we provide the result here for comparison. From Lemma 11.1, taking $\psi = 1$, $\lambda = \gamma = e^{-rT}$, and setting $\mu = r$,

$$
\begin{aligned}
C_{BS} &= e^{-rT} \{ e^{rT} S(0) \Phi \left(\frac{\log(S(0)/X) + [r + (\sigma^2/2)]T}{\sigma \sqrt{T}} \right) \quad (11.5) \\
&\quad - X \Phi \left(\frac{\log(S(0)/X) + [r - (\sigma^2/2)]T}{\sigma \sqrt{T}} \right) \}.
\end{aligned}
$$

---

[3]

$$
\Phi(x) = \frac{1}{\sqrt{2\pi}} \int_{-\infty}^{x} \exp(-z^2/2) dz
$$

The formula for the pricing of an option which gives the purchaser the right to buy a stock for X dollars at future time T, as formulated above, in the case of C_A and C_B depends on knowledge of three parameters, μ, σ, and r while C_{BS} employs only σ and r. Ostensibly, r should be easy to determine. It is frequently argued to be simply the riskless interest rate (i.e., that of a Treasury bill). A bit later, we will question whether a universal r value is reasonable. Knowledge of μ and σ would appear to be more uncertain. They can be estimated quite handily from past stock records, but we are not talking about the μ and σ values of the past. Rather we want to use those which are appropriate from now until the time T (when the option is exercised or is allowed to expire). One circular way to proceed (and one of particular appeal to believers in efficient market theory) is to use what might be called (politely) "answer analysis." We know what call options for strike price X with execution time T are selling for on the open market. So, we simply "plug in" values of μ and/or σ which give the price. This would give us "implied" estimates of μ and/or σ. Here, we have assumed a kind of "deus ex machina" character of the market. The market is assumed to be so efficient that the aggregated wisdom of buyers and sellers gives us option prices which appropriately incorporate all relevant information. We recall the old story about the efficient market analyst who, seeing a $100 bill lying on the street, disdains to pick it up, for he is sure that if a $100 bill had been lying on the street, it would have been picked up instantly after it was dropped there. The reality of the $100 bill is discarded in favor of a doctrinal belief in perfect market efficiency.

The reality of the "right" values to use for μ and σ (for C_A or C_B)is even more complicated. The Upbeat Mutual Fund, on the basis of careful analysis, has made the judgment that, contrary to the general expectation, Blahblahbiotech will be allowed to bring its AIDS drug to market within one year. Accordingly, UMF might decide to use the formula for C_B with a μ value well above the historical estimate. On the other hand, the Bad Bear Mutual Fund, on the basis of careful analysis, has determined that the early clinical results will make Blahblahbiotech's entry a write-off within one year. BBMF might well decide to use C_B (or C_A) with a μ value well below that of the historical estimate. The market, efficient or not, will reflect the views of UMF and BBMF, together with a host of other investor inputs. Moreover, there will be a fair amount of dynamism in the market. Today's market prices may well be quite different to those of next week. If there were no such dynamism, the believers in efficient market theory would have a more realistic case than they do. Suppose that we saw on the street a deposit slip dated May of 1962 to which was clipped a $100 bill. Then, indeed, we might disdain to believe our eyes.

If we could find a way to determine by a mathematical formula the "fair market value" of an option to buy 1,000 shares of Blahblahbiotech one year from today at strike price X, that would be simply incredible, worthy of a Nobel Prize at the very least. Before the work of Black and Scholes, there was a folk theorem to the effect that perhaps the answer would be to use Method A in (11.3) with the growth μ replaced by the interest rate r. The analogy was to the story told at the beginning of this chapter to the effect that the fair price to take delivery of a share of stock S at time T was $S(0)\exp(rt)$, where the interest rate was used rather than the growth of the stock.

11.2.3   The Black–Scholes Hedge Using Binomial Trees

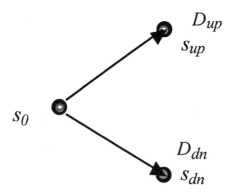

Figure 11.2. Simple Binomial Tree with Options.

Suppose, however, we were to ask another question ("another" because options are not precisely analogized by payoffs in horse races). Namely, can we come up with a way in which a brokerage firm could match the value of a call option by continually readjusting a portfolio consisting of a mixture of shares of the stock on which the option is based and Treasury bills, in such a way that there would be no risk to the brokerage firm? If we neglect transaction costs to the firm, the answer is "Yes". The brokerage firm, like a bookie at a race track, makes its money on service fees built into the cost of the commodity being sold. The brokerage firm does not care whether the stock goes up or down, provided that it can determine a "risk neutral" price. Given that the goal has changed from one of forecasting to one of accounting, we will not be surprised if the stochastic effects are modest in the solution of the problem of achieving "risk neutrality."

Let us return to the situation in Figure 11.2 but with the addition of a stock (with outcomes in the next increment of time s_{up} and s_{dn}) to which the prizes D_{up} and D_{dn} are linked. We will refer now to the prizes D_{up} and D_{dn} as *derivatives*.

A woman approaches the dealer and states she would like to play a game in which she receives D_{up} if the stock goes up in the next step and D_{dn} if it goes down. How much should the dealer charge her to play this game?

The dealer decides to emulate the derivative by forming a portfolio of the u units of the stock on which the derivative is based plus a position of v in a bond paying interest rate r. Using Figure 11.2, let us determine u and v.

$$us_{up} + ve^{r\Delta(t)} = D_{up} \text{ when the stock goes up}$$
$$vs_{dn} + ve^{r\Delta(t)} = D_{dn} \text{ when the stock goes down.} \qquad (11.6)$$

Solving for u and v, we have

$$u = \frac{D_{up} - D_{dn}}{s_{up} - s_{dn}}$$
$$v = e^{-r\Delta(t)}(D_{up} - us_{up}).$$

At time zero, the value of the portfolio (and therefore of the emulated derivative) is given by

$$
\begin{aligned}
D_0 &= s_u + v \\
&= \exp(-r\Delta(t))\left(\frac{s_0 e^{r\Delta(t)} - s_{dn}}{s_{up} - s_{dn}} D_{up} + \left(1 - \frac{s_0 e^{r\Delta(t)} - s_{dn}}{s_{up} - s_{dn}}\right) D_{dn}\right) \\
&= qD_{up} + (1-q)D_{dn}, \qquad (11.7)
\end{aligned}
$$

where

$$q = \frac{s_0 e^{\Delta(t)} - s_{dn}}{s_{up} - s_{dn}}. \qquad (11.8)$$

(Verify that $0 \le q \le 1$.)

1. Now, at time zero, the lady who wishes to play the game gives the dealer D_0 in cash.

2. The dealer buys u shares of the stock at a cost of us_0.

3. So, at time zero, the dealer has u shares of stock, and cash equal to $D_0 - us_0 = v$.

4. If the stock moves up at time $\Delta(t)$, the dealer sells his u shares at price s_u. He then has cash equal to

$$us_{up} + ve^{r\Delta(t)} = \frac{D_{up} - D_{dn}}{s_{up} - s_{dn}}s_{up} + D_{up} - \frac{D_{up} - D_{dn}}{s_{up} - s_{dn}}s_{up} = D_{up}.$$

5. If the stock moves down at time $\Delta(t)$, the dealer sells his u shares at price s_u. He then has cash equal to

$$us_{dn} + v = \frac{D_{up} - D_{dn}}{s_{up} - s_{dn}}s_{dn} + D_{up} - \frac{D_{up} - D_{dn}}{s_{up} - s_{dn}}s_{up} = D_{dn}.$$

So, then, the dealer has achieved a strategy of buying a portfolio which gives him zero profit and zero loss whether the stock moves up or down. By imposing a discipline of buying the right amount of stocks and bonds at each step, he has achieved, apparently, a situation where he controls the "state of nature." We note that

$$q = \frac{s_0 e^{r\Delta(t)} - s_{dn}}{s_{up} - s_{dn}}$$

has the formal properties of a probability. That is, q is between zero and one. We have seen that the value of the dealer's portfolio at time zero is given by

$$D_0 = e^{-r\Delta(t)}[qD_{up} + (1-q)D_{dn}].$$

For a price received from the lady D_0, he agrees to pay the lady D_{up} at time $\Delta(t)$ if the stock moves to s_{up} and D_{dn} if it moves down to s_{dn}. At time $\Delta(t)$, the dealer has a portfolio of value D_{up} if the stock goes up and D_{dn} if it goes down. The dealer has, it would appear, created his own reality. Maintaining his strategy of updating the portfolio at each tick of time, he will have achieved the ability to have assets, absent his obligations, of zero. This q is, accordingly, frequently referred to as the *martingale* measure: it has "expectation" zero. Like a bookie, the dealer has managed to eliminate his risk and simply live on commissions.

Let us consider a modified version of equation (11.5), namely,

$$
\begin{aligned}
S(t + \Delta(t)) &= S(t)\exp\left[\left(\mu - \frac{1}{2}\sigma^2\right)\Delta(t)\right]\exp[\sigma\sqrt{\Delta(t)}] \text{ with prob } q \\
&= S(t)\exp\left[\left(\mu - \frac{1}{2}\sigma^2\right)\Delta(t)\right]\exp[-\sigma\sqrt{\Delta(t)}]\text{with prob } 1 - q.
\end{aligned}
\tag{11.9}
$$

Note that

$$
\begin{aligned}
q &= \frac{s_0[e^{r\Delta(t)} - e^{\mu\Delta(t)-\sigma\sqrt{\Delta(t)}}]}{s_0 e^{\mu\Delta(t)}[e^{\sigma\sqrt{\Delta(t)}} - e^{-\sigma\sqrt{\Delta(t)}}]} \\
&= \frac{e^{\Delta(t)(r-\mu)} - e^{-\sigma\sqrt{\Delta(t)}}}{e^{\sigma\sqrt{\Delta(t)}} - e^{-\sigma\sqrt{\Delta(t)}}} \\
&= \frac{1 + \Delta(t)(r-\mu) + \Delta(t)^2(r-\mu)^2/2 - 1 + \sigma\sqrt{\Delta(t)} - \sigma^2\Delta(t)/2}{2\sigma\sqrt{\Delta(t)}} \\
&= 1/2\left(1 - \sqrt{\Delta(t)}\frac{[\mu + \sigma^2/2 - r]}{\sigma}\right).
\end{aligned}
$$

Again, we look at

$$S(t) = S(0)\exp\left[\left(\mu - \frac{1}{2}\sigma^2\right)t + \sigma\sqrt{t}\frac{2X_n - n}{\sqrt{n}}\right].
\tag{11.10}$$

But, with the martingale measure, we have

$$
\begin{aligned}
E_Q(2X_n - n) &= 2nq - n \\
&= n\left(1 - \sqrt{\Delta(t)}\frac{[\mu + \sigma^2/2 - r]}{\sigma}\right) - n \\
&= -n\sqrt{t/n}\frac{[\mu + \sigma^2/2 - r]}{\sigma}.
\end{aligned}
\tag{11.11}
$$

$$
\begin{aligned}
Var(2X_n) &= 4nq(1 - q) \\
&= n\left(1 - \sqrt{t/n}[\mu + \frac{\sigma^2}{2} - r]/\sigma\right)\left(1 + \sqrt{\Delta(t)}[\mu + \frac{\sigma^2}{2} - r]/\sigma\right) \\
&= n\left(1 - (t/n)((\mu + \sigma^2/2 - r)/\sigma)^2\right) \\
&\to n + \text{ a constant as } n \text{ becomes large, i.e., as } \Delta(t) \to 0.
\end{aligned}
\tag{11.12}
$$

Therefore,

$$
\frac{2X_n - n}{\sqrt{n}} \to \mathcal{N}\left(-\sqrt{t}[\mu + \sigma^2/2 - r]/\sigma, 1\right).
\tag{11.13}
$$

Let

$$
Z = \frac{2X_n - n}{\sqrt{n}} + \frac{\sqrt{t}[\mu + \sigma^2/2 - r]}{\sigma}.
\tag{11.14}
$$

Z is normal with mean zero and unit variance. Then

$$
\begin{aligned}
S(t) &= S(0)\exp\left[\left(\mu - \frac{1}{2}\sigma^2\right)t + \sigma\sqrt{t}\frac{2X_n - n}{\sqrt{n}}\right] \\
&\to S(0)\exp\left[\left(\mu - \frac{1}{2}\sigma^2\right)t - \sigma\sqrt{t}\frac{\sqrt{t}[\mu + \sigma^2/2 - r]}{\sigma} + \sigma\sqrt{t}Z\right].
\end{aligned}
\tag{11.15}
$$

Then, under the risk neutral Q measure,

$$
\begin{aligned}
E_Q(S(t)) &= S(0)\exp\left[\left(r - \frac{1}{2}\sigma^2\right)t\right]E_Q\left(\exp[\sigma\sqrt{t}Z]\right) \\
&= S(0)e^{rt}.
\end{aligned}
\tag{11.16}
$$

Not surprisingly, considering the way the portfolio is constantly rebalanced, its return is the same as that of a riskless bond paying the rate r. In other words, although under the probability model, $E(S(t)) = S(0)\exp(\mu t)$, the risk neutral hedging model buys and sells the stock in such a way that the expected value of the stock under the hedging discipline is only $E(S(t)) = S(0)\exp(rt)$ *regardless of* the stock's growth rate μ. In essence, we have taken a risky stock and split it into riskless debt (which the dealer holds) and a risky call (which is sold to the lady).

11.2.4   A Game Independent of the Odds

In their provocative book *Financial Calculus*, Baxter and Rennie [1, p. 7] state that for the purpose of pricing derivatives, "... seductive though the strong law [of large numbers] is, it is also completely useless." They show how serious they are about this view when they consider a wager which is based on the progress of a stock [1, p. 15]. At present, the stock is worth $1. In the next tick of time, it will either move to $2.00 or to $0.50. A wager is offered which will pay $1.00 if the stock goes up and $0.00 if the stock goes down. The authors form a portfolio consisting of 2/3 of a unit of stock and a borrowing of 1/3 of a $1.00 riskless bond. The cost of this portfolio is $0.33 at time zero. After the tick, it will either be worth 2/3 × $2.00 − 1/3 × $1.00 = $1.00 or 2/3 × $ 0.50 - 1/3 × $1.00 = $0.00. From this, they infer that "the portfolio's initial value of $0.33 is also the bet's initial value." They call this exercise the whole [Black-Scholes] story in one step. Certainly, if the seller gets $0.33 for selling the option, he must break even whether the stock advances or declines. Would the buyer be as willing to pay $0.33 for the option if the chance of the stock going up were .001 as he would if the chance were .999? Probably not. Would the seller be willing to sell the option for $0.33 if he know the probability the stock would go up next tick were .999? Probably not. However, concatenating the Baxter and Rennie one step many times, one can indeed emulate a manner in which the seller could, by continuous rebalancing of his portfolio, maintain a risk neutral position in the case of an option. [4]

The risk neutral discipline has essentially overridden nature. We have taken a stock, any stock, and by constantly readjusting our position in it,

---

[4]Some might argue as follows: In fact, several things are occurring in the example which are often confused and combined. Observe that there are two immediate future states, up and down. The portfolio exactly replicates the payoff of the bet in each state and involves a net investment of $.33. Hence, unless the bet also sold for $.33, an arbitrage profit could be earned by buying the cheaper and selling the dearer of the two. In that no risk would be involved in such a transaction, it would be invariant to (and devoid of information about) risk preferences. Even if the market for the stock were in equilibrium (which would imply $.33 was an equilibrium price for the bet), including general agreement about the states, outcomes, probabilities and risk preferences create a jointness which cannot be untangled from the information provided. For example, if the market is employing risk neutral pricing, then prob(up)=1/3, prob(down)=2/3, the expected value of the stock, portfolio and bet are their prices and the expected return on each upon the revelation is 0. Suppose, however, the market is engaging in risk averse pricing, such that the probabilities are 50-50. The expected value of the stock is $1.25 and of the portfolio and bet is $.50. The instantaneous expected return on the first is 25% and on the latter two is 50%. Further, there is no way to earn an arbitrage profit between the portfolio and the bet. In particular, there is no way to arbitrage this result back to risk neutral pricing. If the only concern is the derivation of the hedged, no arbitrage price of the bet, a shortcut may be employed. In that preferences do not enter this computation, the simplest approach is to assume the stock price is risk-neutral, extract the implied probabilities (i.e., 1/3, 2/3), and price the bet (i.e., $1/3 × $1 + 2/3 × 0 = $.33$). By using this approach, it is not even necessary to derive the hedge portfolio to price the bet. As discussed above, this procedure in no way implies that the price of the stock or the bet is *actually* risk neutral.

apparently obtained freedom from risk but at the cost of realizing only the returns available in a risk free bond. Are there hitches in all this risk neutrality? Of course, lots of them. Here are a few

- Transaction costs are not really free. The closer the hedge gets to being riskless, the more frequently one must rebalance (and this results in material transaction costs).

- The realistic value of r will be significantly higher than that of a Treasury bill.

- Historical records show that the Black-Scholes formula [2], which we develop below, generally does not give the actual market price of a call option. To correct this imperfection in nature, it is customary for some traders to plug in whatever value is necessary for σ to give the market price for the option. We may recall in a chemistry or physics lab when we did not get the answer demanded by the science, there was some temptation to plug in whatever would conform to the established physical model. Such a procedure was called "dry labing" and generally regarded as cheating. Amongst believers in EMH, such a plug-in approach to a σ so obtained has a much more respectable name, *implied volatility*.

- Stock prices may jump (with substantial discontinuities), and this may defeat the hedging strategy. Stock price evolution is not just a smooth function of time.

- In the case of horse betting, there is an arbitrary mechanism which sets payoffs at the instant the race starts. The bookmaker is allowed to set payoffs with his profit margin locked into the payoff. Suppose that the bet is placed a week before the instant the race starts. What costs must the bookmaker incur in the intervening time period to rebalance the payoffs for the bets he has covered? The answer is "zero." And this is the reason the analogy between bookmaking and selling options is flawed.

The Black-Scholes Theorem can be described as a proof of a result suspected by many to be true. Perhaps the best introduction to the result comes from asking the old question of how much an investor should pay at time T for a stock which today has price $S(0)$. The answer is deemed (by many) to be obvious. The seller of the stock future buys the stock today at a price $S(0)$ using money he borrows at interest rate r. If the agreed upon price is $S(0) \exp(rT)$, then when the buyer pays it at time T, the seller can pay back the loan he entered upon to buy the stock. So, the story goes, it really makes no difference what buyer or seller believes the growth rate of the stock is (that has already been incorporated magically into the current price of the stock). The broker of the future naturally will add a commission to the cost. The only market aspect of the deal will be the possible competition between brokers to lower their commissions.

Let us turn to the "risk free" purchase of a European call option. Recall that here we pay at time zero for the right to buy the stock at time T for strike price X.

$$C_{BS} = e^{-rT}E[\text{Max}(S(T) - X, 0)] \tag{11.1}$$

$$= e^{-rt}\frac{1}{\sqrt{2\pi\sigma^2 t}}\int_{\ln[\frac{X}{S(0)}]}^{\infty}(S(0)e^z - X)\exp\left(-\frac{1}{2\sigma^2 t}\left(z - (r - \frac{\sigma^2}{2})t\right)^2\right).$$

where we note that the formula is the same as that for Method B except that the growth rate is r rather than μ.

The risk neutral determination value for a call option was first given by Black and Scholes [2]. The Black–Scholes model has had enormous impact on the trading of options. Consequently, it has itself changed the mechanism of the market. However, if one looks at the actual market price of an option at a given time, it is seldom the case that it is the same or even close to the Black-Scholes value if one uses historical measures of volatility. But since the true believers know the Black-Scholes value is correct, they take the actual market price of the option and calculate backwards to determine the *implied volatility*. That this *implied volatility* will be different for different time horizons is taken care of by noting that the appropriate average volatility will naturally be different for longer and shorter time epochs. That the *implied volatility* can also be different for the same stock using the same time horizon but different strike prices is the kind of rude remark that is best left unsaid in polite financial circles.

11.3   The Black–Scholes Derivation Using Differential Equations

We recall the Brownian model for stock progression:

$$dS = \mu S dt + \sigma S dz. \tag{11.18}$$

Let f be a derivative security (i.e., one which is contingent on S). Then, from Ito's lemma, we have:

$$df = \left(\frac{\partial f}{\partial S}\mu S + \frac{\partial f}{\partial t} + \frac{1}{2}\frac{\partial^2 f}{\partial S^2}(\sigma S)^2\right)dt + \frac{\partial f}{\partial S}\sigma S dz. \tag{11.19}$$

Multiplying (8.51) by $\partial f/\partial S$, and isolating $\partial f/\partial S\sigma S dz$ on the left side in both (8.51) and (8.52), we have:

$$\frac{\partial f}{\partial S}\sigma S dz = \frac{\partial f}{\partial S}dS - \frac{\partial f}{\partial S}\mu S dt$$

$$\frac{\partial f}{\partial S}\sigma S dz = df - \left(\frac{\partial f}{\partial S}\mu S + \frac{\partial f}{\partial t} + \frac{1}{2}\frac{\partial^2 f}{\partial S^2}(\sigma S)^2\right)dt.$$

Setting the two right-hand sides equal (stochastic though they be), we have:

$$df - \frac{\partial f}{\partial S} dS = \left(\frac{\partial f}{\partial t} + \frac{1}{2} \frac{\partial^2 f}{\partial S^2} (\sigma S)^2 \right) dt. \qquad (11.20)$$

Let us consider a portfolio which consists of one unit of the derivative security and $-\partial f / \partial S$ units of the stock. The instantaneous value of the portfolio is then

$$\mathcal{P} = f - \frac{\partial f}{\partial S} S. \qquad (11.21)$$

Over a short interval of time, the change in the value of the portfolio is given by

$$d\mathcal{P} = df - \frac{\partial f}{\partial S} dS = \left(\frac{\partial f}{\partial t} + \frac{1}{2} \frac{\partial^2 f}{\partial S^2} (\sigma S)^2 \right) dt. \qquad (11.22)$$

Now since (11.22) has no dz term, the portfolio is riskless during the time interval dt. We note that the portfolio consists both in buying an option and selling the stock. Since, over an infinitessimal time interval, the Black–Scholes portfolio is a riskless hedge, it could be argued that the portfolio should pay at the rate r of a risk free security, such as a Treasury short-term bill. That means that

$$d\mathcal{P} = r\mathcal{P}dt = r \left(f - \frac{\partial f}{\partial S} S \right) dt = \left(\frac{\partial f}{\partial t} + \frac{1}{2} \frac{\partial^2 f}{\partial S^2} (\sigma S)^2 \right) dt. \qquad (11.23)$$

Finally, that gives us the Black–Scholes differential equation

$$rf = \left(rS \frac{\partial f}{\partial S} + \frac{\partial f}{\partial t} + \frac{1}{2} \frac{\partial^2 f}{\partial S^2} (\sigma S)^2 \right). \qquad (11.24)$$

It is rather amazing that the Black–Scholes formulation has eliminated both the Wiener term and the stock growth factor μ. Interestingly, however, the stock's *volatility* σ remains. Essentially, the Black–Scholes evaluation of a stock is simply driven by its volatility, with high volatility being prized. We note that μ has been replaced by the growth rate r of a riskless security. Over a short period of time, the portfolio will be riskless. (We recall how, in the Black–Scholes solution, we used a hedge where we bought options and sold stock simultaneously.) This risklessness will not be maintained at the level of noninfinitessimal time. However, if one readjusts the portfolio, say, daily, then (making the huge assumption that sudden jumps cannot happen within a short period of time), it could be argued that assuming one knew the current values of r and σ, a profit could be obtained by purchasing options when the market value was below the Black–Scholes valuation and selling them when the market value was above that of the Black–Scholes valuation (assuming no transaction costs). (Such a fact, it could be argued, in which all traders acted on the Black–Scholes valuation, would drive the market. In reality, if the Treasury Bill rate is used for r

and historical estimates are used for μ and σ, the actual value for which an option is traded is generally significantly different from the Black–Scholes value.)

Now, we recall that a *European call option* is an instrument which gives the owner the right to purchase a share of stock at the *exercise price* X, T time units from the date of purchase. Naturally, should the stock actually be priced less than X at time T, the bearer will not exercise the option to buy at price X. Although we get to exercise the option only at time T, we must pay for it today. Hence, we must discount the value of an option by the factor $\exp(-rt)$. Since we have seen that the Black–Scholes equation involves no noise term, it is tempting to conjecture that the fair evaluation of an option to purchase a share of stock at exercise price X is given by

$$
\begin{aligned}
C_{BS} &= e^{-rT} E[\mathrm{Max}(S(T) - X, 0)] \qquad\qquad (11.25)\\
&= e^{-rt} \frac{1}{\sqrt{2\pi\sigma^2 t}} \int_{\ln[\frac{X}{S(0)}]}^{\infty} (S(0)e^z - X) \exp\left[-\frac{1}{2\sigma^2 t}(z - (r - \frac{\sigma^2}{2})t)^2\right] dz,
\end{aligned}
$$

where we note that the growth rate is r rather than μ.

11.4   Black–Scholes: Some Limiting Cases

Consider, in Tables 11.1 and 11.2, the Black–Scholes pricing model compared to Model A and Model B in the case where a stock has a rather high growth rate $\mu = .15$ with a fixed riskless interest rate of 5% and a variety of volatilities and strike prices X. (Of course, we are looking at a case where the option buyer's estimates of the growth rate μ and volatility σ were correct. From the standpoint of the buyer, who decides to buy the call option, standing at time zero, he probably *believes* his estimate for (μ, σ) is correct. One question we should be examine is the value of the option to the buyer, given his current state of information.) We shall assume the option is for an exercise time of six months in the future, and that the price of the stock at the present time is $100. The purpose of this exercise is simply to look at Black-Scholes in comparison to two older pricing models in the very optimistic case where the person using the model knows μ and σ.

Table 11.1. Six-Month Options: $\sigma = .20, \mu = .15$									
X	102	104	106	108	110	112	114	116	118	120
C_{BS}	5.89	4.99	4.20	3.51	2.91	2.40	1.98	1.63	1.36	1.16
C_B	8.58	7.47	6.46	5.55	4.73	4.00	3.37	2.82	2.35	1.95
C_A	9.02	7.85	6.79	5.83	4.97	4.21	3.54	2.96	2.47	2.05

Table 11.2. Six-Month Options: $\sigma = .40, \mu = .15$									
X	102	104	106	108	110	112	114	116	118	120
C_{BS}	11.48	10.63	9.82	9.07	8.37	7.72	7.11	6.54	6.01	5.53
C_B	13.84	12.89	12.00	11.16	10.37	9.62	8.92	8.26	7.64	7.06
C_A	14.55	13.55	12.62	11.73	10.90	10.11	9.37	8.68	8.03	7.43

We note that as the volatility σ increases, the three strategies become more similar. To note the effect of increasing and decreasing σ, we show in Tables 11.3 and 11.4 results for very low σ (.001) and very high σ (2.00).

Table 11.3. Six-Month Options: $\sigma = .001, \mu = .15$									
X	102	104	106	108	110	112	114	116	118	120
C_{BS}	0.52	0.00	0.00	0.00	0.00	0.00	0.00	0.00	0.00	0.00
C_B	5.37	3.51	1.66	0.00	0.00	0.00	0.00	0.00	0.00	0.00
C_A	5.65	3.69	1.74	0.00	0.00	0.00	0.00	0.00	0.00	0.00

Table 11.4 . Six-Month Options: $\sigma = 2.00, \mu = .15$									
EP	102	104	106	108	110	112	114	116	118	120
C_{BS}	52.17	51.71	51.25	50.80	50.36	49.93	49.50	49.08	48.67	48.26
C_B	53.37	52.91	52.45	52.00	51.56	51.13	50.70	50.28	49.87	49.47
C_A	56.11	55.62	55.14	54.67	54.21	53.75	53.30	52.86	52.43	52.00

Let us consider limiting behavior as the volatility goes first to infinity and then to zero. Suppose that a stock is currently selling for $S(0)$. We wish to buy an option T time units in the future with strike price X. As the volatility of the stock goes to infinity, then we note that both Black–Scholes (11.17) and Method B (11.3) tell us that the option is so valuable that its fair price is simply the current value of the stock, namely $S(0)$, irrespective of the value of μ.

On the other hand, let us suppose that the value of the volatility is zero. Then the Black–Scholes price is

$$
\begin{aligned}
C_{BS} &= S(0) - e^{-rT}X \text{ if } S(0)e^{rT} \geq X \\
&= 0 \text{ otherwise.}
\end{aligned}
\tag{11.26}
$$

Next, let us consider the situation where the growth rate of the stock is actually negative ($\mu = -.15$) in Tables 11.5 and 11.6, respectively.

Table 11.5. Six-Month Options: $\sigma = .001, \mu = -.15$									
X	102	104	106	108	110	112	114	116	118	120
C_{BS}	0.52	0.00	0.00	0.00	0.00	0.00	0.00	0.00	0.00	0.00
C_B	0.00	0.00	0.00	0.00	0.00	0.00	0.00	0.00	0.00	0.00
C_A	0.00	0.00	0.00	0.00	0.00	0.00	0.00	0.00	0.00	0.00

Table 11.6. Six-Month Options: $\sigma = 2.00, \mu = -.15$									
X	102	104	106	108	110	112	114	116	118	120
C_{BS}	52.17	51.71	51.25	50.80	50.36	49.93	49.50	49.08	48.67	48.26
C_B	49.77	49.30	48.84	48.39	47.95	47.51	47.08	46.66	46.25	45.84
C_A	45.03	44.61	44.19	43.78	43.38	42.99	42.60	42.22	41.85	41.48

We note that Black–Scholes values a call option at, say $102, equally whether the growth rate of the stock is +.15 or −.15. Naturally, it is unfair to note that Method A and Method B are more accurate than Black-Scholes. That would be true if we really knew μ, but generally we have only noisy estimates for this parameter. Nevertheless, markets are made, in large measure, by differences in information (opinion).

Table 11.7. Computed Values of Six-Month Options (with Bear Jumps).
$\sigma = .20, \mu = .15$

Ex. Pr.	102	104	106	108	110	112	114	116	118	120
C_{BS}	5.89	4.99	4.20	3.51	2.91	2.40	1.98	1.63	1.36	1.16
C_B	8.58	7.47	6.46	5.55	4.73	4.00	3.37	2.82	2.35	1.95
C_A	9.02	7.85	6.79	5.83	4.97	4.21	3.54	2.96	2.47	2.05
Sim.	5.97	5.13	4.38	3.72	3.13	2.62	2.18	1.80	1.48	1.21

Table 11.8. Six-Month Options (with Bear Jumps).
$\sigma = .40, \mu = .15$

EP	102	104	106	108	110	112	114	116	118	120
C_{BS}	11.48	10.63	9.82	9.07	8.37	7.72	7.11	6.54	6.01	5.53
C_B	13.84	12.89	12.00	11.16	10.37	9.62	8.92	8.26	7.64	7.06
C_A	14.55	13.55	12.62	11.73	10.90	10.11	9.37	8.68	8.03	7.43
Sim.	10.89	10.08	9.33	8.62	7.96	7.34	6.76	6.23	5.73	5.28

Now, based on Tables 11.7 and 11.8, the pricing of the Black–Scholes model appears inspired. Of course, we have simply added on the kind of unexpected downward turn which is not accounted for by the geometric Brownian walk unmodified. On the other hand, our imposition of bear jumps has depressed the expected growth rate of the stock to essentially 1%, and most of the value of the option is due to volatility.

The vendor of the option typically has no strong views about a particular stock. He or she is selling options in many stocks and is only interested that he or she retrieves his or her supposed opportunity cost rate η. Accordingly, the vendor of the option might use use Black–Scholes with r replaced by η.

$$C_{\text{vendor}} = e^{-\eta T}\{e^{\eta T}S(0)\Phi\left(\frac{\log(S(0)/X) + [\eta + (\sigma^2/2)]T}{\sigma\sqrt{T}}\right)$$
$$-X\Phi\left(\frac{\log(S(0)/X) + [\eta - (\sigma^2/2)]T}{\sigma\sqrt{T}}\right)\}. \qquad (11.27)$$

On the other hand, the buyer of the option will have fairly strong views about the stock and its upside potential. The buyer could use Black–Scholes replacing r by μ, where, typically, $\mu > \eta > r$. Thus,

$$C_{\text{buyer}} = e^{-\mu T}\{e^{\mu T}S(0)\Phi\left(\frac{\log(S(0)/X) + [\mu + (\sigma^2/2)]T}{\sigma\sqrt{T}}\right)$$
$$-X\Phi\left(\frac{\log(S(0)/X) + [\mu - (\sigma^2/2)]T}{\sigma\sqrt{T}}\right)\}. \qquad (11.28)$$

We note that typically, in the mind of the call option buyer, μ is rather large. Perhaps some investors will buy solely on the basis of a large stock volatility, but this is unusual. Option buying is frequently a leveraging device whereby an investor can realize a very large gain by buying call options rather than stocks. The seller of the option is probably expecting an $\eta < \mu$ value as the reasonable rate of return on his/her investments overall. It is observed [3] that the arithmetic mean annual return on U.S. common stocks from 1926 on is over 10%. Let us suppose we are dealing with an initial stock price of $100 and that the vendor uses $\eta = .10$ and the buyer believes $\mu = .15$. In Tables 11.9 and 11.10, we show the values of $C_{(vendor)}$ and $C_{(buyer)}$, respectively. This may appear confusing, for we have arrived at a price for the vendor and one for the buyer, and they are generally not the same. *Pareto efficiency* is the situation where all parties are better off by undertaking a transaction. Clearly, at least from their respective viewpoints, we do have Pareto efficiency (assuming that the commission is not so high as to swamp the anticipated profit to the buyer). The difference between the price the buyer is willing to pay and that for which the vendor is willing to sell must be positive, or there will be no trade.

Suppose that an investor believes the rate of growth of a stock is .15 overall, bear jumps included. Then, if we are to include the bear jumps, we need to increase the value of the Brownian growth to $.15 + .14 = .29$. So, let us now compute the simulated buyer's price, with discount to present value rate being $\mu = .15$. We also compute the vendor's price using the Black—Scholes formula with riskless rate $\eta = .10$ (we will assume that the vendor will use the nominal volatility values of .20 and .40, as shown in Tables 11.9 and 11.10).

Table 11.9. Six-Month Options (with Bear Jumps): $\sigma = .20, \mu = .15$

Ex Pr	102	104	106	108	110	112	114	116	118	120
C_{BS}	5.89	4.99	4.20	3.51	2.91	2.40	1.98	1.63	1.36	1.16
C_{vendor}	7.17	6.16	5.26	4.45	3.74	3.13	2.59	2.13	1.75	1.42
C_{buyer}	10.53	9.34	8.24	7.23	6.30	5.46	4.71	4.04	3.44	2.91

Table 11.10. Six-Month Options (with Bear Jumps): $\sigma = .40, \mu = .15$

EP	102	104	106	108	110	112	114	116	118	120
C_{BS}	11.48	10.63	9.82	9.07	8.37	7.72	7.11	6.54	6.01	5.53
C_{ven}	12.63	11.73	10.88	10.08	9.34	8.63	7.98	7.36	6.79	6.25
C_{buy}	15.56	14.54	13.58	12.67	11.80	10.98	10.21	9.49	8.81	8.17

It is unlikely that a buyer will be able to acquire options at the orthodox Black—Scholes rate (i.e., the one using Treasury bill interest rates of .05). But suppose she can. Suppose she correctly guesses $\sigma = .2$. The Black—Scholes price for a strike price of $108 six months in the future is $3.5357. Suppose she gets lucky and the growth rate over the next six months is $\mu = .15$. However, this is the aggregate growth (including Poissonian bear jumps of size 10% once a year and of size 20% once every five years). Simulation allows us to see what she can expect. We display the results in the simugram in Figure 11.3. The expected value of the option is $7.23. However, she should realize that around 55% of the time she will have lost her purchase price of the option. There are many other things the prospective buyer might choose to try before making the decision as to whether or not the option should be bought. Most of these are rather easy to achieve with simulation. The point here is that she should view the purchase of the call option as something risky. Mathematics has not secured for her a free lunch.

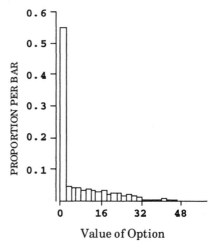

Figure 11.3. Simugram of Option (Present) Values.

11.5   Conclusions

In actuality, a buyer of options has many to choose from. Let us suppose that a rational buyer has computed the value of options (using his own estimates for the growth rates and the volatilities). Suppose he has confined himself to stocks with roughly the same volatilities. Suppose further that he is interested in European call options maturing one year from today. If he has no emotional attachment to one stock or the other (big and frequently false assumption), he would then be well advised to purchase options in the stock which give him the largest positive difference between what his computations give him for the value of the option and the actual market price of that option. It will be unusual for the buyer to have the Black-Scholes price as his personal valuation of the option, and he will seldom see the stock offered for the Black-Scholes price on the market.

So, our hypothetical buyer puts down his money for a one year option in the stock which seems, for a given risk measure, to give him the greatest expected (or median or twenty percentile) return. When he and others buy the option, if the volume of sales over a given time interval appears high, this will encourage vendors of the stock to take note that perhaps they can raise the option's price. Similarly, vendors of options which are not selling at expected volume levels, may decide to lower the prices of their options.

And, as all this is going on, we are aware that there exists a wide diversity of ways for each buyer to consider balancing expected gain against volatility.

And there are many time horizons for which he can buy an option. And there are many other possible ways he can invest his capital: in real estate, bonds, wheat, etc. It is the concatenation of all these opportunities for buying and selling. viewed from the standpoints of each buyer and seller, which make up the market. The dynamism of the world in which the market exists is such that any notion of reaching equilibrium for most potential investments is generally unlikely so long as the market is allowed to "work its will." The possibilities for finding undervalued (from the standpoint of the buyer) stocks and derivatives to purchase exist day by day, hour by hour, minute by minute.

Finally, we cannot emphasize too strongly that the time indexed profile of the probability distribution of the proposed investment is much more reliable than simply looking at expected values. It is frequently the case that the purchase of an option with a high expected value of gain will, most of the time, be a losing proposition.

At the end of the day, an option based on a stock is simply a security to be purchased or not depending on its risk profile and the way an investor views that risk profile in comparison to those of other investments. It might seem absurd that one could come up with a formula which would give, at a given time, strike price and execution time, the value of an option on a stock which is following a random trajectory. Indeed, it is absurd. If transaction costs are truly zero, then the Black-Scholes evaluation of an option is correct *if one is basing evaluation on the expected value of the option.* And if cold fusion were a reality and anti-gravitational devices existed, that would be nice too.

The reality is that transaction costs are not free, and looking only at the expected value of an option (as opposed to its entire risk profile) will frequently lead to disaster. In many ways, option trading has become a useful surrogate for margin buying. Before the Crash of 1929, an investor could use his portfolio to leverage purchasing of stocks by a factor of ten to one. As the market started downward, the broker would dump stocks in the portfolio to meet margin calls. This put downward pressure on the prices of the stocks being sold. This kind of feedback mechanism led to the Crash of 1929. These days, in the United States, margin leveraging is a more modest two to one.

The purchasing of options appears to be a relatively benign alternative to margin leveraging. If the buyer purchases an option for a strike price which the stock does not realize, then he loses the purchase price of the option without any direct negative pressure on the stock. Indeed, if the stock price is rising, then vendors of uncovered options will have to go into the market and purchase shares to cover the calls, thus putting an upward pressure on the stock price. Though some argue that the availability of options exerts a stabilizing effect on the market, one has to question this judgment. The ability of large companies to sell uncovered options can have disastrous consequences.

The selling of covered call options may be very desirable for a fund manager who is trying to maintain a somewhat steady rate of return in markets good and bad. If the market moves into a phase of low or negative growth, the selling of the call options will bring in some income even though the purchasers will not exercise the options. If the market moves into a good growth phase, then the selling of the call options limits the upside profit from the stock to that of the strike price minus the original cost of the stock to the fund. Suppose that, by very clever balancing, the manager of such a fund managed to obtain a return of, say, 3% in bear markets and 10% in strong markets. That might be the basis for an attractive alternative to bonds. In other words, the vendor might be able to use call options as a way of trading high gains for low risk.

From the standpoint of the buyer of call options, it is frequently the case that the purpose of the trade is to assume high risk in the hopes of substantial gains. There are situations where this can make a great deal of sense. But the authors wonder how often the purchasers of call options bother to crank out and examine a risk profile such as that shown in Figure 11.3. The purchaser of a call option ought not believe that the equation of Black, Scholes and Merton will bring determinism into what is, in fact, the very risky business of using options for leveraging purposes.

The one sure thing about the stock market continues to be that it will fluctuate. And this fluctuation produces risk. However, the investor has two weapons at his/her disposal to reduce risk: portfolio diversification and time. The investor also has another weapon: reason and the knowledge that if a deal appears too good to be true, it probably is.

Problems[5]

Problem 11.1. Using equations (11.2) and (11.3), find the option values for Method A and Method B for a stock with present value 100, $\mu = .10$, $r = .06$, $\sigma = .1$, strike price $X = 110$ and time horizon one year.

Problem 11.2. Verify Lemma 11.1.

Problem 11.3. Consider the Black–Scholes differential equation (11.24) with the boundary condition that

$$
\begin{aligned}
f(S,T) &= S(T) - X \text{ if } S(T) - X > 0 \\
&= 0, \text{ otherwise.}
\end{aligned}
$$

---

[5]For those uncomfortable with simulations, the reader will find assistance in the statistical appendix at the end of this book.

Prove that

$$f(S,t) = S\Phi\left(\frac{\log(S/X) + [r + (\sigma^2/2)](T-t)}{\sigma\sqrt{T-t}}\right)$$
$$-Xe^{[-r(T-t)]}\Phi\left(\frac{\log(S/X) + [r - (\sigma^2/2)](T-t)}{\sigma\sqrt{T-t}}\right).$$

Problem 11.4. A group of investors is considering the possibility of creating a European option-based mutual fund. As a first step in a feasibility study, they decide to compare investments of $1 million in a portfolio of 20 stocks as opposed to a portfolio of 20 options. They want to obtain histograms of investment results after one year. Let us suppose that all stocks in the portfolios are bought at a cost of $100 per share. Let us assume the usual model of stock growth,

$$S(t) = S(0)\exp(\mu t + \sigma\sqrt{t}\epsilon),$$

where ϵ is normally distributed with mean 0 and variance 1. Let us take two values of μ, namely, .10 and .15. Also, let us consider two values of σ, namely, .15 and .30. Consider several strike prices for the options: for example, the expected value of the stock at the end of one year, and various multiples thereof. Assume that the options are completely fungible. Thus, at the end of the year, if a stock is $10 over the strike price, the option purchased for the Black–Scholes price is worth $10 (i.e., one does not have to save capital to buy the stock; one can sell the option). For the "riskless" interest rate, use two values: .06 and .08. Clearly, then, we are considering a leveraged portfolio and seeing its performance in relationship to a traditional one. Carry out the study assuming that there is no correlation between the stocks.

Problem 11.5. Carry out the study in Problem 11.4 with the following modification. Use the μ value of .24 and the two σ values of .15 and .30. Then assume that there is an across-the-board bear jump mechanism whereby a sudden drop of 10% happens, on the average, once a year and a sudden drop of 20% happens on the average once every five years. The overall growth is still roughly .10. Use the Black–Scholes riskless price as before without adding in the effect of the Poisson jumps downward.

References

[1] Baxter, M. and Rennie, A. (1996). *Financial Calculus: An Introduction to Derivative Pricing.* New York: Cambridge University Press.

[2] Black, F. and Scholes, M. (1973). "The pricing of options and corporate liabilities," *Journal of Political Economy*, Vol. 81, 637–659.

[3] Bogle, J. C. (1999). *Common Sense and Mutual Funds: New Imperatives for the Intelligent Investor.* New York: John Wiley & Sons.

[4] Campbell, J. Y., Lo, A. W. and MacKinlay, A. C. (1997). *The Econometrics of Financial Markets.* Princeton, NJ: Princeton University Press.

[5] Hull, J. C., (1993). *Options, Futures, and Other Derivative Securities.* Englewood Cliffs, NJ: Prentice Hall.

[6] Smith, C. W. (1976). "Option pricing: A review," *Journal of Financial Economics*, 3, 3–51.

[7] Thompson, J. R. (1999). *Simulation: A Modeler's Approach.* New York: John Wiley& Sons, 115–142.

Chapter 12

Summary, Some Unsettled (Unsettling) Questions, and Conclusions

12.1 Summary

We now summarize briefly the contents of this book. In Chapter 1, we outlined some of the basic institutional factors associated with how stock markets operate and raised questions about whether these markets are efficient. We suggested that, in an efficient market, there would be very little that an investor or securities analyst could do to provide superior (above average) returns for his or her own accounts or for those of clients. We implied that, if markets are perfectly efficient, it makes no sense to study a particular security or, for that matter, read the typical 800-page "Investments" textbook required in college undergraduate or graduate classes. Further, it makes little sense for professionals to acquire such designations as the CFA (Chartered Financial Analyst). All securities should be correctly priced at any time, and it should make no difference when or what stocks are purchased.

In Chapter 2, the reader was introduced to the basic concepts of utility theory. The essence of utility analysis rests on the basic economic premise that people prefer more good things to less, and that a continued flow of good things will have increasingly reduced marginal value as larger quantities are enjoyed. The law of diminishing marginal utility underlies this conceptualization, and it gives shape to utility of income (wealth) functions that have positive first and negative second derivatives. Utility theory forms the cornerstone of portfolio theory, and is the underlying explanation for why most individual investors are presumed to be risk averse.

Chapter 3 illustrated why diversification is an important fundamental requirement of rational investment choice. The chapter began by examining the traditional Markowitz constrained optimization model, which depends for its solution on notions of mean returns from individual stocks, variances of those returns, and covariance among the returns. We ultimately concluded that what is essentially a sound procedure (diversification) applied to a concept (risk) that is, at best, difficult to quantify, all put together with historical data may have no relationship at all to the underlying facts in the future! Chapter 3 provided a better way of looking at risk through simugrams of future stock prices.

The materials in Chapters 4 and 5 summarized the current theory as generally accepted in financial economics along with an extensive critique of that theory. Chapter 4 detailed an explicit discussion of the capital asset pricing model (CAPM) and the arbitrage pricing model (APM). The chapter outlined various methods to make the CAPM and the APM operational, but left for Chapter 5 to provide a specific critique of both. We found that neither model is much accepted any place but in textbooks but there is nothing left to occupy the vacuum that remains if either (or both) models is (are) rejected. Chapter 5 developed a stinging critique of modern financial economic theory. After one has read this chapter, it is clear that what is done in many investments textbooks really has no meaning.

In Chapter 6, we examined the basic valuation equations derived by J. B. Williams seventy years ago. In this very simple model, the "value" of a security is simply the present value of the future income stream expected to be generated by the security in question. Next, we amplified these equations by assuming various probability distributions for the underlying variables (sales, expenses, earnings, dividends, etc.), and computed means and variances for them. We suggested that this sort of methodology could be used to generate price or "value" distributions where a given market (or individual) rate of return, r, is required. Variations on this methodology were seen to be still used today by many analysts where the market rate of return, r, is taken from the capital asset pricing model (see discussion in Chapters 4 and 5).

In Chapters 7 and 8, we provided a detailed procedure for analyzing securities. Borrowing from concepts developed by Benjamin Graham and others from the 1930s to the 1960s, we examined concepts that were often only verbally constructed and built a more quantitative framework. In Chapter 9, we examined the importance of compound interest and observed that security growth might be viewed as noisy compound interest. Chapter 10 continued to develop our conceptualization of risk profiling and applied the idea to bundles of securities (called portfolios). We argued that looking at the expected (present) values of an investment in and of itself is seldom a good idea.

Finally, we introduce and analyze the options market in Chapter 11 as a way of considering the notion of risk. It was our view that a simple application of the Williams equations will not allow an investor to make a rational choice. This is an extremely important point since many modern day analysts and investors try to make decisions using variations of these equations by discounting essentially an uncertain (in the sense used in Chapter 5) income stream by means of taking a discount rate from the CAPM. Since the notion of "risk" is specifically incorporated into the evaluation by assuming that risk can be determined by means, variances, and covariances of historical stock returns (see discussion of the CAPM in Chapter 4), we see why the procedure is inappropriate. In Chapter 11, we also reproduced the Black-Scholes-Merton option pricing equation and suggest some serious problems associated with its "search" for risk neutrality. Having summarized the contents of the book, we are left with several unsettled (unsettling) conclusions. These are addressed below.

12.2   What Is Uncertainty Aversion?

In Chapter 2, we discuss the concept of the utility of wealth, including the arguments advanced for diminishing marginal utility of wealth. The latter implies risk aversion. In Chapter 5, we reviewed the Knightian/Keynesian distinction between risk and uncertainty and argued that the real world is characterized by the latter. The question arises as to whether the concepts of risk aversion can be extended into a world of uncertainty (e.g., are there such things as uncertainty premia?).

Taking the latter first, in an opportunity cost framework, risk premia are purported to exist because risk adverse investors demand them in order to forego a less risky (or riskless) alternative investment. Hence, a continuum of (measurable) risk is presumed to exist, grounded in a riskless alternative (e.g., as in the CML and SML in Chapter 4). This is basically a world of stochastic certainty (i.e., we may not know the unique outcome of all future events, but we at least know the parameters of the process generating the outcomes). This can also be viewed as a world of certainty to which lotteries have been added but where one has the option of not purchasing tickets (and, thus, in some sense remaining in a world of certainty).

Much of this book has been devoted to a more realistic way of looking at such a world. As demonstrated in Chapter 9, efforts to remain at or near the world of certainty (e.g., government bonds) offer relatively low returns which are further pounded by inflation and taxes. It would seem that there is little "near certainty" to go around, and such as there is sells at substantial premiums. Furthermore, in the unlikely event that the government debt is effectively repudiated (e.g., hyperinflation), those who have been living off coupon payments are poorly positioned to survive in the resulting environment. Hence, near certainty may not be near enough.

If the alternatives being foregone are (1) earn little on your money while hoping the government does not fall or (2) consume everything in the next instant and die or (3) seek some other universe governed by different rules, then "taking a chance" suddenly does not seem so onerous as the normal risk premium literature would imply. There is effectively no alternative to forego (leaving the philosophical issues in 2 and 3 above for another day) in the first place and, hence, nothing with respect to which one can be averse.

12.3   Can There Still Be Differential Uncertainty Premia?

Premia is an *ex-ante* concept. It requires some sort of ability to form expectations about the future. Perhaps this need not be a mathematical expectation (which Keynes indicated could not be done in an uncertain world), where all possible outcomes and probabilities could be known (i.e., a world of risk) or the associated distributions. A fair amount of discussion in this book has been devoted to a demonstration that the latter contains more potentially useful information than the former. The point being raised here is the reliability of estimates of either, especially since the latter involves a more complex estimate.

Of course, one can (and must do) use *ex-post* data of some sort to provide the *ex-ante* estimates. This gets us to the question of whether there is a (return generating) process. Factually, of course, there is not (a point often forgotten by neoclassical economists). But may we model it as though there were and, in particular, as though the last 75 years of American experience provides the correct parameters? Of course, we *can*, as a good bit of this book has demonstrated. Whether or not we *may* contains all of the elements of a test of faith (e.g., we can neither prove it nor disprove it and we will not know until it is too late).

If we may say so, a broadly diversified portfolio of stocks held over long periods approaches a free lunch. Yet even the neoclassical theorists (cf [1]) are now contending that much of this differential return was not expected (and, hence, was not a premium). From the perspective of uncertainty, all we can say is that stocks outperformed money market instruments by substantial amounts over that time interval. Anyone building a stock portfolio over the decades, especially with the decline in dividend yields, will still retire dependant upon the kindness of strangers (i.e., whatever the market is prepared to pay for the pieces of paper — which are not even in a physical form anymore — he or she has accumulated).

The analysis to this point implies some changes to the casino analogy which has been employed at several points in this book. From Section 12.1 above, we no longer enter the casino because it is the only game in town or, in the alternative, demand a premium to do so. Now, the town, the countryside, and everything else is within the casino; we cannot get out because there is no place else and we are charged to be there (e.g., consumption expenditures). Attempting to meet the charges with near certain strategies becomes somewhat similar to placing simultaneous bets on black and red at roulette (while hoping green does not appear).

In trying to meet the charges from other games, we encounter the arguments of section 12.2 above. There is no guarantee of odds nor payoffs, nor that the wheel is not rigged nor the dice loaded. All of these can be changed at any time, even during the game, without our knowledge. If we seek to place the same bets as a player who has had a winning run, we have no idea if he is lucky, playing with dice loaded in his favor, or about to be given dice loaded against him nor whether the house will change the payoffs in the middle of his throw. Neoclassical financial economists solve this problem by assuming everything is known, constant, and fair. We are dubious.

12.4   Is Equity Premia Belief Sufficient to (at least) Index?

The notion here is that if one believes stocks will do better than debt in the future, even though one may have no idea about the relative performance of individual stocks, does this justify holding a large, diversified, passive portfolio such as an index fund? And the answer is: not quite. Although the belief transforms the uncertainty to something approaching risk, Grossman and Stiglitz [2] have demonstrated some problems even in a world of risk.

Suppose everyone sought to hold an index fund? If everybody is free-riding, there is no gas money and the bus does not move. There is no incentive for information production or analysis, so that prices become ever less informed. This is a world where everyone is buying his or her wealth-production of existing (and new) shares at existing prices. IPO prices could be set anywhere and the issue would sell out. If a price goes up or down for any reason, everyone is content to continue holding (because these shares are now a greater (lesser) proportion of a greater (lesser) wealth. There is no rational speculator or any other equilibrating price setting mechanism in this scenario.

The quote in the Preface to this book identified risk with the possibility of long run stock returns being terrible. In the context of this discussion, it is probably better identified with uncertainty. Whatever the label, it is a problem. And it can only be addressed by faith or belief. As we demonstrated in Chapter 5, the neoclassical efforts to "solve" it by "science" actually eliminated it from view by assumptions which are really strongly held beliefs. The point here is that, even with faith, a totally passive stock selection strategy can have problems. Somebody must mind the store (or drive the bus). What we are discussing here would also be the result if everyone invested as though he or she believe the EMH. If the price is right, one is content to hold at that price. Unfortunately, this dictum applies irrespective of the level of the price and substantial instability can result.

It might appear that the type of analysis advanced in this book would allow an individual to win at this game. We must advise that there are no guarantees here, either. If the system has no foundation in fundamental valuation, there is no intrinsic value to which it can converge (in the sense of Chapter 4). If we add to this potential instability the endogenous volatility discussed in Chapter 5, dot com bubbles become easier to explain. The scary part is the possibility that the whole market is like that (consider the Tokyo exchange experience over the last two decades).

Nor is this all. During the 2000–2002 market decline, it became known that many companies had liabilities that were tied more closely to their stock prices than had been previously realized. One example was provided by off-balance sheet entities whose obligations were guaranteed by stock of the parent, with price declines requiring more stock to be posted. Others involved debt which became due and payable if the stock price fell below a particular level. Even the more traditional debt which accelerated upon a ratings decline came into play as ratings agencies increasingly used stock price declines as signals for downgrades. To the extent stock price declines also trigger debt acceleration, volatility and instability can lead to contagion.

12.5   Does This Mean We Are Advocating Behavioral Finance?

We almost made it to the end of the book without having to address behavioral finance, which is a recent "scientific" alternative to the neoclassical fantasy. The basic idea is that real people have limited abilities to process information, may have cognitive biases, make mistakes, and so forth such that the idealized neoclassical results are not attained in real life. There is still an implied, underlying process, except that it is now imperfectly perceived and reacted to by economic agents. Notions from psychology are then employed on the market.

Just as we cannot put the market on the witness stand to testify about what "its" expectations were, we cannot put "it" on the couch either. The neoclassical argument is that if these human errors are systematic, they will be detected, exploited, and eliminated; if they are not systematic, then they simply contribute to the noise in prices. On this point, we finally have common ground with the neoclassical financial economists. Combining psychology with the neoclassical fantasy appears to raise error rather than to reduce it. Philosophically, it can clearly do nothing to establish the existence of an underlying process.

12.6 What Strategies Might Hold Promise?

A basic message of this book has been diversification across securities and time. Another has been to try to avoid overpriced stocks. The problem with this advice is that most stocks, most of the time, seem overpriced. The good news is that there are close to 10,000 widely traded stocks and, depending upon the type of portfolio, only 20-100 good ones need be found.

The overpriced stock issue is not (entirely) a value versus growth distinction. Many "value" stocks are cheap for good reason. Growth at a reasonable price is Warren Buffett's strategy, and one that we have previously endorsed. But what about numerous growth stocks where the most optimistic scenario would need to recur year after year simply to provide a fair return at the current price? It is hard to find an upside in such a situation. Such stocks may provide fine trading opportunities, but that is not the focus of this book.

Another message of this book has been that if the data are poor and/or the process is not well understood, complex modeling will only tend to make the problem worse. Even assuming there is a process, we would suggest that one or both of these problems is present in most stock analyses. This was further illustrated by the case study of Ameritape. After extensive analysis in Chapters 7, 8, and 10, we were able to demonstrate the basic conclusions in terms of the reduced form valuation models from Chapter 6. The insider secret here is that if there are enough reliable data to estimate the present value models, there is probably no need to do so. Inserting a growth rate into the reduced form models will generally work as well. And, of course, if there are not enough data for the models, why pretend to estimate the parameters characterizing the models?

Finally, the world of "stochastic certainty" (or risk) seems to be one of uniform (and high) confidence. One "knows" the true distributions. The experience of an outlier is good or bad luck, but one either places the bet again or, if necessary, performs a Bayesian adjustment, and then places the next bet. Apparent arbitrage opportunities are undertaken on a massive scale, with substantial leverage (and, thus, little cushion for continued divergence prior to the hoped for eventual convergence), for what are often small potential margins of profit.

Sooner or later (and generally sooner), some of these "big bets" go bad. The divergence may get worse, leading to margin calls which cannot be met. This appears to have been the case with Long-Term Capital Management (see further discussion below). Or the profit margins may go away while the overhead remains, which may have been at least the energy trading part of the Enron story. Or the party on the other side of the trade which was to produce the profit defaults, leaving the loss trade and/or the debt uncovered. This appears to be the story for the third parties in the off balance sheet Enron entities. The successful strategy here is: DO NOT PLUNGE (you may be diving into the shallow end of the pool).

Another problem with uniform confidence is that assumed distributions for new businesses are not as good as the track record of established businesses. This led not only to the dot com bubble but also the tech bust and all the talk about first mover advantage. In fact, especially in regard to infrastructure, the investment experience of first movers has been rather poor for 200 years. Railroads in Argentina, power in Mexico, first canals and then railroads in the United States, were all built with bonds sold to Europeans which went into default. International entities have taken up this slack in recent years. The old adage that "pioneers are those with arrows in their backs" seems to hold today as well as over the years.

Note that IBM was fairly late both to business and to personal computing. The automobile was invented decades before Ford. RCA tied the inventor of television up in court until his patents expired. In contrast to first mover advantage, an equally compelling argument (especially with ease of entry, e.g., no intellectual property) can be made to let somebody else make the mistakes and build excess capacity (which appears to be the Global Crossing story). A lot of deals which make little sense the first time at 100 cents on the dollar may become attractive the second time at five to ten cents on the dollar.

12.7 Conclusions

Persons who indulge in market purchases based on fine notions of a Brownian regularity and rapidly attained equilibria unmoderated by the modifications which are almost always necessary if we are to look at real stocks from the varying standpoints of real people and real institutions, should not be surprised when they lose their (or, more commonly, other people's) shirts.

The Long-Term Capital Management Limited Partnership (which we discussed at some length throughout this book) is a case in point. The fund was conceived and managed by John Meriweather, a former Salomon Brothers vice chairman with advice from Myron Scholes and Robert Merton, the 1997 winners of the Nobel Prize in Economics for developing the Black–Scholes options pricing model with the late Fischer Black (discussed in Chapter 11).

Andrei Schleifer [4] has provided an intriguing argument for the failure of the fund:

> Royal Dutch and Shell are independently incorporated in the Netherlands and England, respectively. In 1907, they formed an alliance agreeing to merge their interests on a 60-40 basis while remaining separate and distinct entities. All their profits, adjusting for corporate taxes and control rights, are effectively split into these proportions.

Information clarifying the linkages between the two companies is widely available. This makes for an easy prediction for the efficient markets theory: If prices are right, the market value of Royal Dutch should always equal 1.5 times the market value of Shell.

. . .

In the early 1990s, Royal Dutch traded at a 7% discount from parity, while in the late 1990s it traded at up to a 20% premium. A shrewd investor who noticed, for example, that in the summer of 1997, Royal Dutch traded at an 8% to 10% premium relative to Shell, would have sold short the expensive Royal Dutch shares and hedged his position with the cheaper Shell shares. Sadly for this investor, the deviation from the 60-40 parity only widened in 1998, reaching nearly 20% in the autumn crisis. This bet against market inefficiency lost money, and a lot of money if leveraged.

In this case, it is said that when Long Term Capital Management collapsed, during the Russian crisis, it unwound a large position in the Royal Dutch and Shell trade. Smart investors can lose a lot of money at the times when an inefficient market becomes even less efficient.

. . .

The inefficiency in the pricing of Royal Dutch and Shell is a fantastic embarrassment for the efficient markets hypothesis because the setting is the best case for that theory.

. . .

But if markets fail to achieve efficiency in this near-textbook case, what should we expect in more complicated situations, when the risks of arbitrage are greater? [1]

Because the enormous losses of the LTCM could well have caused a chain reaction leading to a market crash, Federal Reserve Chairman Allan Greenspan, in an almost unprecedented move, put together a $3.5 billion bailout. The portfolio, at the time of the bailout, was priced at two cents on the dollar.

One of Benjamin Franklin's maxims was "Experience keeps a hard school, and a fool will learn by no other." In the two decades between the publication of the Black–Scholes equation and the formation of LTCM, it might have been better if investigators had stressed the model by massive simulations, using both historical data and "what if?" scenarios. The cost would have been perhaps a few million dollars for salaries and computer time. Now, by actual utilization of real money in a real market, we have

---

[1] To paraphrase a Keynesian observation, markets can trade away from intrinsic value longer than most investors can remain solvent!

learned that the assumptions of the Black–Scholes options pricing model are flawed. But at enormous cost.

Given the power of modern computers, the problem approached by Markowitz is nowadays quite manageable. Markowitz dealt with the problem of finding maximum expected growth of a portfolio subject to an acceptable level of volatility. We recall, however, one of the maxims of John Tukey [7, p. 13]:

> Far better an approximate answer to the right question, which is often vague, than an exact answer to the wrong question, which can always be made precise.

Or, as Professor Paul Davidson (a leading post Keynesian theorist) used to say, "It's better to be roughly right than precisely wrong!" With all due respect to Markowitz's pathbreaking work in portfolio theory, we have to observe that it is indeed the mathematically neat solution to a problem selected for solution tractability rather than concordance with reality. The truth is, we will seldom be able to come up with a model in the security arena which is very close to the reality being modeled. But we can do much better than did Markowitz in approaching reality. In this book, we have included bear jumps in our model to take account of the fact that portfolios are not very good at hedging against broad market declines. Moreover, instead of showing expected yield for a given level of volatility, we have shown a broad spectrum of probabilities for various portfolio values x years out. These are bits of information worth knowing to the responsible investor or investment counsellor. Simply look at what could happen and make decisions accordingly. It is unacceptable for anyone to claim, "I picked a portfolio which maximized expected yield for the allowable volatility given me by my client. Things just did not work out as we had hoped, but we did the best we could." Wrong. The client has a poor feel for what "allowable volatility" might be (so do most of us). What he or she really wants to see is what could happen and what are the probabilities of the various possibilities happening. The simugrams in Chapters 9 and 10 do this. The Markowitz methodology does not.

Although the closed form representation of the system in Chapter 9 is not easily obtained, we get the simugram rather easily by simulation. We believe that the simugram representation of end results is greatly superior than looking at expected gains plotted versus variabilities. And, since we are unconcerned about mathematical tractability, we can revise our model as much as we like to bring our model as close to reality as our prior insights and the data will permit.

Those who believe in the Efficient Market Hypothesis frequently like to point out that managed mutual funds tend to exhibit average returns 3 percent annually below the Wilshire 5000 index. Does this imply that random selection of stocks from the overall market is better than the selections made by experts? Actually, if one looks at the performance of managed funds, overall they do about as well as random selection (this is not including their management costs). It is the management fees and the activity fees caused by extensive buying and selling of securities in the funds which brings their performance below that of an index fund. (In a way, it is a pity that mutual fund managers do not perform worse than random selection, for if they did, one could construct contrarian strategies which exhibited performance better than those of the random selection of the index strategy.)

Some EMF enthusiasts argue that the fact that managed mutual funds do, on the average (even with management and transaction fees neglected) no better than Wilshire or S&P index funds (which historically give essentially the same results) shows that the market is efficient. So quickly does the market respond to news, they say, that even honest and competent technical analysts cannot squeeze (in the long run) profitable information from all the fundamental and technical data available. So, the argument goes, it must be the case that one can do no better than buying stocks randomly at the prices available today.

The generally poor performance of managed funds relative to that of index funds is real enough. However, it does not follow that the market is efficient. Indeed, the example quoted from Schleifer above shows there are clear cases where the pricing of the market is not only not efficient but apparently may remain so over extended periods of time. The fact that something has not been done does not mean that it cannot be done.

Let us suppose, for the sake of argument, that there are being created all the time better and better procedures for obtaining returns above random selection. Let us suppose there is developed by Kryptonite Investment Services a new computer model for buying stocks in the retail food sector. Suppose that the KIS model really works. The KIS clients would do well for a time. Then, after a while, other analysts would reverse engineer the KIS product, and KIS would lose its edge. KIS would be in a situation where it would have to build new and better models all the time. Otherwise, it would be a flash in the pan with outstanding results for a brief period of time followed by mediocre (or even poor, since general knowledge of the model might thwart its use) performance. If such situations existed, then, indeed, we might have no improvement by managed funds in the aggregate even though better and better technical analysis were being developed all the time. The market would indeed be incorporating improved model information more or less continually, and, thus, would be efficient.

Thousands of person-hours are being spent daily by technical analysts to find patterns in the historical record which might have been exploited to find better returns than those obtained by an index investment. Some of these have enabled short to moderate term advantage for those who employ them. Most, however, have simply shown how difficult it is to extrapolate models from the past. At the level of fundamental analysis, the seeking of companies which have a good product to vend, makes a great deal of sense for the long term investor. Simply having a good product is not enough, however. The management of the firm with a good product is also important. Does the firm have a good plan of action to bring the product to market? Is the product so good that substitutability of another product by another firm is not a serious consideration in the short to medium term? Does the firm have high quality of production as a major consideration? Are the finances of the firm sound enough to endure short to medium term difficulties? These are very important considerations. We doubt very much whether the market efficiently incorporates such information into the day-to-day pricing of a stock. But how to capture this information usefully has been and continues to remain a serious problem for investors. And, of course, we must not neglect Keynes's "beauty contest" considerations. Whether the firm has a good product and is efficiently and effectively bringing it to market is important. But do other investors perceive that the firm is desirable? This psychological component of investing has been mastered by some, such as George Soros. But it too always breaks down sooner or later. Later, however, can be years down the road. Buying well managed firms with competitive products to sell would seem to be a reasonable consideration for any investor.

What if we had not only market efficiency but complete foreknowledge of the earnings future of every company on the exchange? Then every stock would operate like a riskless bond. Of course, we have no such foreknowledge. However, if there were no general bear jumps downward, a big diversified portfolio would act somewhat like a riskless bond. But there are bear jumps downward. Operating, as with classical Markowitz theory, as though everything were modelable as a geometric Brownian walk, without allowances for sudden broad downturns, is not a very good idea. The best strategy we can offer is the building of portfolios in the light of models of returns based on the best data and insights available. Perhaps the display of the stochastic profile of anticipated performance of the portfolio, as given in this book via the simugram, is about the best we can do. If investors focus, not on the expected value of the return and the associated volatility, but rather on the stochastic profile of what might happen, then that would be a big step forward.

The whole area of modeling markets is an exciting one. Anyone investing or anyone advising an investor may make ruinous mistakes if they do not ask questions such as "what happens if ...". Such questions, typically, are not easy to answer if we demand closed-form solutions. Simulation enables us to ask the questions we need to ask rather than restricting ourselves to looking at possibly irrelevant questions for which we have formulas at hand.

Assignment

Problem 12.1. Go forth and seek thy treasure!

References

[1] Fama, E. and French, K. (2002). "The equity premium," *Journal of Finance*, April, 637–659.

[2] Grossman, S.J. and Stiglitz, J. (1980). The impossibility of informationally efficient markets," *American Economic Review*, June, 393–408.

[3] Markowitz, H. (1959). *Portfolio Selection*. New York: John Wiley & Sons.

[4] Shleifer, A (2000). "Arbitrage is inherently risky," *The Wall Street Journal*, December 28, A10.

[5] Siconolfi, M., Raghavan, A., and Pacelle, M. (1998). "All bets are off: How the salesmanship and brainpower failed at Long-Term Capital," *The Wall Street Journal*, November 16, A1, A18–A19.

[6] Thompson, J. R. (1999). *Simulation: A Modeler's Approach*. New York: John Wiley & Sons.

[7] Tukey, J. W. (1962),"The future of data analysis," *Annals of Mathematical Statistics*, Vol. 33, 1–67.

Appendix A

A Brief Introduction to Probability and Statistics

A.1   Craps: An Intuitive Introduction to Probability

In this game, played with great gusto by millions, the player throws two six-sided dice. We shall assume that one of these dice is white and the other is black. If, in the first throw, the player throws a seven (W1B6, W2B5, W3B4, W4B3, W5B2, W6B1) or an eleven (W5B6, W6B5), he wins the game. We note that there are 36 possible results of the throw: (W1B1, W1B2, W1B3, W1B4, W1B5, W1B6; W2B1, W2B2, W2B3, W2B4, W2B5, W2B6;W3B1, W3B2, W3B3, W3B4, W3B5, W3B6; W4B1, W4B2, W4B3, W4B4, W4B5, W4B6; W5B1, W5B2, W5B3, W5B4, W5B5, W5B6; W6B1, W6B2, W6B3, W6B4, W6B5, W6B6).

This collection would be looked upon as the *sample space S*. Clearly

$$P(S) = 1.$$

We have to get one of the 36 results. Intuitively, each of these 36 results has the same chance of occurring: 1/36. These 36 elements represent the basic primitive events of the probability space.

We now give an example of a *random variable*. For a toss of the dice:

Let X = number of white pips + number of black pips.

A.1.1   Random Variables, Their Means and Variances

What is the probability of winning on the first throw? We need to find $P(7 \text{ or } 11)$. How shall do this? We look at the primitive elements which map under the random variable to 7 or 11. These are W1B6, W2B5, W3B4,

W4B3, W5B2, W6B1, W5B6, W6B5. Now, we know the probability of each of these primitive events: each has probability $1/36$. So, then the concept of a *random variable* is a mapping from the space of primitive events to some other space (here to the integers 2, 3, 4, 5, 6, 7, 8, 9, 10, 11, 12) in such a way that the inverse map gets us to the primitive events on which the probability is naturally defined. A random variable has the property that the probability that the random variable is equal to a particular value can be computed from the space of original events:

$$P(7 \text{ or } 11 \text{ on first toss}) =$$
$$P \text{ (W1B6, W2B5, W3B4, W4B3, W5B2, W6B1, W5B6, W6B5)} =$$
$$P(\text{W1B6})+ P(\text{W2B5})+ P(\text{W3B4})+ P(\text{W4B3})+ P(\text{W5B2})+ P(\text{W6B1})$$
$$+P(\text{W5B6})+P(\text{W6B5}) = \frac{8}{36} = .222222.$$

By simply looking at the primitive events that map into 2,3,4,5,6,7,8,9,10,11,1 we note that

$$P(2) = P(12) = \frac{1}{36}$$
$$P(3) = P(11) = \frac{2}{36}$$
$$P(4) = P(10) = \frac{3}{36}$$
$$P(5) = P(9) = \frac{4}{36}$$
$$P(6) = P(8) = \frac{5}{36}$$
$$P(7) = \frac{6}{36}.$$

The *expected value* of a random variable X is its average value. Here

$$\mu = E(X) = \sum_{x=2}^{x=12} XP(X) = 2\frac{1}{36} + 3\frac{2}{36} + 4\frac{3}{36} + 5\frac{4}{36} + \ldots$$
$$= 7. \tag{A.1}$$

The *variance* of a random variable is given by

$$\sigma^2 = E(X-\mu)^2 = E(X^2)-2\mu E(X)+\mu^2 = E(X^2)-\mu^2 = 54.833-7^2 = 5.833. \tag{A.2}$$

Now, the rules of the game of craps tell us that the player loses if he gets a 2, 3, or 12 on the first throw. What is the probability of this?

$$\text{Prob (2, 3, or 12 on first throw)} = P(\text{W1B1, W1B2, W2B1, W6B6})$$
$$= \frac{1}{36} + \frac{1}{36} + \frac{1}{36} + \frac{1}{36}$$
$$= \frac{4}{36}.$$

Now, the player may throw some number other than 2, 3, 12, 7 or 11. Suppose the number is 4. The probability of this is

$$P(4) = P(W1B3) + P(W2B2) + P(W3B1) = \frac{3}{36}.$$

The rule is that the player wins if he throws a second 4 before throwing a 7. We have seen already that the probability of a 7 is 6/36. So, the probability of getting a 4 before a 7 is

$$\frac{3}{3+6} = \frac{1}{3}.$$

The probability of the player winning the game by throwing a 4 on the first round and then, on subsequent throws, getting a 4 before rolling a 7 is:

$$\frac{3}{36} \times \frac{1}{3} = \frac{1}{36} = .027778.$$

Another way the player could win is to throw a 10 on the first round and then throw a second 10 before throwing a 7. Now the probability of throwing a 10 is

$$P(10) = P(W4B6, W5B5, W6B4) = \frac{3}{36},$$

the same as the probability of throwing a 4. Thus, the probability of winning by throwing first a 10 and then throwing another 10 before throwing a 7 is

$$\frac{3}{36} \times \frac{1}{3} = \frac{1}{36} = .027778.$$

We then note that the probability of getting a 5 on the first toss is

$$P(5) = P(W1B4, W2B3, W3B2, W4B1) = \frac{4}{36}.$$

So, the probability of winning by throwing first a 5 and then throwing another 5 before getting a 7 is

$$\frac{4}{36} \times \frac{4}{10} = \frac{4}{90} = .0444444.$$

By symmetry, we see that this is the same probability as throwing first a 9 and then getting a second 9 before throwing a 7.

Finally, the probability of throwing first a 6 is

$$P(5) = P(W1B5, W2B4, W3B3, W4B2, W5B1) = \frac{5}{36}.$$

Then, the probability of winning by throwing first a 6 and then a second 6 before throwing a 7 is

$$\frac{5}{36} \times \frac{5}{11} = .0631313.$$

By symmetry, this is the same as the probability of throwing first an 8 and then throwing another 8 before throwing a 7.

In summary, the probability of winning the game of craps is given by

$$P(\text{Winning}) = P(7 \text{ or } 11 \text{ on first toss})$$
$$+2 \times [P(4 \text{ on first toss})P(4 \text{ before } 7) + P(5 \text{ on first toss})P(5 \text{ before } 7)$$
$$+P(6 \text{ on first toss})P(6 \text{ before } 7)] = .492928.$$

Craps is interesting in that it can be used to capture at the level of intuition the key concepts of probability theory. We notice, for example, that all the probabilities in the game are actually generated from the 36 elementary events. Each of these has probability 1/36. But the game itself pays off on the basis of the sum total of the two dice (without regard to their color). The sum total of the two faces is a random variable. For any value of the random variable, we can find the elementary events which form the basis for the necessary computation.

Here, as we have mentioned, there are only eleven throws of interest to us (2,3,4,5,6,7,8,9,10,11,12). Suppose we hear that someone has won at a game of craps, just that he has won. We then want to compute the probability that the person won on the first round. This could only have been done if he had thrown a 7 or an 11. That probability we know is 8/36=.222222. We have, however, the additional information that the player has won. Common sense might lead us to the following formula

$$P(\text{win}) \, P(\text{win on first round}|\text{win}) \;=\; P(\text{win and win on first round})$$
$$=\; P(\text{win} \cap \text{win first round}). \;\; (A.3)$$

Here $P(\text{win on first round}|$ win) is termed the "*conditional probability* that he won on the first toss given that he won at all*." We can then solve for this conditional probability by using a bit of algebra:

$$P(\text{win on first round}| \text{ win}) = \frac{P(\text{win and win on first round})}{P(\text{win})}. \;\; (A.4)$$

Now, here, we note that the event that he won on the first round implies that he won at all. Hence the solution to (A.2) is given by

$$P(\text{win on first round}| \text{ win}) = \frac{.222222}{.492928} = .450816. \;\; (A.5)$$

This discussion should be used as a template any time one needs reminding what a random variable is and how we compute the probability that a random variable has a particular set of values.

Let us get more abstract and ask how many sets could we construct from the 36 primitive elements (without replacement)? Let us construct one such set. We can put any number of primitive elements into the set. The first element W1B1 can either be in our set or out of it. That means two choices for inclusion of W1B1. And the same choice is available as to whether to include W1B2 or not. Aggregating across each of the primitive elements, we find that the number of sets is

$$2 \times 2 \times \ldots \times 2 = 2^{36} = 6,871,947,674.$$

We can easily compute the probability P of each of these sets by looking at the primitive events included in each of them and adding up their probabilities (each equal to $1/36$). [1]

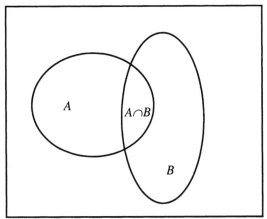

Figure A.1. Sets and Their Intersections.

---

[1]We can do more and take arbitrarily many unions of these 6,871,947,674 sets and intersections, and complements and so forth and so on *ad infinitum*. The class of sets so obtained is called the *sigma field* generated by the primitive events. Now, as noted, we can find the probability of each of these sets by simply looking to see which primitive events are included.

For example, the case where none of the elementary sets is included is called *the empty set* or ∅. We have to get one of the elementary events; something must happen; so the probability that nothing happens is zero. Thus,

$$P(\emptyset) = 0.$$

Now the entire sample space S of all the elementary events (36 of them here) must have probability one, for one of the elementary tosses must happen:

$$P(S) = 1.$$

Now, in the set of real numbers, we can consider the connected intervals as the basic building blocks of a sigma field. And then we can ad infinitum look at unions, intersections, and complements for the real numbers. The resulting sigma field is called the *Borel field*. In order for us to compute the probability of a set B of real numbers, we need to be able to assure ourselves that the inverse $X^{-1}(B)$ is a member of the sigma field in our primitive probability space (where we know what the probabilities are). In such a case,

$$P'(B) = P(X^{-1}(B)).$$

So, a random variable is a real variable such that the inverse image of the Borel sets is a member of the sigma field in the primitive probability space. Thus, the probability measure P and the random variable X will induce a probability measure P' on the Borel sets.

As we see from Figure A.1,

$$P(A \cup B) = P(A) + P(B) - P(A \cap B). \tag{A.6}$$

Suppose that we have two sets A and B which have the property that

$$P(A \text{ and } B) = P(A \cap B) = P(A) \times P(B). \tag{A.7}$$

Then, we say that A and B are *stochastically independent* under the probability measure P.

If A and B are *disjoint* (have no points in common), then we can write

$$P(A \text{ and/or } B) = P(A \cup B) = P(A) + P(B). \tag{A.8}$$

For any set A on which a probability measure is defined, we must have

$$0 \leq P(A) \leq 1. \tag{A.9}$$

Let us return briefly to the game of craps. Suppose we have a second random variable, Y, which is equal to 1 if the sum of the two dice is odd and 2 if it is even. Let us view the two random variables in the light of the eleven possible outcomes of X and their probabilities

Table A.1. The Game of Craps.				
Primitive Events	X	Y	XY	$P(X,Y)$
W1B1	2	2	4	1/36
W1B2,W2B1	3	1	3	2/36
W1B3,W2B2,W3B1	4	2	8	3/36
W1B4,W2B3,W3B2,W4B1	5	1	5	4/36
W1B5,W2B4,W3B3,W4B2,W5B1	6	2	12	5/36
W1B6,W2B5,W3B4,W4B3,W5B2,W6B1	7	1	7	6/36
W2B6,W3B5,W4B4,W5B3,W6B2	8	2	16	5/36
W3B6,W4B5,W5B4,W6B3	9	1	9	4/36
W4B6,W5B5,W6B4	10	2	20	3/36
W5B6,W6B5	11	1	11	2/36
W6B6	12	2	24	1/36

Firstly, we can easily compute the mean and variance of Y.

$$
\begin{aligned}
E(Y) &= 2 \times 1/36 + 3 \times 2/36 + \ldots = 1.5 \\
E(Y^2) &= 2^2 \times 1/36 + 3^2 \times 2/36 + \ldots = 2.5 \\
Var(Y) &= E(Y^2) - [E(Y)]^2 = 2.5 - 1.5^2 = .25.
\end{aligned}
$$

Generally, we may try to find a simple measure of the apparent interaction between two variables, X and Y. One such is the covariance of X and Y

$$Cov(X,Y) = E[(X - \mu_X)(Y - \mu_Y)]. \tag{A.10}$$

A more popular measure of apparent interaction is the *correlation* between X and Y,

$$\rho(X,Y) = \frac{Cov(X,Y)}{\sqrt{\sigma_X^2 \sigma_Y^2}}. \qquad (A.11)$$

Now, although covariances can take values from $-\infty$ to $+\infty$, the correlation can only take values between -1 and $+1$. To prove this fact, we note that:

$$0 \leq E[a(X - \mu_X) - (Y - \mu_Y)]^2 = a^2\sigma_X^2 + \sigma_Y^2 - 2aCov(X,Y), \quad (A.12)$$

where a is an arbitrary real constant which we elect to be $Cov(X,Y)/\sigma_X^2$. This gives us immediately a version of *Cauchy's Inequality*,

$$\rho^2 \leq 1. \qquad (A.13)$$

Now, the reader should verify that if Y is simply a positive multiple of X, then $\rho = 1$. If Y is a negative multiple of X, then $\rho = -1$. We are interested, in portfolio design, in looking at the correlation between two securities. To the extent that this correlation is close to 1, the diversification benefit of including both stocks in the portfolio is marginal.

A.2 Combinatorics Basics

Let us first compute the number of ways that we can arrange in a distinctive order k objects selected without replacement from n, $n \geq k$, distinct objects. We easily see that there are n ways of selecting the first object, $n - 1$ ways of selecting the second object, and so on until we select $k - 1$ objects and note that the kth object can be selected in $n - k + 1$ ways. The total number of ways is called the *permutation* of n objects taken k at a time, $P(n,k)$, and is seen to be given by

$$P(n,k) = n(n-1)(n-2)\cdots(n-k+1) = \frac{n!}{(n-k)!}, \qquad (A.14)$$

where $m! = m(m-1)(m-2)\cdots2 \times 1$, $0! = 1$. In particular, there are $n!$ ways that we can arrange n objects in a distinctive order. Next, let us compute in how many ways we can select k objects from n objects when we are not concerned with the distinctive order of selection. This number of ways is called the *combination* of n objects taken k at a time, and is denoted by $C(n,k)$. We can find it by noting that $P(n,k)$ could be first computed by finding $C(n,k)$ and then multiplying it by the number of ways k objects could be distinctly arranged (i.e., $k!$). So we have

$$P(n,k) = C(n,k)P(k,k) = C(n,k)k!$$

and thus

$$\binom{n}{k} = C(n,k) = \frac{n!}{(n-k)!k!}. \qquad (A.15)$$

For example, the game of stud poker consists in the drawing of 5 cards from a 52 card deck (4 suits, 13 denominations). The number of possible hands is given by

$$C(52,5) = \frac{52!}{47!5!} = 2,598,960.$$

We are now in a position to compute some basic probabilities which are slightly harder to obtain than, say, those concerning tossing a die. Each of the 2,598,960 possible poker hands is equally likely. To compute the probability of a particular hand, we simply evaluate

$$P(\text{hand}) = \frac{\text{number of ways of getting the hand}}{\text{number of possible hands}}.$$

Suppose we wish to find the probability of cards all of the same suit. There are $C(4,1)$ of choosing a suit. Then there are $C(13,5)$ ways of selecting 5 cards (without regard to their order) out of the 13 cards of that suit. Hence,

$$P(\text{a flush hand}) = \frac{C(4,1)C(13,5)}{C(52,5)}$$

$$= \frac{4 \times (9)(10)(11)(12)(13)}{(5!)(2,598,960)} = .0019808.$$

Finding the probability of getting four cards of a kind (e.g., four aces, four kings) is a bit more complicated. There are $C(13,1)$ ways of picking a denomination, $C(4,4)$ ways of selecting all the four cards of the same denomination, and $C(48,1)$ of selecting the remaining card. Thus,

$$P(\text{four of a kind}) = \frac{C(13,1)C(4,4)C(48,1)}{C(52,5)}$$

$$= \frac{(13)(1)(48)}{2,598,960} = .00024.$$

Similarly, to find the probability of getting two pairs, we have

$$P(\text{two pairs}) = \frac{C(13,2)C(4,2)C(4,2)C(44,1)}{C(52,5)}$$

$$= \frac{(78)(6)(6)(44)}{2,598,960} = .0475.$$

A.3   Bayesian Statistics

A.3.1   Bayes' Theorem

Suppose that the sample space S can be written as the union of disjoint sets: $S = A_1 \cup A_2 \cup \cdots \cup A_n$. Let the event H be a subset of S which has non-empty intersections with some of the A_i's. Then

$$P(A_i|H) = \frac{P(H|A_i)P(A_i)}{P(H|A_1)P(A_1) + P(H|A_2)P(A_2) + \cdots + P(H|A_n)P(A_n)}.$$
(A.16)

To explain the conditional probability given by equation (A.16), consider a diagram of the sample space, S. Consider that the A_i's represent n disjoint states of nature. The event H intersects some of the A_i's.

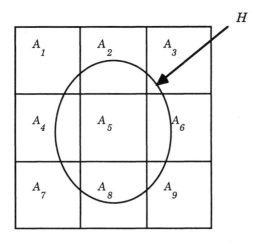

Figure A.2. Bayes Venn Diagram.

Then,

$$P(H|A_1) = \frac{P(H \cap A_1)}{P(A_1)} = \frac{P(A_1|H)P(H)}{P(A_1)}.$$

Solving for $P(A_1|H)$, we get

$$P(A_1|H) = \frac{P(H|A_1)P(A_1)}{P(H)},$$

and in general,

$$P(A_i|H) = \frac{P(H|A_i)P(A_i)}{P(H)}.$$
(A.17)

Now,

$$
\begin{aligned}
P(H) &= P((H \cap A_1) \cup (H \cap A_2) \cup \cdot \cup (H \cap A_n)) \\
&= \sum P(H \cap A_i), \text{ since the intersections } (H \cap A_i) \text{ are disjoint} \\
&= \sum P(H|A_i)P(A_i), \text{ where } j = 1, 2, \ldots, n.
\end{aligned}
$$

Thus, with (A.17) and $P(H)$ given as above, we get (A.16).

The formula (A.16) finds the probability that the true state of nature is A_i given that H is observed. Notice that the probabilities $P(A_i)$ must be known to find $P(A_i|H)$. These probabilities are called *prior* probabilities because they represent information prior to experimental data. The $P(A_i|H)$ are then *posterior* probabilities. For each $i = 1, 2, \ldots, n$, $P(A_i|H)$ is the probability that A_i was the state of nature in light of the occurrence of the event H.

A.3.2   A Diagnostic Example

Consider patients being tested for a particular disease. It is known from historical data that 5% of the patients tested have the disease, further, that 10% of the patients that have the disease test negative for the disease, and that 20% of the patients who do not have the disease test positive for the disease. Denote by D^+ the event that the patient has the disease, by D^- the event that the patient does not, and denote by T^+ the event the patient tests positive for the disease, and by T^- the event the patient tests negative.

If a patient tests positive for the disease, what is the probability that the patient actually has the disease? We seek the conditional probability, $P(D^+|T^+)$. Here, T^+ is the observed event, and D^+ may be the true state of nature that exists prior to the test. (We trust that the test does not cause the disease.) Using Bayes' theorem,

$$
\begin{aligned}
P(D^+|T^+) &= \frac{P(T^+|D^+)P(D^+)}{P(T^+)} \qquad\qquad\qquad \text{(A.18)} \\[2mm]
&= \frac{P(T^+|D^+)P(D^+)}{P(T^+|D^+)P(D^+) + P(T^+|D^-)P(D^-)} \\[2mm]
&= \frac{.9 \times .05}{.9 \times .05 + .2 \times .95} = 0.1915.
\end{aligned}
$$

Thus, there is nearly a 20% chance given a positive test result that the patient has the disease. This probability is the posterior probability, and if the patient is tested again, we can use it as the new prior probability. If the patient tests positive once more, we use equation (A.18) with an updated version of $P(D^+)$, namely, .1915.
The posterior probability now is:

$$
\begin{aligned}
P(D^+|T^+) &= \frac{P(T^+|D^+)P(D^+)}{P(T^+)} \\[2mm]
&= \frac{P(T^+|D^+)P(D^+)}{P(T^+|D^+)P(D^+) + P(T^+|D^-)P(D^-)} \\[2mm]
&= \frac{.9 \times .1915}{.9 \times .1915 + .2 \times .8085} = 0.5159.
\end{aligned}
$$

Twice the patient tests positive for the disease and the posterior probability that the patient has the disease is now much higher. As we gather more and more information with further tests, our posterior probabilities will better and better describe the true state of nature.

In order to find the posterior probabilities as we have done, we needed to know the prior probabilities. A major concern in a Bayes application is the choice of priors, a choice which must be made sometimes with very little prior information. One suggestion made by Bayes is to assume that the n states of nature are equally likely (Bayes' Axiom). If we make this assumption in the example above, that is, that $P(D^+) = P(D^-) = .5$, then

$$P(D^+|T^+) = \frac{P(T^+|D^+)P(D^+)}{P(T^+|D^+)P(D^+) + P(T^+|D^-)P(D^-)}$$

$P(D^+)$ and $P(D^-)$ cancel, giving

$$P(D^+|T^+) = \frac{P(T^+|D^+)}{P(T^+|D^+) + P(T^+|D^-)}$$
$$= \frac{.9}{.9 + .2}$$
$$= 0.8182.$$

This is much higher than the accurate probability, .1912. Depending upon the type of decisions an analyst has to make, a discrepancy of this magnitude may be very serious indeed. As more information is obtained, however, the effect of the initial choice of priors will become less severe.

A.4   The Binomial Distribution

Let us suppose we are selecting from a very large—effectively infinite—population of black and white balls. Suppose the probability a ball is black is p and that the probability a ball is white is $q = 1 - p$. Suppose that out of n draws, the first x are black and the next $n - x$ are white. Suppose that out of n draws we get x black balls and $n - x$ white balls. The probability is given by

$$\underbrace{pp \cdots p}_{x} \underbrace{(1-p)(1-p) \cdots (1-p)}_{n-x \text{ times}} = p^x (1-p)^{n-x}.$$

But suppose that we are not interested in the order in which the black balls appear, just their total number x out of n draws. Then we have the *binomial probability function*

$$P(X = x) = \binom{n}{x} p^x (1-p)^{n-x}, \quad x = 0, 1, 2, \ldots, n. \tag{A.19}$$

The binomial distribution may be viewed as the sum of n independent *Bernoulli variables*. A Bernoulli variable, Y takes the value 1 with probability p and the value 0 with probability $q = 1 - p$. Thus, the expected value of $Y = p \times 1 + q \times 0 = p$. Similarly, the expected value of $Y^2 = p \times 1^2$ $= p$. Then the variance of $Y = E(Y^2) - [E(Y)]^2 = p - p^2 = p(1-p)$.

Returning to the binomial variable x, we have

$$X = y_1 + y_2 + \ldots + y_n. \tag{A.20}$$

So, for the binomial distribution,

$$\begin{aligned} E(X) &= E(y_1) + E(y_2) + \ldots + E(y_n) \\ &= p + p + \ldots + p \\ &= np. \tag{A.21} \end{aligned}$$

$$E(X^2) = E(y_1^2 + y_2^2 + \ldots + y_n^2 + n^2 - n \text{ terms like } y_i y_j \text{ where } i \neq j). \tag{A.22}$$

Thus

$$E(X^2) = np + n(n-1)p^2. \tag{A.23}$$

And

$$Var(X) = E(X^2) - [E(X)]^2 = np + n(n-1)p^2 - (np)^2 = np(1-p). \tag{A.24}$$

We now bring in a concept which is used extensively throughout this book, that of the *cumulative probability distribution function*.

$$F(x) = P(X \leq x). \tag{A.25}$$

For the binomial distribution, we have

$$F(x) = \sum_{j=0}^{j \leq x} \binom{n}{j} p^j (1-p)^{n-j}, \quad j = 0, 1, 2, \ldots, n. \tag{A.26}$$

We note that the binomial distribution is discrete. $F(x)$ is described by step functions. We show the (cumulative) distribution function of a binomial variate when $p = .7$ and $n = 3$ in Figure A.3.

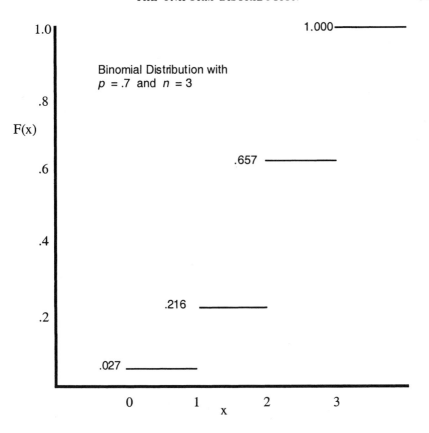

Figure A.3. CDF of the Binomial Distribution.

A.5   The Uniform Distribution

We now look at a random variable X which is characterized by its cdf

$$
\begin{aligned}
F(x) \;&=\; 0, \text{ if } x < 0 \\
&=\; x, \text{ if } 0 \le x \le 1 \\
&=\; 1, \text{ if } x > 1 \,.
\end{aligned}
\qquad (A.27)
$$

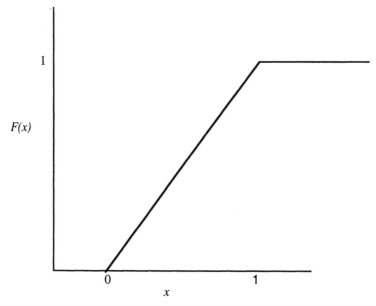

Figure A.4. CDF of the Uniform Distribution.

The uniform distribution has no jumps in the cdf. A uniform random variable is, therefore, an example of a *continuous random variable*. For a continuous random variable, we may desire to look at its derivative, $f(x) = dF(x)/dx$. $f(x)$ is called the *probability density functions* of the random variable X. In the case of the uniform distribution, we note that

$$f(x) = \frac{dF(x)}{dx} \quad = \quad 1 \text{ for } 0 \leq x \leq 1;$$
$$= \quad 0 \text{ otherwise .}$$

For simulation purposes, the uniform distribution is of particular importance. Suppose that we have another continuous random variable X with cdf $F(.)$. We will now consider F as itself a random variable, $Y = F(x)$. Its cdf $G(y)$ is easily found by the following argument:

$$G(y) = P(Y \leq y) = P(F(x) \leq y) = P(x \leq F^{-1}(y)) = y. \qquad \text{(A.28)}$$

The consequence is that if we know the cdf $F(x)$, then we may obtain a simulated value from the distribution by finding u randomly distributed on the unit interval. Then

$$x = F^{-1}(u). \qquad \text{(A.29)}$$

Practically every numerical computer compiler has a uniform random number generator. So (A.29) will generally yield an easy way for us to generate a simulated observation of the random variable with cdf F.

A.6   Moment–Generating Functions

We now consider the joint density function of n random independent and identically distributed random variables of a continuous random variable X having density function $f(.)$. Then a natural definition of the joint density of $(x_1, x_2, \ldots, x_n)$ is

$$
\begin{aligned}
f(x_1, x_2, \ldots, x_n) &= \\
&= \lim_{\epsilon_1, \epsilon_2, \ldots, \to 0} \frac{P[x_1 < X_1 < x_1 + \epsilon_1]}{\epsilon_1} \frac{P[x_2 < X_2 < x_2 + \epsilon_2]}{\epsilon_2} \cdots \\
&= \lim_{\epsilon_1, \epsilon_2, \ldots, \to 0} \frac{f(x_1)\epsilon_1}{\epsilon_1} \frac{f(x_2)\epsilon_2}{\epsilon_2} \cdots \frac{f(x_n)\epsilon_n}{\epsilon_n} \\
&= f(x_1)f(x_2)\ldots f(x_n).
\end{aligned}
\tag{A.30}
$$

The late Salomon Bochner once mentioned the rather modest result in (A.30) as being R.A. Fisher's greatest contribution to statistics. Note that it enables us to write the density of an n–dimensional random variable as the product of n one-dimensional densities.

Next, let X be a random variable with cumulative distribution function $F(x)$. The *moment-generating function $M_X(t)$* via

$$
M_X(t) = E(e^{tX}),
\tag{A.31}
$$

where t is an arbitrary real variable.

Assuming that differentiation with respect to t commutes with expectation operator E, we have

$$
\begin{aligned}
M'_X(t) &= E(Xe^{tX}) \\
M''_X(t) &= E(X^2 e^{tX})
\end{aligned}
$$

$$
\vdots
$$

$$
M_X^{(k)}(t) = E(X^{(k)} e^{tX}).
$$

Setting t equal to zero, we see that

$$
M_X^{(k)}(0) = E(X^k).
\tag{A.32}
$$

Thus, we see immediately the reason for the name moment-generating function (m.g.f.). Once we have obtained $M_X(t)$, we can compute moments of arbitrary order (assuming they exist) by successively differentiating the m.g.f. and setting the argument t equal to zero. As an example of this application, let us consider an r.v. distributed according to the binomial distribution with parameters n and p. Then,

$$
\begin{aligned}
M_X(t) &= \sum_0^n e^{tx} \binom{n}{x} p^x (1-p)^{n-x} \\
&= \sum_0^n \binom{n}{x} (pe^t)^x (1-p)^{n-x}.
\end{aligned}
$$

Now recalling the binomial identity

$$\sum_{0}^{n} \binom{n}{x} a^x b^{n-x} = (a+b)^n,$$

we have

$$M_X(t) = [pe^t + (1-p)]^n. \tag{A.33}$$

Next, differentiating with respect to t, we have

$$M_X'(t) = npe^t[pe^t + (1-p)]^{n-1}. \tag{A.34}$$

Then, setting t equal to zero, we have

$$E(X) = M_X'(0) = np. \tag{A.35}$$

Differentiating again with respect to t and setting t equal to zero, we have

$$E(X^2) = M_X''(0) = np + n(n-1)p^2. \tag{A.36}$$

In order to calculate the variance, it suffices to recall that for any r.v. X we have

$$\text{Var}(X) = E(X^2) - [E(X)]^2. \tag{A.37}$$

Thus, for the binomial X

$$\text{Var}(X) = np(1-p). \tag{A.38}$$

Of course we have already found the mean and variance of the binomial distribution via (A.20)−(A.24). Generally speaking the moment generating function is an easier way to compute moments than the direct approach. However, we shall shortly see an even more important use of the moment generating function.

A.7   The Normal (Gaussian) Distribution

Consider the normal density function

$$f(x) = \frac{1}{\sqrt{2\pi\sigma^2}}\exp\left(-\frac{1}{2\sigma^2}(x-\mu)^2\right), \quad -\infty < x < \infty. \tag{A.39}$$

We would like to satisfy ourselves that we have a true density function. We note, first of all, that $f(x) > 0$ for all $-\infty < x < \infty$. Next, we need to show that $F(\infty) = 1$, that is, that

$$\int_{-\infty}^{\infty} \frac{1}{\sqrt{2\pi\sigma^2}}\exp\left(-\frac{1}{2\sigma^2}(x-\mu)^2\right) dx = 1. \tag{A.40}$$

Let us make the transformation

$$z = \frac{x - \mu}{\sigma}. \tag{A.41}$$

The left-hand side of (A.40) becomes

$$\int_{-\infty}^{\infty} \frac{1}{\sqrt{2\pi}} e^{-z^2/2} dz = A. \tag{A.42}$$

Clearly, A is non-negative. Hence it will suffice to show that $A^2 = 1$. Now

$$A^2 = \frac{1}{2\pi} \int_{-\infty}^{\infty} \int_{-\infty}^{\infty} \exp[-\frac{1}{2}(z^2 + w^2)] dz dw. \tag{A.43}$$

Let us transform to polar coordinates, with

$$r^2 = z^2 + w^2; \tan(\theta) = w/z.$$

Thus,

$$A^2 = \frac{1}{2\pi} \int_0^{\infty} \int_0^{2\pi} e^{-r^2/2} r dr d\theta = \frac{1}{2\pi} 2\pi \int_0^{\infty} e^{-r^2/2} r dr = 1. \tag{A.44}$$

The moment generating function of a normal random variable can be found via

$$
\begin{aligned}
M_X(t) &= \frac{1}{\sqrt{2\pi}\sigma} \int_{-\infty}^{\infty} e^{tx} \exp\left(-\frac{1}{2\sigma^2}(x - \mu)^2\right) dx \\
&= \frac{1}{\sqrt{2\pi}\sigma} \int_{-\infty}^{\infty} \exp\left(-\frac{1}{2\sigma^2}(x^2 - 2\mu x - 2\sigma^2 tx + \mu^2)\right) dx \\
&= \frac{1}{\sqrt{2\pi}\sigma} \int_{-\infty}^{\infty} \exp\left(-\frac{1}{2\sigma^2}(x^2 - 2x(\mu + t\sigma^2) + \mu^2)\right) dx \\
&= \frac{1}{\sqrt{2\pi}\sigma} \int_{-\infty}^{\infty} \exp\left(-\frac{1}{2\sigma^2}(x - \mu^*)^2\right) dx \exp\left(t\mu + \frac{t^2\sigma^2}{2}\right),
\end{aligned}
$$

where $\mu^* = \mu + t\sigma^2$. But recognizing that the integral is simply equal to $\sqrt{2\pi}\sigma$, we see that the m.g.f. of the normal distribution is given by

$$M_X(t) = \exp\left(t\mu + \frac{t^2\sigma^2}{2}\right). \tag{A.45}$$

By evaluating the first two derivatives of $M_X(t)$ at 0, the reader may now easily verify that the mean and variance of the normal distribution are μ and σ^2, respectively. [2]

---

[2] A related distribution, of particular interest to persons involved with market models, is the *lognormal distribution*. Suppose that we have a random variable X such that its logarithm is normally distributed with mean μ and variance σ. Then we say that X has the lognormal distribution with density:

$$f(x) = \frac{1}{\sqrt{2\pi}\sigma x} \exp\left(-\frac{1}{2\sigma^2}(\ln x - \mu)^2\right) \quad \text{for } x > 0.$$

The possible mechanical advantages of the m.g.f. are clear. One integration (summation) operation plus k differentiations yield the first k moments of a random variable. However, the moment-generating aspect of the m.g.f. pales in importance to some of its properties relating to the summation of independent random variables. Let us suppose, for example, that we have n independently distributed r.v.'s $X_1, X_2, \ldots, X_n$ with m.g.f.'s $M_1, M_2, \ldots, M_n$, respectively. Suppose that we wish to investigate the distribution of the r.v.

$$Y = c_1 X_1 + c_2 X_2 + \cdots + c_n X_n,$$

where $c_1, c_2, \ldots, c_n$ are fixed constants. Let us consider using the moment-generating functions to achieve this task. We have

$$M_Y(t) = E[\exp t(c_1 X_1 + c_2 X_2 + \cdots + c_n X_n)].$$

Using the independence of $X_1, X_2, \ldots, X_n$ we may write

$$
\begin{aligned}
M_Y(t) &= E[\exp t c_1 X_1] E[\exp t c_2 X_2] \cdots E[\exp t c_n X_n] \\
&= M_1(c_1 t) M_2(c_2 t) \cdots M_n(c_n t).
\end{aligned}
\tag{A.46}
$$

Given the density (or probability) function, we know what the m.g.f. will be. But it turns out that, under very general conditions, the same is true in the reverse direction; namely, if we know $M_X(t)$, we can compute a unique density (probability) function that corresponds to it. The practical implication is that if we find a random variable with an m.g.f. we recognize as corresponding to a particular density (probability) function, we know immediately that the random variable has the corresponding density (probability) function. Thus, in many cases, we are able to use (A.46) to give ourselves immediately the distribution of Y. Consider, for example, the sum

$$Y = X_1 + X_2 + \cdots + X_n$$

of n independent binomially distributed r.v.'s with the same probability of success p and the other parameter being equal to $n_1, n_2, \ldots, n_n$, respectively. Thus, the moment-generating function for Y is

$$
\begin{aligned}
M_Y(t) &= [pe^t + (1-p)]^{n_1} [pe^t + (1-p)]^{n_2} \cdots [pe^t + (1-p)]^{n_n} \\
&= [pe^t + (1-p)]^{n_1 + n_2 + \cdots + n_n}.
\end{aligned}
$$

We note that, not unexpectedly, this is the m.g.f. of a binomial r.v. with parameters $N = n_1 + n_2 + \cdots + n_n$ and p.

Next, we note that the moment generating function for

$$Z = c_1 X_1 + c_2 X_2 + \ldots c_n X_n,$$

where the X_j are independent normal variables with parameters μ_j and σ_j^2, is simply

$$M_Z(t) = \exp\left(t \sum_{i=1}^{n} c_i \mu_i + \frac{t^2}{2} \sum_{i=1}^{n} (c_i \sigma)^2 \right).
\tag{A.47}$$

From (A.45), we recognize that Z must be a normal random variable with mean $\sum_{i=1}^{n} c_i \mu_i$ and variance $\sum_{i=1}^{n} c_i^2 \sigma_i^2$.

A.8   The Central Limit Theorem

We are now in a position to derive one version of the *central limit theorem*. Let us suppose we have a sample $X_1, X_2, \ldots, X_n$ of independently and identically distributed random variables with mean μ and variance σ^2. We wish to determine, for n large, the approximate distribution of the sample mean

$$\bar{X} = \frac{X_1 + X_2 + \cdots + X_n}{n}.$$

We shall examine the distribution of the sample mean when put into the standard form. Let

$$Z = \frac{\bar{X} - \mu}{\sigma/\sqrt{n}} = \frac{X_1 - \mu}{\sigma\sqrt{n}} + \frac{X_2 - \mu}{\sigma\sqrt{n}} + \cdots + \frac{X_n - \mu}{\sigma\sqrt{n}}.$$

Now, utilizing the independence of the X_i's and the fact that they are identically distributed with the same mean and variance, we can write

$$
\begin{aligned}
M_Z(t) &= E(e^{tZ}) = \prod_{i=1}^{n} E\left[\exp\left(t\frac{X_i - \mu}{\sigma\sqrt{n}}\right)\right] \\
&= \left\{E\left[\exp\left(t\frac{X_1 - \mu}{\sigma\sqrt{n}}\right)\right]\right\}^n \\
&= \left\{E\left[1 + t\frac{X_1 - \mu}{\sigma\sqrt{n}} + \frac{t^2}{2}\frac{(X_1 - \mu)^2}{\sigma^2 n} + o\left(\frac{1}{n}\right)\right]\right\}^n \\
&= \left(1 + \frac{t^2}{2n}\right)^n \to e^{t^2/2} \text{ as } n \to \infty.
\end{aligned}
\qquad (A.48)
$$

But (A.48) is the m.g.f. of a normal distribution with mean zero and variance one. Thus, we have been able to show that the distribution of the sample mean of a random sample of n i.i.d. random variables with mean μ and variance σ^2 becomes "close" to the normal distribution with mean μ and variance σ^2/n as n becomes large.

Perhaps the easiest method of remembering the CLT is that if a statistic is the result of a summing process, then

$$Z = \frac{\text{statistic - E(statistic)}}{\sqrt{\text{Var(statistic)}}} \qquad (A.49)$$

is approximately normally distributed with mean 0 and variance 1.

A.9   The Gamma Distribution

Consider the *gamma function*:

$$\Gamma(\alpha) = \int_0^\infty x^{\alpha-1}e^{-x}dx \text{ for } \alpha > 0. \tag{A.50}$$

Integrating by parts, we obtain

$$\Gamma(\alpha) = (\alpha - 1)\Gamma(\alpha - 1) \text{ for } \alpha > 1. \tag{A.51}$$

When $\alpha = n$, with n a positive integer, repeating (A.51) $n - 1$ times yields

$$\Gamma(n) = (n - 1)!, \tag{A.52}$$

since $\Gamma(1) = \int_0^\infty e^{-x}dx = 1$.The random variable X has the *gamma distribution* with parameters α and β, if its p.d.f. is

$$f(x) = \frac{1}{\beta^\alpha\Gamma(\alpha)}x^{\alpha-1}e^{-x/\beta} \text{ for } x > 0 \tag{A.53}$$

and zero elsewhere, where both constants, α and β, are positive. The mean of X is $\alpha\beta$ and the variance of X is $\alpha\beta^2$.

The gamma distribution with parameter $\alpha = 1$ is called the *(negative) exponential distribution* with parameter β. That is, the exponential r.v. X has the p.d.f.

$$f(x) = \frac{1}{\beta}e^{-x/\beta} \text{ for } x > 0 \tag{A.54}$$

and zero elsewhere, where $\beta > 0$. It also follows from the above that X has the mean β and variance β^2.

The gamma distribution with parameters $\alpha = \nu/2$ and $\beta = 2$, where ν is a positive integer, is called the *chi-square* (χ_ν^2 for short) distribution with ν degrees of freedom. The chi-square r.v. X has the p.d.f.

$$f(x) = \frac{1}{2^{\nu/2}\Gamma(\nu/2)}x^{\nu/2-1}e^{-x/2} \text{ for } x > 0 \tag{A.55}$$

and zero elsewhere. The r.v. X has the mean ν and variance 2ν.

The m.g.f. of a gamma variate with parameters α and β can be computed in the following way:

$$\begin{aligned} M(t) &= \frac{1}{\Gamma(\alpha)\beta^\alpha}\int_0^\infty e^{tx}x^{\alpha-1}e^{-x/\beta}dx \\ &= \frac{1}{\Gamma(\alpha)\beta^\alpha}\int_0^\infty x^{\alpha-1}e^{-x(1-\beta t)/\beta}dx; \end{aligned}$$

now this integral is finite only for $t < 1/\beta$ and substituting $y = x(1 - \beta t)/\beta$ yields

$$
\begin{aligned}
M(t) &= \frac{1}{\Gamma(\alpha)} \int_0^\infty y^{\alpha-1} e^{-y} dy \left(\frac{1}{1 - \beta t}\right)^\alpha \\
&= \left(\frac{1}{1 - \beta t}\right)^\alpha \quad for \ t < \frac{1}{\beta},
\end{aligned}
\tag{A.56}
$$

where the last equality follows from the form of the p.d.f. of the gamma distribution with parameters α and 1. In particular, the m.g.f. for a chi-square r.v. with ν degrees of freedom has the form

$$
M(t) = \left(\frac{1}{1 - 2t}\right)^{\nu/2} = (1 - 2t)^{-\nu/2}, \ t < 1/2.
\tag{A.57}
$$

Suppose we consider the moment generating function of the square of a $\mathcal{N}(0, 1)$ random variable.

$$
\begin{aligned}
M_{Z^2}(t) &= \frac{1}{\sqrt{2\pi}} \int_{-\infty}^\infty e^{tz^2} e^{(-z^2/2)} dz \\
&= \frac{1}{\sqrt{2\pi}} \int_{-\infty}^\infty \exp\left(-\frac{1}{2} z^2(1 - 2t)\right) dz \\
&= \frac{1}{\sqrt{2\pi}} \int_{-\infty}^\infty e^{-\frac{w^2}{2}} dw (1 - 2t)^{\frac{1}{2}} \\
&= (1 - 2t)^{\frac{1}{2}},
\end{aligned}
\tag{A.58}
$$

where $w = z\sqrt{1 - 2t}$.

Next, let us consider the sum of squares of n random variables independently distributed as $\mathcal{N}(0, 1)$. That is, we wish to consider the moment generating function of

$$
\chi^2 = \sum_{i=1}^n z_I^2.
$$

Then, from (A.45) and (A.46), we have

$$
M_{\chi^2}(t) = \left(\frac{1}{1 - 2t}\right)^{n/2}.
\tag{A.59}
$$

We recognize this to be a χ^2 variable with n degrees of freedom.

A.10 Conditional Density Functions

Let us return to questions of interdependence between random variables and consider briefly conditional distribution of one random variable given that another random variable has assumed a fixed value.

If two random variables X and Y are discrete and have a joint probability function $f(x,y)$, then, the *conditional probability function* of the r.v. X, given that $Y = y$, has the form

$$f(x_i|y) = P(X = x_i|Y = y) = \frac{f(x_i,y)}{f_Y(y)}, \tag{A.60}$$

where

$$f_Y(y) = \sum_{\text{all values of } x_j} f(x_j,y). \tag{A.61}$$

Next, let us now suppose that r.v.'s X and Y are continuous and have joint c.d.f.

$$F(x,y) = \int_{-\infty}^{x}\int_{-\infty}^{y} f(u,v)dudv. \tag{A.62}$$

We can obtain the *marginal density function* of Y via

$$f_Y(y) = \int_{-\infty}^{\infty} f(x,y)dx. \tag{A.63}$$

Writing the statement of joint probability for small intervals in X and Y, we have

$$P(x < X \le x + \varepsilon \cap y < X \le y + \delta)$$
$$P(y < Y \le y + \delta)P(x < X \le x + \varepsilon|y < Y \le y + \delta).$$

Now, exploiting the assumption of continuity of the density function, we can write

$$\int_{x}^{x+\varepsilon}\int_{y}^{y+\delta} f(x,y)dydx = \int_{y}^{y+\delta} f_Y(y)dy \int_{x}^{x+\varepsilon} f_{X|y}(x)dx$$
$$= \varepsilon\delta f(x,y) = \delta f_Y(y)\varepsilon f_{X|y}(x).$$

Here, we have used the terms f_Y and $f_{X|y}$ to denote the *marginal density function* of Y, and the *conditional density function* of X given $Y = y$, respectively. This gives us immediately

$$f_{X|y}(x) = \frac{f(x,y)}{f_Y(y)}. \tag{A.64}$$

Note that this is a function of the argument x, whereas y is fixed; y is the value assumed by the random variable Y.

A.11   The Weak Law of Large Numbers

Let us now consider the set of n data drawn from some probability distribution. Prior to the experiment which yields the data, they can be treated as a sequence of n independent and identically distributed (i.i.d.) random variables $X_1, X_2, \ldots, X_n$. Such sequence will be labeled as a *random sample* of size n. Suppose that the mean and variance of the underlying probability distribution are μ and σ^2, respectively. Otherwise, the probability distribution is unknown. We shall find the mean and variance of the sample mean of the random sample.

It is easy to see that

$$\begin{aligned} \mu_{\bar{x}} &= \frac{E(X_1 + X_2 + \cdots + X_n)}{n} = \frac{E(X_1) + E(X_2) + \cdots + E(X_n)}{n} \\ &= \frac{\mu + \mu + \cdots + \mu}{n} = \mu. \end{aligned}$$

In this derivation, we have not used independence or the fact that all the r.v.'s have the same distribution, only the fact that they all have the same (finite) mean. We say that $\bar{X}$ is an *unbiased* estimator of μ.

Next we shall derive the variance of $\bar{X}$:

$$\begin{aligned} \sigma_{\bar{x}}^2 &= E[(\bar{X} - \mu)^2] \\ &= E\left[\left(\frac{(X_1 - \mu)}{n} + \frac{(X_2 - \mu)}{n} + \cdots + \frac{(X_n - \mu)}{n} \right)^2 \right] \\ &= \sum_{i=1}^{n} \frac{E[(X_i - \mu)^2]}{n^2} + \text{terms like } E\left[\frac{(X_1 - \mu)(X_2 - \mu)}{n^2} \right]. \end{aligned}$$

Now, by independence, the expectation of the cross-product terms is zero:

$$\begin{aligned} E[(X_1 - \mu)(X_2 - \mu)] &= \int_{-\infty}^{\infty} (x_1 - \mu)(x_2 - \mu) f(x_1) f(x_2) dx_1 dx_2 \\ &= E(X_1 - \mu) E(X_2 - \mu) = 0 \end{aligned}$$

(the argument for discrete distributions is analogous). Thus, we have

$$\sigma_{\bar{x}}^2 = \frac{\sigma^2}{n}.$$

We note that in the above derivation the fact that the X_i's are identically distributed has been superfluous. Only the facts that the r.v.'s are independent and have the same μ and σ^2 have been needed. The property that the variability of $\bar{X}$ about the true mean μ decreases as n increases is of key importance in experimental science. We shall develop this notion further below.

APPENDIX A

Let us begin by stating the celebrated *Chebyshev's inequality*. If Y is any r.v. with mean μ_y and variance σ_y^2, then for any $\varepsilon > 0$

$$P(|Y - \mu_y| > \varepsilon) \leq \frac{\sigma_y^2}{\varepsilon^2}. \tag{A.65}$$

As a practical approximation device, it is not a particularly useful inequality. However, as an asymptotic device, it is invaluable. Let us consider the case where $Y = \bar{X}$. Then, we have

$$P(|\bar{X} - \mu| > \varepsilon) \leq \frac{\sigma^2}{n\varepsilon^2}, \tag{A.66}$$

or equivalently

$$P(|\bar{X} - \mu| \leq \varepsilon) > 1 - \frac{\sigma^2}{n\varepsilon^2}. \tag{A.67}$$

Equation (A.67) is a form of the *weak law of large numbers*. The WLLN tells us that if we are willing to take a sufficiently large sample, then we can obtain an arbitrarily large probability that $\bar{X}$ will be arbitrarily close to μ [3].

---

[3] In fact, even a more powerful result, the *strong law of large numbers*, is available. In order to make the difference between the WLLN and SLLN more transparent, let us denote the sample mean based on a sample of size n by $\bar{X}_n$, so that the dependence of $\bar{X}$ on n be emphasized. Now we can write the WLLN in the following way:

$$\lim_{n \to \infty} P(|\bar{X}_n - \mu| \leq \varepsilon) = 1 \tag{A.68}$$

for each positive ε. On the other hand, the SLLN states that

$$P(\lim_{n \to \infty} |\bar{X}_n - \mu| = 0) = 1. \tag{A.69}$$

Loosely speaking, in the WLLN, the probability of $\bar{X}_n$ being close to μ for only one n at a time is claimed, whereas in the SLLN, the closeness of $\bar{X}_n$ to μ for all large n simultaneously is asserted with probability one. The rather practical advantage of the SLLN is that if $g(x)$ is some function, then

$$P(\lim_{n \to \infty} |g(\bar{X}_n) - g(\mu)| = 0) = 1. \tag{A.70}$$

The WLLN and the SLLN are particular cases of convergence in probability and almost sure convergence of a sequence of r.v.'s, respectively. Let $Y_1, Y_2, \ldots, Y_n, \ldots$ be an infinite sequence of r.v.'s. We say that this sequence of r.v.'s converges *in probability* or *stochastically* to a random variable Y if

$$\lim_{n \to \infty} P(|Y_n - Y| > \varepsilon) = 0$$

for each positive ε. We say that the sequence $Y_1, Y_2, \ldots, Y_n, \ldots$ converges *almost surely* or converges *with probability one* if

$$P(\lim_{n \to \infty} |Y_n - Y| = 0) = 1.$$

A.12   The Multivariate Normal Distribution

The random vector $\mathbf{X}$ of dimension p is said to have *multivariate normal* (or *p-dimensional multinormal* or *p-variate normal*) distribution if its p.d.f. is given by

$$f(\mathbf{x}) = |2\pi\Sigma|^{-1/2} \exp\left[-\frac{1}{2}(\mathbf{x}-\boldsymbol{\mu})'\Sigma^{-1}(\mathbf{x}-\boldsymbol{\mu})\right], \qquad (A.71)$$

where $\boldsymbol{\mu}$ is a constant vector and Σ is a constant positive definite matrix. It can be shown that $\boldsymbol{\mu}$ and Σ are the mean vector and covariance matrix of the random vector $\mathbf{X}$, respectively. For short, we write that $\mathbf{X}$ is $\mathcal{N}(\boldsymbol{\mu},\Sigma)$ distributed. Now if the covariance matrix Σ is diagonal, $\Sigma = diag(\sigma_{11}, \sigma_{22}, \ldots, \sigma_{pp})$, the density function can be written as:

$$f(\mathbf{x}) = \prod_{i=1}^{p}(2\pi\sigma_{ii})^{-1/2}\exp\left[-\frac{1}{2}(x_i-\mu_i)\sigma_{ii}^{-1}(x_i-\mu_i)\right]. \qquad (A.72)$$

Thus, the elements of $\mathbf{X}$ are then mutually independent normal random variables with means μ_i and variances σ_{ii}, $i = 1, 2, \ldots, p$, respectively. If the random vector $\mathbf{X}$ is multivariate normal, then the property that its elements are uncorrelated one with another (i.e., that $Cov(X_i, X_j) = 0, i \neq j$) implies their mutual independence.

A.13   The Wiener Process

A *stochastic process* $\{X_t\}$ is a collection of random variables indexed on the real variable t. Typically, t is time. Let us suppose the $\{X_t\}$ process has the property that for any collection of t values $t_1 < t_2 < \ldots < t_n$, the vector random variable $(X_{t_1}, X_{t_2}, \ldots, X_{t_n})$ is an n dimensional normal distribution. Then $\{X_t\}$ is a *Gaussian process*.

Next, suppose that a stochastic process $W(t)$ has the following properties:

- $W(0) = 0$;

- For any t, $W(t)$ is normal with mean zero and variance t;

- If the intervals $[t_1, t_2]$ and $[t_3, t_4]$ do not overlap, then the random variables $W(t_2) - W(t_1)$ and $W(t_4) - W(t_3)$ are stochastically independent.

Then $W(t)$ is called a *Wiener process*.
   We define a *Brownian process* $S(t)$ as

$$S(t) = \mu t + \sigma W(t), \qquad (A.73)$$

where $W(t)$ is a Wiener process. We write this as the stochastic differential equation

$$dS(t) = \mu t + \sigma dW(t). \qquad (A.74)$$

If the logarithm of $S(t)$ is a Brownian process, then we say that $S(t)$ is a *geometric Brownian process*. We may write this as the stochastic differential equation

$$\frac{dS(t)}{Sdt} = \mu t + \sigma dW(t). \qquad (A.75)$$

Then, we have

$$S(t) = S(0) \exp\left[\left(\mu - \frac{1}{2}\sigma^2\right)t + \sigma\sqrt{t}Z\right]$$

$$= S(0) \exp(\mu t) \times \exp\left(\sigma\sqrt{t}Z - (\frac{1}{2}\sigma^2 t)\right) \qquad (A.76)$$

where the Z are independently distributed as $\mathcal{N}(0,1)$. To find the percentile values of a security at time t, we use the percentile values from

$Z_{critical}$	$P(Z > Z_{critical} = P(Z < -Z_{critical}$
2.3263	.01
1.6449	.05
1.2816	.10
1.0364	.15
.8416	.20
.6745	.25
.5244	.30
.2533	.40
0	.50

A.14   The Poisson Process and the Poisson Distribution

Let us consider a counting process described by Poisson's Four Axioms:

1. $P(1$ occurrence in $[t, t + \epsilon]) = \lambda\epsilon$;

2. $P($ more than 1 occurrence in $[t, t + \epsilon]) = o(\epsilon)$
   where $\lim_{\epsilon \to 0} o(\epsilon)/\epsilon = 0$;

3. $P(k$ in $[t_1, t_2]$ and m in $[t_3, t_4]), = P(k$ in $[t_1, t_2])P(m$ in $[t_3, t_4])$
   if $[t_1, t_2] \bigcap [t_3, t_4] = \emptyset$;

4. $P(k$ in $[t_1, t_1 + s]) = P(k$ in $[t_2, t_2 + s])$ for all $t_1, t_2,$ and s.

Then we may write

$$P(k + 1 \text{ in } [0, t + \epsilon] = P(k + 1 \text{ in } [0, t])P(0 \text{ in } [t, t + \epsilon]) \qquad (A.77)$$

$$+ P(k \text{ in } [0, t])P(1 \text{ in } [t, t + \epsilon]) + o(\epsilon)$$

$$= P(k + 1, t)(1 - \lambda\epsilon) + P(k, t)\lambda\epsilon + o(\epsilon),$$

where $P(k,t) = P(k \text{ in } [0,t])$.

Then we have

$$\frac{P(k+1,t+\epsilon) - P(k+1,t)}{\epsilon} = \lambda[P(k,t) - P(k+1)] + \frac{o(\epsilon)}{\epsilon}. \qquad \text{(A.78)}$$

Taking the limit as $\epsilon \to 0$, we have the differential-difference equation:

$$\frac{dP(k+1,t)}{dt} = \lambda[P(k,t) - P(k+1,t)]. \qquad \text{(A.79)}$$

Taking $k = -1$, since we know that it is impossible for a negative number of events to occur, we have

$$\frac{dP(0,t)}{dt} = -\lambda P(0,t). \qquad \text{(A.80)}$$

So,

$$P(0,t) = \exp(-\lambda t). \qquad \text{(A.81)}$$

Next, we for for $P(1,t)$ via

$$\frac{P(1,t)}{dt} = \lambda[\exp(-\lambda t) - P(1,t)], \qquad \text{(A.82)}$$

with solution,

$$P(1,t) = \exp(-\lambda t)(\lambda t). \qquad \text{(A.83)}$$

Continuing in this fashion, we quickly conjecture that the general solution of (A.78) is given by

$$P(k,t) = \frac{e^{-\lambda t}(\lambda t)^k}{k!}. \qquad \text{(A.84)}$$

(A.83) defines the *Poisson distribution*. We can quickly compute the mean and variance both to be λt. If there are N happenings in a time interval of length T, then a natural estimate for λ is found by solving

$$\hat{\lambda} T = N. \qquad \text{(A.85)}$$

This gives us

$$\hat{\lambda} = \frac{N}{T}. \qquad \text{(A.86)}$$

Of special interest to us is the probability that no shock (event) occurs in the interval from t to $t+s$. Clearly, that is given by $P(0,s) = \exp(-\lambda s)$. This immediately enables us to write the cumulative distribution function of the time it takes to reach an event, namely,

$$F(t) = 1 - \exp(-\lambda t). \qquad \text{(A.87)}$$

Now, we know that the cdf of a continuous random variable is distributed as a uniform random variable on the interval from 0 to 1. This gives us a ready means for generating a simulated time until first occurrence of an event.

1. Generate u from $U(0,1)$.

2. Set $u = 1 - \exp(-\lambda t)$.

3. $t = \log(1 - u)/\lambda$.

Once we observe an occurrence time t_1, we start over with the new time origin set at t_1.

Note that we now have a stochastic process. At any given time t, the probability an event happens in the interval $(t, t + \epsilon)$ is given by $\lambda \times \epsilon$. The probability is stochastically independent of the history prior to t.

A.14.1   Simulating Bear Jumps

Let us suppose we have a bank account of one million pesos which is growing at the rate of 20% per year. Unfortunately, a random devaluation of currency in the amount of 10% occurs on the average of once a year. A random devaluation of currency in the amount of 20% occurs on the average of once every five years. What will be the value of the bank account six months in the future in present day pesos? If there are no devaluations, the answer is

$$S(10) = 1,000,000 \exp(.5 \times .2) = 1,105,171.$$

On the other hand, we have to deal with the concatenation of our "sure thing" bank account with the cancatenation of two Poisson bear jump mechanisms. The 10% jumps have $\lambda_{10\%}$ given by

$$\lambda_{10\%} = \frac{1}{1} = 1. \tag{A.88}$$

The 20% jumps have $\lambda_{20\%}$ given by

$$\lambda_{20\%} = \frac{1}{5} = .2. \tag{A.89}$$

Returning to (A.84), to handle the 10% jumps, we generate a uniform random variate u_1 on [0,1] we have the following downjump multiplier table:

1. If $u_1 < \exp(-1 \times .5) = .60653$, use multiplier 1.00.

2. If $.60653 \le u_1 < .90980$, use multiplier .9.

3. If $.909080 \le u_1 < .98561$, use multiplier .81.

4. If $.98561 \le u_1 < .99824$, use multiplier .729.

5. If $.99824 \le u_1 < .99982$, use multiplier .6561.

6. If $.99824 \le u_1$, use multiplier .59049.

To handle the 20% bear jumps, we generate a uniform random variate u_2. We then have the following downjump multiplier table:

1. If $u_1 < \exp(-.2 \times .5) = .90484$, use multiplier 1.00

2. If $.90484 \leq u_2 < .99532$, use multiplier .8

3. If $.99532 \leq u_2 < .99985$, use multiplier .64.

4. If $.99985 \leq u_2$, use multiplier .512.

Many standard software packages have automatic Poisson generators. To use such a routine, one simply enters λT and a random number of bear jumps, from 0 to infinity, is generated.

A.15 Parametric Simulation

For the standard lognormal model for stock growth, we have

$$S(t) = S(0) \exp\left[\left(\mu - \frac{1}{2}\sigma^2\right)t + \sigma\sqrt{t}Z\right]. \qquad (A.90)$$

Then, from (A.90), we have, for all t and Δt

$$r(t + \Delta t, t) = \frac{S(t + \Delta t)}{S(t)} = \exp\left[\left(\mu - \frac{\sigma^2}{2}\right)\Delta t + Z\sigma\sqrt{\Delta t}\right]. \qquad (A.91)$$

Defining $R(t + \Delta t, t) = \log(r(t + \Delta t, t))$, we have

$$R(t + \Delta t, t) = \left(\mu - \frac{\sigma^2}{2}\right)\Delta t + \epsilon\sigma\sqrt{\Delta t}.$$

Then

$$E[R(t + \Delta t, t)] = \left(\mu - \frac{\sigma^2}{2}\right)\Delta t. \qquad (A.92)$$

We will take μs to be given on an annual basis. Then, if the data are taken at N points separated by Δt, let the sample mean $\bar{R}$ be defined by

$$\bar{R} = \frac{1}{N}\sum_{i=1}^{N} R(i). \qquad (A.93)$$

By the strong law of large numbers, the sample mean $\bar{R}$ converges almost surely to its expectation $(\mu - \sigma^2/2)\Delta t$. Next, we note that

$$[R(t + \Delta t, t) - E(R(t + \Delta t, t))]^2 = \epsilon^2\sigma^2\Delta t, \qquad (A.94)$$

so

$$\text{Var}[R(t + \Delta t, t)] = E[R(t + \Delta t, t) - \left(\mu - \frac{\sigma^2}{2}\right)\Delta t]^2 = \sigma^2\Delta t. \qquad (A.95)$$

$$s_R^2 = \frac{1}{N-1} \sum_{i=1}^{N} (R(i) - \bar{R})^2. \tag{A.96}$$

The most utilized estimation technique in statistics is the *method of moments*. By this procedure, we replace the mean by the sample mean, the variance by the sample variance, etc. Doing this in (A.94), we have

$$\hat{\sigma}^2 = \frac{s_R^2}{\Delta t}. \tag{A.97}$$

Then, from (A.92) we have

$$\hat{\mu} = \frac{\bar{R}}{\Delta t} + \frac{\hat{\sigma}^2}{2}. \tag{A.98}$$

Having estimated from historical data μ and σ, we can now simulate the value of our stock using (A.90) for any desired time horizon.

A.15.1   Simulating a Geometric Brownian Walk

First we start with a simple geometric Brownian case. The program is incredibly simple. We need only assume available a uniform generator, a normal generator, and a sorting routine.

Simugram

1. Enter $S(0)$,T, $\hat{\mu}$, and $\hat{\sigma}$.

2. Repeat 10,000 times.

3. Generate normal observation Z with mean zero and variance 1 .

4. Set $S(T) = S(0) \exp[(\mu - \frac{1}{2}\sigma^2)T + \sigma\sqrt{T}Z]$.

5. End repeat.

6. Sort the 10,000 end values.

7. Then $F(v) = \dfrac{\text{Number of sorted values} \leq v}{10,000}$.

We will now add on the possibility of two types of bear jumps: a 10% downturn on the average of once every year ($\lambda_1 = 1$) and a 20% downturn on the average of once every five years ($\lambda_2 = .2$).

SimugramwithJumps

1. Enter $S(0)$, T, $\hat{\mu}$, $\hat{\sigma}$, λ_1, λ_2.

2. Repeat 10,000 times.

3. Generate normal observation Z with mean zero and variance 1 .

4. Set $S(T) = S(0) \exp[(\mu - \frac{1}{2}\sigma^2)T + \sigma\sqrt{T}Z]$.

5. Generate a Poisson variate m_1 from $Po(\lambda_1 \times T)$.

6. Replace $S(T)$ by $.9^{m_1} \times S(T)$

7. Generate a Poisson variate m_2 from $Po(\lambda_2 \times T)$.

8. Replace $S(T)$ by $.8^{m_2} \times S(T)$

9. End repeat.

10. Sort the 10,000 end values.

11. Then $F(v) = \dfrac{\text{Number of sorted values} \leq v}{10,000}$.

A.15.2 The Multivariate Case

Now, we assume that we have p securities to be considered in a portfolio with weights $\{c_j\}$ which are non-negative and sum to one. We estimate the $\{\mu_j\}$ and the $\{\sigma\}$ precisely as in the one security case. Accordingly, for the jth security, we let

$$R_j(i) = \log\left(\frac{S_j(i)}{S_j(i-1)}\right).$$

Then

$$\bar{R}_j = \frac{1}{N}\sum_{i=1}^{N} R_j(i)$$

and

$$s_{R_j}^2 = \frac{1}{N-1}\sum_{i=1}^{N}(R_j(i) - \bar{R}_j)^2.$$

So we have an estimates for σ_j^2 and μ_j, namely,

$$\hat{\sigma}_j{}^2 = \frac{s_{R_j}^2}{\Delta t} \tag{A.99}$$

and

$$\hat{\mu}_j = \frac{\bar{R}_j}{\Delta t} + \frac{\hat{\sigma}_j^2}{2}. \tag{A.100}$$

Now, we must also account for the correlation between the growths of stocks:

$$\hat{\sigma_{jm}} = \frac{1}{N-1}\sum_{i=1}^{N}(R_j(i) - \bar{R}_j)(R_m(i) - \bar{R}_m). \tag{A.101}$$

Then, we have as our estimated covariance matrix

$$\hat{\Sigma} = \begin{pmatrix} \hat{\sigma}_{11} & \hat{\sigma}_{12} & \cdots & \hat{\sigma}_{1p} \\ \hat{\sigma}_{12} & \hat{\sigma}_{22} & \cdots & \hat{\sigma}_{2p} \\ \cdots & \cdots & \cdots & \cdots \\ \hat{\sigma}_{1p} & \hat{\sigma}_{2p} & \cdots & \hat{\sigma}_{pp} \end{pmatrix}, \tag{A.102}$$

For a covariance matrix, we may obtain a *Cholesky decomposition*

$$\hat{\Sigma} = \mathbf{LL}^T, \tag{A.103}$$

where $\mathbf{L}$ is a lower triangular matrix, frequently referred to as the matrix square root of $\hat{\Sigma}$. Subroutines for obtaining the matrix square root are available in most standard matrix compatible software such as MATLAB, Splus, and SAS.

Let us generate p independent normal variates with mean 0 and variance one, putting them into a row vector, $\mathbf{Z} = (z_1, z_2, \ldots, z_p)$. Then, we compute the row vector

$$\mathbf{V} = \mathbf{ZL}^T = (v_1, v_2, \ldots, v_p).$$

Then, the price of the jth stock in the joint simulation at time T is given by

$$S_j(t) = S_j(0)\exp\left[\left(\mu_j - \frac{1}{2}\sigma_j^2\right)T + v_j\sqrt{T}\right]. \tag{A.104}$$

Then, for the p-dimensional case:

Simugram Multivariate

1. Enter $\{S_j(0)\}_{j=1}^p$, T, $\{\hat{\mu}_j\}$, and $\mathbf{L}$.

2. Repeat 10,000 times.

3. Generate p independent normal variates with mean 0 and variance one, putting them into a row vector, $\mathbf{Z} = (z_1, z_2, \ldots, z_p)$.

4. Compute the row vector

$$\mathbf{V} = \mathbf{ZL}^T = (v_1, v_2, \ldots, v_p).$$

5. For each of the p stocks, compute

$$S_j(t) = S_j(0)\exp\left[\left(\mu_j - \frac{1}{2}\sigma_j^2\right)T + v_j\sqrt{T}\right].$$

6. For the ith repeat save the $(S_1(T), S_2(T), \ldots, S_p(T))$ as the row vector $\mathbf{S_i}$.

7. End repeat.

The multivariate simulation with the two bear jump processes added becomes:

Simugram Multivariate with Jumps

1. Enter $\{S_j(0)\}_{j=1}^{p}$, T, $\{\hat{\mu}_j\}$, $\mathbf{L}$, λ_1 and λ_2.

2. Repeat 10,000 times.

3. Generate a Poisson variate m_1 from $Po(\lambda_1 \times T)$.

4. Replace $S_j(0)$ by $.9^{m_1} \times S_j(T)$.

5. Generate a Poisson variate m_2 from $Po(\lambda_2 \times T)$.

6. Replace $S_j(0)$ by $.8^{m_2} \times S_j(T)$.

7. Generate p independent normal variates with mean 0 and variance one, putting them into a row vector, $\mathbf{Z} = (z_1, z_2, \ldots, z_p)$.

8. Compute the row vector

$$\mathbf{V} = \mathbf{Z}\mathbf{L}^T = (v_1, v_2, \ldots, v_p).$$

9. For each of the p stocks, compute

$$S_j(T) = S_j(0)\exp\left[\left(\mu_j - \frac{1}{2}\sigma_j^{2}\right)T + v_j\sqrt{T}\right].$$

10. For the ith repeat save the $(S_1(T), S_2(T), \ldots, S_p(T))$ as the row vector $\mathbf{S_i}(T)$.

11. End repeat.

Suppose we have a portfolio consisting of p stocks each weighted by $c_{\geq}0$ such that $A\sum c_i = 1$. Then to obtain the simugram of the portfolio, we must look at the 10,000 values from the above simulations, of the form

$$P_i(T) = \sum_{j=1}^{p} c_j S_{i,j}(T). \tag{A.105}$$

Here, the i refers to the number of the simulation. We can then form the portfolio simugram by sorting the $P_i(T)$ and obtaining the cumulative distribution function. This gives us the algorithm:

Simugram Portfolio

1. Enter $\mathbf{S}(T)$, $\{c_j\}$.

2. For all i from 1 to 10,000, find $P_i(T) = \sum_{j=1}^{p} c_j S_{i,j}(T)$.

3. Sort the $P_i(T)$.

4. $F(v) = \dfrac{\text{Number of sorted values } P_i(t) \leq v}{10,000}$.

The *Portfolio Simugram* may then be used for a host of purposes. We might use it as an assist, for example, in deciding whether one wished to replace one stock in the portfolio by another. Also, we might use the Portfolio Simugram as an input into a program for optimizing the allocation of weights according to a particular Portfolio Simugram percentile (or a combination of several percentiles) for a particular time horizon (or several time horizons).

We wish to make it clear that, due to the fact that one is using historical estimates of growth and volatility rather than the actual values, rather than the true ones (which we cannot know), the portfolio optimization based on simugrams should only be used as an exploratory and speculative tool. It is not a magic bullet, nor do we claim it to be.

A.16   Resampling Simulation

Let us suppose we have a data base showing the year to year change in a stock or a stock index. We can then obtain a data base of terms like

$$R_i = \log\left(\frac{S(t_i)}{S(t_{i-1})}\right).$$

In other words, we know that

$$S(t_i) = S(t_{i-1}) \times \exp(R_i).$$

Suppose we have a data base of n such terms, $\{R_1, R_2, \ldots, R_n\}$. Let us make the (frequently reasonable) assumption that the ups and downs of the stock or the index in the past are a good guide to the ups and downs in the future. It would not be a good idea, if we wished to forecast the value of the stock five years in advance, randomly to sample (with replacement) five of the R_i's, say, $\{R_3, R_{17}, R_{20}, R_{20}, R_{31}\}$ and use

$$\hat{S}(5) = S(0) \times \exp[R_3 + R_{17} + R_{20} + R_{20} + R_{31}].$$

On the other hand, if we wished to obtain, not a point estimate for $S(T)$, but an estimate for the distribution of possible values of $S(5)$, experience shows that this frequently can be done as follows:

Simugram Resampling

1. Enter $S(0)$, T, and the $\{R_i\}$.

2. Repeat 10,000 times

3. For pass i, randomly sample with replacement T values from $\{R_1, R_2, \ldots, R_n\}$, say, $\{R_{i,1}, R_{i,2}, R_{i,3}, R_{i,4}, \ldots, R_{i,T}\}$.

4. Compute

$$SS(T) = S(0) \times \exp[R_{i,1} + R_{i,2} + R_{i,3} + R_{i,4} + \ldots + R_{i,T}].$$

5. Obtain the simugram (empirical cumulative distribution function) from the resulting 10,000 values of $SS(T)$. That is, compute

$$F_T(v) = \frac{\text{Number of sorted values } \{SS(T)\} \leq v}{10,000}. \qquad \text{(A.106)}$$

A.16.1   The Multivariate Case

Here, for each of p stocks, we compute, from the historical record (taking, for all i, $t_i - t_{i-1} = h$ where $T = nh$ and the entire historical record is of length $TT = Nh$), we compute the N by p matrix whose elements are

$$R_{i,j} = \log\left(\frac{S_j(t_i)}{S_j(t_{i-1})}\right).$$

Simugram Multivariate Resampling

We then proceed very much as in the parametric case:

1. Enter $\{S_j(0)\}_{j=1}^p$, T, and the $\{R_{i,j}\}_{i,j}$.

2. Repeat 10,000 times.

3. For pass i, randomly sample with replacement n integer values from the length of the historical list (N), say, $l_{i,1}, l_{i,2}, \ldots, l_{i,n}$.

4. For each stock j

$$SS_{i,j}(T) = S_j(0) \times \exp[h \times (R_{l_{(i,1)},j} + R_{l_{(i,2)},j} + \ldots + R_{l_{(i,n)},j})].$$

5. Store $(SS_{i,1}(T), SS_{i,2}(T), \ldots, SS_{i,p}(T))$ as a row vector $\mathbf{SS}_i(T)$.

6. End Repeat.

Now for a portfolio of p stocks, balanced according to

$$P(t) = \sum_{j=1}^{p} c_j S_j(t), \qquad (A.107)$$

where the weights are non-negative and sum to one, we simply use the following algorithm:

Simugram Portfolio Resampling

1. Enter $SS(T)$, $\{c_j\}$.

2. For all i from 1 to 10,000, find $P_i(T) = \sum_{j=1}^{p} c_j SS_{i,j}(T)$.

3. Sort the $P_i(T)$.

4. $F(v) = \dfrac{\text{Number of } P_i(T) \leq v}{10,000}$.

We may then proceed to obtain simugrams of the portfolio value at a given time. Because of the correlations between stock values, it is essential that, when we randomly select a year, we sample the annual growth factors of the stocks in the portfolio for that year.

As a cautionary note, we recommend that the reader generate his or her set of 10,000 stock combinations only once for each time horizon. Obtaining a new random sample of stocks for every iteration of an optimization algorithm tends to increase dramatically the time to convergence of algorithms, such as the polytope algorithm of Nelder and Mead which we present later in this Appendix.

A.17   A Portfolio Case Study

Next we take the 90 stocks in the S&P 100 that were in business prior to 1991. The data base we shall use will utilize the 12 years 1990–2001, utilizing monthly data, both for the estimation of the parameters characterizing the simple geometric Brownian model parameters (without Poissonian jumps), namely the $\{\mu_j\}_{j=1}^{j=80}$ and the covariance matrix $\{\sigma_{i,j}\}$, and for obtaining resampling months.

We look below at the result of one of many possible optimization criteria that might have been considered. We find the allocation of an investment in the portfolio amongst the 90 stocks maximizing the one year lower 20 percentile, with the constraint that no stock has more than 5% of the portfolio share.

id	permno	ticker	μ	σ	par alloc	npar alloc

Table A.2. Portfolio Allocation from S&P 100.
Maximizing One Year 20 Percentile
with Max 5% in Any Stock.

id	permno	ticker	μ	σ	par alloc	npar alloc
1	10104	ORCL	0.44	0.65	0.05	0.05
2	10107	MSFT	0.39	0.48	0.05	0.05
3	10145	HON	0.12	0.44	0.00	0.00
4	10147	EMC	0.48	0.61	0.00	0.01
5	10401	T	0.05	0.40	0.00	0.00
6	10890	UIS	0.36	0.68	0.00	0.01
7	11308	KO	0.07	0.31	0.00	0.00
8	11703	DD	0.05	0.28	0.00	0.00
9	11754	EK	−0.11	0.35	0.00	0.00
10	11850	XOM	0.12	0.17	0.00	0.00
11	12052	GD	0.19	0.25	0.05	0.05
12	12060	GE	0.23	0.26	0.00	0.00
13	12079	GM	0.08	0.36	0.00	0.00
14	12490	IBM	0.32	0.34	0.05	0.00
15	13100	MAY	0.07	0.29	0.00	0.00
16	13856	PEP	0.13	0.28	0.00	0.00
17	13901	MO	0.13	0.32	0.00	0.00
18	14008	AMGN	0.32	0.39	0.05	0.05
19	14277	SLB	0.12	0.36	0.00	0.00
20	14322	S	0.05	0.36	0.00	0.00
21	15560	RSH	0.27	0.49	0.05	0.05
22	15579	TXN	0.40	0.56	0.00	0.01
23	16424	G	0.09	0.33	0.00	0.00
24	17830	UTX	0.21	0.35	0.00	0.00
25	18163	PG	0.16	0.31	0.04	0.00
26	18382	PHA	0.12	0.30	0.00	0.00
27	18411	SO	0.14	0.24	0.05	0.05
28	18729	CL	0.25	0.33	0.05	0.05
29	19393	BMY	0.20	0.26	0.01	0.03
30	19561	BA	0.05	0.35	0.00	0.00
31	20220	BDK	0.06	0.38	0.00	0.00
32	20626	DOW	0.07	0.30	0.00	0.00
33	21573	IP	0.07	0.36	0.00	0.00
34	21776	EXC	0.17	0.33	0.05	0.05
35	21936	PFE	0.26	0.28	0.05	0.05

id	permno	ticker	μ	σ	par alloc	npar alloc
		Table A.2. Portfolio Allocation from S&P 100. Maximizing One-Year 20 Percentile with Max 5% in Any Stock (continued).				
36	22111	JNJ	0.20	0.26	0.00	0.02
37	22592	MMM	0.14	0.25	0.00	0.00
38	22752	MRK	0.17	0.31	0.00	0.00
39	22840	SLE	0.11	0.31	0.00	0.00
40	23077	HNZ	0.06	0.25	0.00	0.00
41	23819	HAL	−0.03	0.47	0.00	0.00
42	24010	ETR	0.11	0.29	0.00	0.01
43	24046	CCU	0.27	0.38	0.00	0.00
44	24109	AEP	0.04	0.22	0.00	0.00
45	24643	AA	0.21	0.37	0.00	0.00
46	24942	RTN	0.02	0.44	0.00	0.00
47	25320	CPB	0.04	0.29	0.00	0.00
48	26112	DAL	0.00	0.33	0.00	0.00
49	26403	DIS	0.05	0.33	0.00	0.00
50	27828	HWP	0.11	0.48	0.00	0.00
51	27887	BAX	0.21	0.24	0.05	0.05
52	27983	XRX	0.04	0.61	0.00	0.00
53	38156	WMB	0.14	0.33	0.00	0.00
54	38703	WFC	0.20	0.31	0.00	0.00
55	39917	WY	0.07	0.32	0.00	0.00
56	40125	CSC	0.17	0.48	0.00	0.00
57	40416	AVP	0.24	0.47	0.00	0.00
58	42024	BCC	0.00	0.33	0.00	0.00
59	43123	ATI	−0.05	0.39	0.00	0.00
60	43449	MCD	0.05	0.26	0.00	0.00
61	45356	TYC	0.37	0.33	0.05	0.05
62	47896	JPM	0.15	0.38	0.00	0.00
63	50227	BNI	0.03	0.27	0.00	0.00
64	51377	NSM	0.35	0.69	0.05	0.05
65	52919	MER	0.31	0.44	0.00	0.01
66	55976	WMT	0.32	0.31	0.05	0.05
67	58640	NT	0.23	0.64	0.00	0.00
68	59176	AXP	0.19	0.31	0.00	0.00
69	59184	BUD	0.20	0.21	0.05	0.05
70	59328	INTC	0.37	0.52	0.00	0.00
71	59408	BAC	0.14	0.34	0.00	0.00
72	60097	MDT	0.28	0.28	0.05	0.05
73	60628	FDX	0.23	0.35	0.00	0.00
74	61065	TOY	0.06	0.48	0.00	0.00
75	64186	CI	0.20	0.28	0.00	0.00
76	64282	LTD	0.16	0.42	0.00	0.00
77	64311	NSC	−0.02	0.34	0.00	0.00
78	65138	ONE	0.10	0.37	0.00	0.00
79	65875	VZ	0.11	0.29	0.00	0.00
80	66093	SBC	0.12	0.29	0.00	0.00
81	66157	USB	0.12	0.37	0.00	0.00
82	66181	HD	0.33	0.32	0.05	0.05
83	66800	AIG	0.26	0.26	0.05	0.05
84	69032	MWD	0.34	0.47	0.00	0.00
85	70519	C	0.35	0.36	0.03	0.05
86	75034	BHI	0.11	0.41	0.00	0.00
87	75104	VIA	0.21	0.37	0.00	0.00
88	76090	HET	0.09	0.39	0.00	0.00
89	82775	HIG	0.24	0.37	0.01	0.00
90	83332	LU	0.10	0.54	0.00	0.00

The agreement between the parametric and nonparametric (resampling) allocations is very close. Naturally, because of the correlation between the various stock growths, the allocation will not be "sharp," i.e., there are many different allocations that may give similar results. For the parametric optimum, our mean portfolio value after one year is 1.32, for the resampling, 1.33. For optimization at the one year horizon, the optimal allocation (using maximization of the lower 20 percentile with constraints of 5% allocation in any one stock) had median portfolio value of 1.30 for the parametric and 1.32 for the nonparametric. (We start at time zero with a portfolio value of 1.00.)

Table A.3. Portfolio Allocation from S&P 100. Maximizing Five Year 20 Percentile with Max 5% in Any Stock.						
id	permno	ticker	μ	σ	par alloc	npar alloc
1	10104	ORCL	0.44	0.65	0.05	0.05
2	10107	MSFT	0.39	0.48	0.05	0.05
3	10145	HON	0.12	0.44	0.00	0.00
4	10147	EMC	0.48	0.61	0.05	0.05
5	10401	T	0.05	0.4	0.00	0.00
6	10890	UIS	0.36	0.68	0.04	0.01
7	11308	KO	0.07	0.31	0.00	0.00
8	11703	DD	0.05	0.28	0.00	0.00
9	11754	EK	−0.11	0.35	0.00	0.00
10	11850	XOM	0.12	0.17	0.00	0.00
11	12052	GD	0.19	0.25	0.00	0.00
12	12060	GE	0.23	0.26	0.00	0.00
13	12079	GM	0.08	0.36	0.00	0.00
14	12490	IBM	0.32	0.34	0.05	0.05
15	13100	MAY	0.07	0.29	0.00	0.00
16	13856	PEP	0.13	0.28	0.00	0.00
17	13901	MO	0.13	0.32	0.00	0.00
18	14008	AMGN	0.32	0.39	0.05	0.05
19	14277	SLB	0.12	0.36	0.00	0.00
20	14322	S	0.05	0.36	0.00	0.00
21	15560	RSH	0.27	0.49	0.05	0.04
22	15579	TXN	0.4	0.56	0.05	0.04
23	16424	G	0.09	0.33	0.00	0.00
24	17830	UTX	0.21	0.35	0.00	0.00
25	18163	PG	0.16	0.31	0.00	0.00
26	18382	PHA	0.12	0.3	0.00	0.00
27	18411	SO	0.14	0.24	0.00	0.00
28	18729	CL	0.25	0.33	0.05	0.05
29	19393	BMY	0.2	0.26	0.00	0.05
30	19561	BA	0.05	0.35	0.00	0.00
31	20220	BDK	0.06	0.38	0.00	0.00
32	20626	DOW	0.07	0.3	0.00	0.00
33	21573	IP	0.07	0.36	0.00	0.00
34	21776	EXC	0.17	0.33	0.00	0.05
35	21936	PFE	0.26	0.28	0.05	0.05

Table A.3. Portfolio Allocation from S&P 100. Maximizing One-Year 20 Percentile with Max 5% in Any Stock (continued).						
id	permno	ticker	μ	σ	par alloc	npar alloc
36	22111	JNJ	0.2	0.26	0.00	0.00
37	22592	MMM	0.14	0.25	0.00	0.00
38	22752	MRK	0.17	0.31	0.00	0.00
39	22840	SLE	0.11	0.31	0.00	0.00
40	23077	HNZ	0.06	0.25	0.00	0.00
41	23819	HAL	-0.03	0.47	0.00	0.00
42	24010	ETR	0.11	0.29	0.00	0.00
43	24046	CCU	0.27	0.38	0.00	0.00
44	24109	AEP	0.04	0.22	0.00	0.00
45	24643	AA	0.21	0.37	0.00	0.00
46	24942	RTN	0.02	0.44	0.00	0.00
47	25320	CPB	0.04	0.29	0.00	0.00
48	26112	DAL	0	0.33	0.00	0.00
49	26403	DIS	0.05	0.33	0.00	0.00
50	27828	HWP	0.11	0.48	0.00	0.00
51	27887	BAX	0.21	0.24	0.01	0.03
52	27983	XRX	0.04	0.61	0.00	0.00
53	38156	WMB	0.14	0.33	0.00	0.00
54	38703	WFC	0.2	0.31	0.00	0.00
55	39917	WY	0.07	0.32	0.00	0.00
56	40125	CSC	0.17	0.48	0.00	0.00
57	40416	AVP	0.24	0.47	0.03	0.00
58	42024	BCC	0	0.33	0.00	0.00
59	43123	ATI	-0.05	0.39	0.00	0.00
60	43449	MCD	0.05	0.26	0.00	0.00
61	45356	TYC	0.37	0.33	0.05	0.05
62	47896	JPM	0.15	0.38	0.00	0.00
63	50227	BNI	0.03	0.27	0.00	0.00
64	51377	NSM	0.35	0.69	0.03	0.02
65	52919	MER	0.31	0.44	0.04	0.00
66	55976	WMT	0.32	0.31	0.05	0.05
67	58640	NT	0.23	0.64	0.00	0.00
68	59176	AXP	0.19	0.31	0.00	0.00
69	59184	BUD	0.2	0.21	0.03	0.05
70	59328	INTC	0.37	0.52	0.02	0.04
71	59408	BAC	0.14	0.34	0.00	0.00
72	60097	MDT	0.28	0.28	0.05	0.05
73	60628	FDX	0.23	0.35	0.00	0.00
74	61065	TOY	0.06	0.48	0.00	0.00
75	64186	CI	0.2	0.28	0.00	0.00
76	64282	LTD	0.16	0.42	0.00	0.00
77	64311	NSC	-0.02	0.34	0.00	0.00
78	65138	ONE	0.1	0.37	0.00	0.00
79	65875	VZ	0.11	0.29	0.00	0.00
80	66093	SBC	0.12	0.29	0.00	0.00
81	66157	USB	0.12	0.37	0.00	0.00
82	66181	HD	0.33	0.32	0.05	0.05
83	66800	AIG	0.26	0.26	0.05	0.05
84	69032	MWD	0.34	0.47	0.00	0.00
85	70519	C	0.35	0.36	0.05	0.05
86	75034	BHI	0.11	0.41	0.00	0.00
87	75104	VIA	0.21	0.37	0.00	0.00
88	76090	HET	0.09	0.39	0.00	0.00
89	82775	HIG	0.24	0.37	0.05	0.01
90	83332	LU	0.1	0.54	0.00	0.00

Next, let us repeat the analysis where we go to maximizing the lower 20 percentile five years out, with a ceiling of 5% for investment in any one of the 90 stocks. The portfolio has not changed very much. The longer the planning horizon, the more the algorithm tends to focus on high growth stocks with time minimizing the effect of high volatility. Unfortunately, the longer we go out into the future, the less confident we are of our estimates of growth and volatility. Also, the twelve year data base employed here is probably too short for good forecasting purposes, particularly if we fail to add on a bear jump subroutine. However, it is also true that in the management of a portfolio, we have the opportunity continually to update our estimates of growth and volatility and to rebalance the portfolio as needed according to the criteria we select.

A.18   A Simple Optimization Algorithm That (Usually) Works

The Nelder-Mead algorithm which we explicate below is available in many software packages. It seems to work reasonably well for optimization of weights in a portfolio. We go through the argument for the two case of two dimensional optimization to give the reader a feel for how the program works. Let us suppose that we are confronted with the task of minimizing a function F which depends upon k variables—say $(X_1, X_2, \ldots, X_k)$. We wish to find the value which approximately minimizes F. To accomplish this task we start out with $k+1$ trial values of $(X_1, X_2, \ldots, X_k)$, of rank k. We evaluate F at each of these values, and identify the three values where F is the minimum (B), the maximum (W) and the second largest $(2W)$. We average all the X points except W to give us the centrum C. We show a pictorial representation for the case where $k = 2$. Note, however, that the flowchart below works for any $k \geq 2$. This is due to the fact that our moves are oriented around the best, the worst, and the second worst points. Note that the algorithm first moves by a succession of expansion steps to a region where F is small. Once this has been achieved, the procedure essentially works by the simplex of vertices surrounding and collapsing in on the optimum (slow in high dimensions). Clearly, the procedure does not make the great leaps characteristic of Newton procedures. It is slow, particularly if k is large. Even, during the expansion steps, the algorithm moves in a zig-zag, rather than in straight thrusts to the target. But the zig-zag and envelopment strategies help avoid charging off on false trails. This algorithm has been presented because of its robustness to noise in the objective function (of particular importance when dealing with such approaches as that delineated in Section A.17), the ease with which it can be programmed for the microprocessor, and the very practical fact that once learned, it is easy to remember.

Nelder-Mead Algorithm Expansion

$P = C + \gamma_R(C - W)$(where typically $\gamma_R = \gamma_E = 1$)
   If$F(P) < F(B)$, then

- $PP = C + \gamma_E(C - W)$ [a]

- If$F(PP) < F(P)$, then

- Replace W with PP as new vertex [c]

- Else

- Accept P as new vertex [b]

- End If

Else
If$F(P) < F(2W)$, Then

- Accept P as new vertex [b]

- Else

Contraction

If $F(W) < F(P)$, then

- $PP = C + \gamma_C(W - B)$ (typically $\gamma_C = 1/2$) [a*]

- If $F(PP) < F(W)$, Then Replace W with PP as new vertex [b*]

- Else Replace W with $(W + B)/2$ and $2W$ with $(2W + B)/2$ (total contraction) [c*]

- End If

Else

Contraction

If$F(2W) < F(P)$, Then

- $PP = C + \gamma_C(P - B)$ [aa]

- If $F(PP) < F(P)$, Then Replace W with PP as new vertex [bb]

- Else Replace W with $(W + B)/2$ and $2W$ with $(2W + B)/2$ (total contraction) [cc]

Else

- Replace W with P

- End If

End If

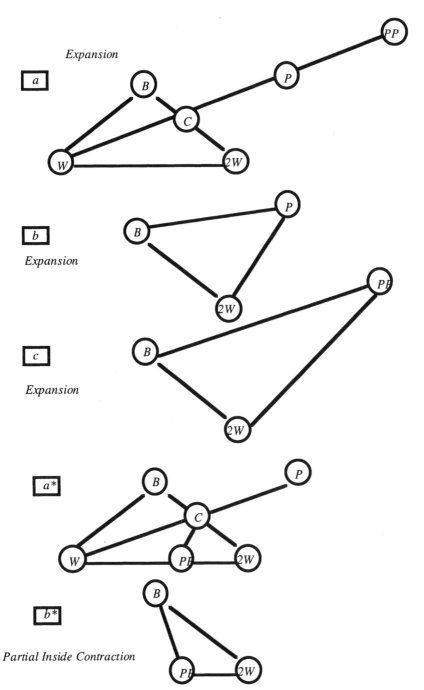

Figure A.5a. Nelder-Mead Polytope Algorithm.

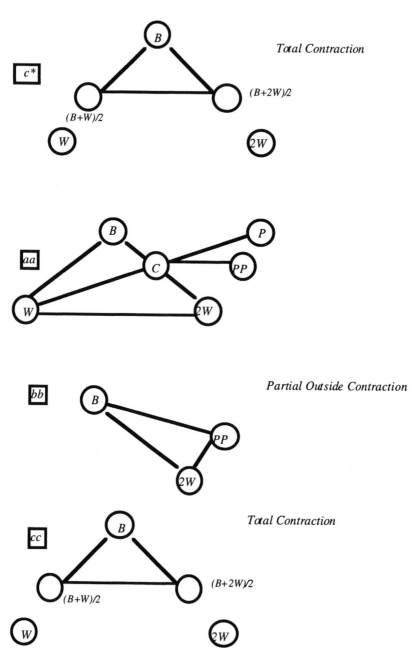

Figure A.5b. Nelder-Mead Polytope Algorithm.

Appendix B

Statistical Tables

B.1. Tables of the Normal Distribution

					Values of $\frac{1}{\sqrt{(2\pi)}} \int_{-\infty}^{z} e^{-\frac{t^2}{2}} dt$					
z	.0	.01	.02	.03	.04	.05	.06	.07	.08	.09
.0	.50000	.50399	.50798	.51197	.51595	.51994	.52392	.52790	.53188	.53586
.1	.53983	.54380	.54776	.55172	.55567	.55962	.56356	.56749	.57142	.57535
.2	.57926	.58317	.58706	.59095	.59483	.59871	.60257	.60642	.61026	.61409
.3	.61791	.62172	.62552	.62930	.63307	.63683	.64058	.64431	.64803	.65173
.4	.65542	.65910	.66276	.66640	.67003	.67364	.67724	.68082	.68439	.68793
.5	.69146	.69497	.69847	.70194	.70540	.70884	.71226	.71566	.71904	.72240
.6	.72575	.72907	.73237	.73565	.73891	.74215	.74537	.74857	.75175	.75490
.7	.75804	.76115	.76424	.76730	.77035	.77337	.77637	.77935	.78230	.78524
.8	.78814	.79103	.79389	.79673	.79955	.80234	.80511	.80785	.81057	.81327
.9	.81594	.81859	.82121	.82381	.82639	.82894	.83147	.83398	.83646	.83891
1.0	.84134	.84375	.84614	.84849	.85083	.85314	.85543	.85769	.85993	.86214
1.1	.86433	.86650	.86864	.87076	.87286	.87493	.87698	.87900	.88100	.88298
1.2	.88493	.88686	.88877	.89065	.89251	.89435	.89617	.89796	.89973	.90147
1.3	.90320	.90490	.90658	.90824	.90988	.91149	.91309	.91466	.91621	.91774
1.4	.91924	.92073	.92220	.92364	.92507	.92647	.92785	.92922	.93056	.93189
1.5	.93319	.93448	.93574	.93699	.93822	.93943	.94062	.94179	.94295	.94408
1.6	.94520	.94630	.94738	.94845	.94950	.95053	.95154	.95254	.95352	.95449
1.7	.95543	.95637	.95728	.95818	.95907	.95994	.96080	.96164	.96246	.96327
1.8	.96407	.96485	.96562	.96638	.96712	.96784	.96856	.96926	.96995	.97062
1.9	.97128	.97193	.97257	.97320	.97381	.97441	.97500	.97558	.97615	.97670
2.0	.97725	.97778	.97831	.97882	.97932	.97982	.98030	.98077	.98124	.98169
2.1	.98214	.98257	.98300	.98341	.98382	.98422	.98461	.98500	.98537	.98574
2.2	.98610	.98645	.98679	.98713	.98745	.98778	.98809	.98840	.98870	.98899
2.3	.98928	.98956	.98983	.99010	.99036	.99061	.99086	.99111	.99134	.99158
2.4	.99180	.99202	.99224	.99245	.99266	.99286	.99305	.99324	.99343	.99361
2.5	.99379	.99396	.99413	.99430	.99446	.99461	.99477	.99492	.99506	.99520
2.6	.99534	.99547	.99560	.99573	.99585	.99598	.99609	.99621	.99632	.99643
2.7	.99653	.99664	.99674	.99683	.99693	.99702	.99711	.99720	.99728	.99736
2.8	.99744	.99752	.99760	.99767	.99774	.99781	.99788	.99795	.99801	.99807
2.9	.99813	.99819	.99825	.99831	.99836	.99841	.99846	.99851	.99856	.99861
3.0	.99865	.99869	.99874	.99878	.99882	.99886	.99889	.99893	.99896	.99900
3.1	.99903	.99906	.99910	.99913	.99916	.99918	.99921	.99924	.99926	.99929
3.2	.99931	.99934	.99936	.99938	.99940	.99942	.99944	.99946	.99948	.99950
3.3	.99952	.99953	.99955	.99957	.99958	.99960	.99961	.99962	.99964	.99965
3.4	.99966	.99968	.99969	.99970	.99971	.99972	.99973	.99974	.99975	.99976
3.5	.99977	.99978	.99978	.99979	.99980	.99981	.99981	.99982	.99983	.99983
3.6	.99984	.99985	.99985	.99986	.99986	.99987	.99987	.99988	.99988	.99989
3.7	.99989	.99990	.99990	.99990	.99991	.99991	.99992	.99992	.99992	.99992
3.8	.99993	.99993	.99993	.99994	.99994	.99994	.99994	.99995	.99995	.99995
3.9	.99995	.99995	.99996	.99996	.99996	.99996	.99996	.99996	.99997	.99997

APPENDIX B

B.2. Tables of the Chi-Square Distribution

Critical Values of $P = 1 - \dfrac{1}{2^{\nu/2}\Gamma(\nu/2)} \displaystyle\int_0^{\chi^2} x^{\nu/2-1}e^{-x/2}dx$

ν	0.100	0.250	0.500	0.750	0.900	0.950	0.975	0.990	0.995	0.998	0
1	0.016	0.102	0.455	1.323	2.706	3.841	5.024	6.635	7.879	9.550	1(
2	0.211	0.575	1.386	2.773	4.605	5.991	7.378	9.210	10.597	12.429	1:
3	0.584	1.213	2.366	4.108	6.251	7.815	9.348	11.345	12.838	14.796	1(
4	1.064	1.923	3.357	5.385	7.779	9.488	11.143	13.277	14.860	16.924	1(
5	1.610	2.675	4.351	6.626	9.236	11.070	12.833	15.086	16.750	18.907	2(
6	2.204	3.455	5.348	7.841	10.645	12.592	14.449	16.812	18.548	20.791	2:
7	2.833	4.255	6.346	9.037	12.017	14.067	16.013	18.475	20.278	22.601	2(
8	3.490	5.071	7.344	10.219	13.362	15.507	17.535	20.090	21.955	24.352	2(
9	4.168	5.899	8.343	11.389	14.684	16.919	19.023	21.666	23.589	26.056	2:
10	4.865	6.737	9.342	12.549	15.987	18.307	20.483	23.209	25.188	27.722	2(
11	5.578	7.584	10.341	13.701	17.275	19.675	21.920	24.725	26.757	29.354	3:
12	6.304	8.438	11.340	14.845	18.549	21.026	23.337	26.217	28.300	30.957	3:
13	7.042	9.299	12.340	15.984	19.812	22.362	24.736	27.688	29.819	32.535	3(
14	7.790	10.165	13.339	17.117	21.064	23.685	26.119	29.141	31.319	34.091	3(
15	8.547	11.037	14.339	18.245	22.307	24.996	27.488	30.578	32.801	35.628	3:
16	9.312	11.912	15.338	19.369	23.542	26.296	28.845	32	34.267	37.146	3(
17	10.085	12.792	16.338	20.489	24.769	27.587	30.191	33.409	35.718	38.648	4(
18	10.865	13.675	17.338	21.605	25.989	28.869	31.526	34.805	37.156	40.136	4:
19	11.651	14.562	18.338	22.718	27.204	30.144	32.852	36.191	38.582	41.610	4:
20	12.443	15.452	19.337	23.828	28.412	31.410	34.170	37.566	39.997	43.072	4(
21	13.240	16.344	20.337	24.935	29.615	32.671	35.479	38.932	41.401	44.522	4(
22	14.041	17.240	21.337	26.039	30.813	33.924	36.781	40.289	42.796	45.962	4(
23	14.848	18.137	22.337	27.141	32.007	35.172	38.076	41.638	44.181	47.391	4(
24	15.659	19.037	23.337	28.241	33.196	36.415	39.364	42.980	45.559	48.812	5:
25	16.473	19.939	24.337	29.339	34.382	37.652	40.646	44.314	46.928	50.223	5:
26	17.292	20.843	25.336	30.435	35.563	38.885	41.923	45.642	48.290	51.627	5(
27	18.114	21.749	26.336	31.528	36.741	40.113	43.195	46.963	49.645	53.023	5(
28	18.939	22.657	27.336	32.620	37.916	41.337	44.461	48.278	50.993	54.411	5(
29	19.768	23.567	28.336	33.711	39.087	42.557	45.722	49.588	52.336	55.792	5(
30	20.599	24.478	29.336	34.800	40.256	43.773	46.979	50.892	53.672	57.167	5(

Index

WILEY SERIES IN PROBABILITY AND STATISTICS
ESTABLISHED BY WALTER A. SHEWHART AND SAMUEL S. WILKS

Editors: *David J. Balding, Peter Bloomfield, Noel A. C. Cressie,*
Nicholas I. Fisher, Iain M. Johnstone, J. B. Kadane, Louise M. Ryan,
David W. Scott, Adrian F. M. Smith, Jozef L. Teugels
Editors Emeriti: *Vic Barnett, J. Stuart Hunter, David G. Kendall*

The **Wiley Series in Probability and Statistics** is well established and authoritative. It covers many topics of current research interest in both pure and applied statistics and probability theory. Written by leading statisticians and institutions, the titles span both state-of-the-art developments in the field and classical methods.

Reflecting the wide range of current research in statistics, the series encompasses applied, methodological and theoretical statistics, ranging from applications and new techniques made possible by advances in computerized practice to rigorous treatment of theoretical approaches.

This series provides essential and invaluable reading for all statisticians, whether in academia, industry, government, or research.

ABRAHAM and LEDOLTER · Statistical Methods for Forecasting
AGRESTI · Analysis of Ordinal Categorical Data
AGRESTI · An Introduction to Categorical Data Analysis
AGRESTI · Categorical Data Analysis, *Second Edition*
ANDĚL · Mathematics of Chance
ANDERSON · An Introduction to Multivariate Statistical Analysis, *Second Edition*
*ANDERSON · The Statistical Analysis of Time Series
ANDERSON, AUQUIER, HAUCK, OAKES, VANDAELE, and WEISBERG ·
    Statistical Methods for Comparative Studies
ANDERSON and LOYNES · The Teaching of Practical Statistics
ARMITAGE and DAVID (editors) · Advances in Biometry
ARNOLD, BALAKRISHNAN, and NAGARAJA · Records
*ARTHANARI and DODGE · Mathematical Programming in Statistics
*BAILEY · The Elements of Stochastic Processes with Applications to the Natural
    Sciences
BALAKRISHNAN and KOUTRAS · Runs and Scans with Applications
BARNETT · Comparative Statistical Inference, *Third Edition*
BARNETT and LEWIS · Outliers in Statistical Data, *Third Edition*
BARTOSZYNSKI and NIEWIADOMSKA-BUGAJ · Probability and Statistical Inference
BASILEVSKY · Statistical Factor Analysis and Related Methods: Theory and
    Applications
BASU and RIGDON · Statistical Methods for the Reliability of Repairable Systems
BATES and WATTS · Nonlinear Regression Analysis and Its Applications
BECHHOFER, SANTNER, and GOLDSMAN · Design and Analysis of Experiments for
    Statistical Selection, Screening, and Multiple Comparisons
BELSLEY · Conditioning Diagnostics: Collinearity and Weak Data in Regression
BELSLEY, KUH, and WELSCH · Regression Diagnostics: Identifying Influential
    Data and Sources of Collinearity
BENDAT and PIERSOL · Random Data: Analysis and Measurement Procedures,
    Third Edition

*Now available in a lower priced paperback edition in the Wiley Classics Library.

BERRY, CHALONER, and GEWEKE · Bayesian Analysis in Statistics and Econometrics: Essays in Honor of Arnold Zellner

BERNARDO and SMITH · Bayesian Theory

BHAT and MILLER · Elements of Applied Stochastic Processes, *Third Edition*

BHATTACHARYA and JOHNSON · Statistical Concepts and Methods

BHATTACHARYA and WAYMIRE · Stochastic Processes with Applications

BILLINGSLEY · Convergence of Probability Measures, *Second Edition*

BILLINGSLEY · Probability and Measure, *Third Edition*

BIRKES and DODGE · Alternative Methods of Regression

BLISCHKE AND MURTHY · Reliability: Modeling, Prediction, and Optimization

BLOOMFIELD · Fourier Analysis of Time Series: An Introduction, *Second Edition*

BOLLEN · Structural Equations with Latent Variables

BOROVKOV · Ergodicity and Stability of Stochastic Processes

BOULEAU · Numerical Methods for Stochastic Processes

BOX · Bayesian Inference in Statistical Analysis

BOX · R. A. Fisher, the Life of a Scientist

BOX and DRAPER · Empirical Model-Building and Response Surfaces

*BOX and DRAPER · Evolutionary Operation: A Statistical Method for Process Improvement

BOX, HUNTER, and HUNTER · Statistics for Experimenters: An Introduction to Design, Data Analysis, and Model Building

BOX and LUCEÑO · Statistical Control by Monitoring and Feedback Adjustment

BRANDIMARTE · Numerical Methods in Finance: A MATLAB-Based Introduction

BROWN and HOLLANDER · Statistics: A Biomedical Introduction

BRUNNER, DOMHOF, and LANGER · Nonparametric Analysis of Longitudinal Data in Factorial Experiments

BUCKLEW · Large Deviation Techniques in Decision, Simulation, and Estimation

CAIROLI and DALANG · Sequential Stochastic Optimization

CHAN · Time Series: Applications to Finance

CHATTERJEE and HADI · Sensitivity Analysis in Linear Regression

CHATTERJEE and PRICE · Regression Analysis by Example, *Third Edition*

CHERNICK · Bootstrap Methods: A Practitioner's Guide

CHILÈS and DELFINER · Geostatistics: Modeling Spatial Uncertainty

CHOW and LIU · Design and Analysis of Clinical Trials: Concepts and Methodologies

CLARKE and DISNEY · Probability and Random Processes: A First Course with Applications, *Second Edition*

*COCHRAN and COX · Experimental Designs, *Second Edition*

CONGDON · Bayesian Statistical Modelling

CONOVER · Practical Nonparametric Statistics, *Second Edition*

COOK · Regression Graphics

COOK and WEISBERG · Applied Regression Including Computing and Graphics

COOK and WEISBERG · An Introduction to Regression Graphics

CORNELL · Experiments with Mixtures, Designs, Models, and the Analysis of Mixture Data, *Third Edition*

COVER and THOMAS · Elements of Information Theory

COX · A Handbook of Introductory Statistical Methods

*COX · Planning of Experiments

CRESSIE · Statistics for Spatial Data, *Revised Edition*

CSÖRGÖ and HORVÁTH · Limit Theorems in Change Point Analysis

DANIEL · Applications of Statistics to Industrial Experimentation

DANIEL · Biostatistics: A Foundation for Analysis in the Health Sciences, *Sixth Edition*

*DANIEL · Fitting Equations to Data: Computer Analysis of Multifactor Data, *Second Edition*

*Now available in a lower priced paperback edition in the Wiley Classics Library.

DAVID · Order Statistics, *Second Edition*
*DEGROOT, FIENBERG, and KADANE · Statistics and the Law
DEL CASTILLO · Statistical Process Adjustment for Quality Control
DETTE and STUDDEN · The Theory of Canonical Moments with Applications in
   Statistics, Probability, and Analysis
DEY and MUKERJEE · Fractional Factorial Plans
DILLON and GOLDSTEIN · Multivariate Analysis: Methods and Applications
DODGE · Alternative Methods of Regression
*DODGE and ROMIG · Sampling Inspection Tables, *Second Edition*
*DOOB · Stochastic Processes
DOWDY and WEARDEN · Statistics for Research, *Second Edition*
DRAPER and SMITH · Applied Regression Analysis, *Third Edition*
DRYDEN and MARDIA · Statistical Shape Analysis
DUDEWICZ and MISHRA · Modern Mathematical Statistics
DUNN and CLARK · Applied Statistics: Analysis of Variance and Regression, *Second
   Edition*
DUNN and CLARK · Basic Statistics: A Primer for the Biomedical Sciences,
   Third Edition
DUPUIS and ELLIS · A Weak Convergence Approach to the Theory of Large Deviations
*ELANDT-JOHNSON and JOHNSON · Survival Models and Data Analysis
ETHIER and KURTZ · Markov Processes: Characterization and Convergence
EVANS, HASTINGS, and PEACOCK · Statistical Distributions, *Third Edition*
FELLER · An Introduction to Probability Theory and Its Applications, Volume I,
   Third Edition, Revised; Volume II, *Second Edition*
FISHER and VAN BELLE · Biostatistics: A Methodology for the Health Sciences
*FLEISS · The Design and Analysis of Clinical Experiments
FLEISS · Statistical Methods for Rates and Proportions, *Second Edition*
FLEMING and HARRINGTON · Counting Processes and Survival Analysis
FULLER · Introduction to Statistical Time Series, *Second Edition*
FULLER · Measurement Error Models
GALLANT · Nonlinear Statistical Models
GHOSH, MUKHOPADHYAY, and SEN · Sequential Estimation
GIFI · Nonlinear Multivariate Analysis
GLASSERMAN and YAO · Monotone Structure in Discrete-Event Systems
GNANADESIKAN · Methods for Statistical Data Analysis of Multivariate Observations,
   Second Edition
GOLDSTEIN and LEWIS · Assessment: Problems, Development, and Statistical Issues
GREENWOOD and NIKULIN · A Guide to Chi-Squared Testing
GROSS and HARRIS · Fundamentals of Queueing Theory, *Third Edition*
*HAHN · Statistical Models in Engineering
HAHN and MEEKER · Statistical Intervals: A Guide for Practitioners
HALD · A History of Probability and Statistics and their Applications Before 1750
HALD · A History of Mathematical Statistics from 1750 to 1930
HAMPEL · Robust Statistics: The Approach Based on Influence Functions
HANNAN and DEISTLER · The Statistical Theory of Linear Systems
HEIBERGER · Computation for the Analysis of Designed Experiments
HEDAYAT and SINHA · Design and Inference in Finite Population Sampling
HELLER · MACSYMA for Statisticians
HINKELMAN and KEMPTHORNE: · Design and Analysis of Experiments, Volume 1:
   Introduction to Experimental Design
HOAGLIN, MOSTELLER, and TUKEY · Exploratory Approach to Analysis
   of Variance
HOAGLIN, MOSTELLER, and TUKEY · Exploring Data Tables, Trends and Shapes

*Now available in a lower priced paperback edition in the Wiley Classics Library.

*Now available in a lower priced paperback edition in the Wiley Classics Library.

KOTZ and JOHNSON (editors) · Encyclopedia of Statistical Sciences: Supplement Volume

KOTZ, READ, and BANKS (editors) · Encyclopedia of Statistical Sciences: Update Volume 1

KOTZ, READ, and BANKS (editors) · Encyclopedia of Statistical Sciences: Update Volume 2

KOVALENKO, KUZNETZOV, and PEGG · Mathematical Theory of Reliability of Time-Dependent Systems with Practical Applications

LACHIN · Biostatistical Methods: The Assessment of Relative Risks

LAD · Operational Subjective Statistical Methods: A Mathematical, Philosophical, and Historical Introduction

LAMPERTI · Probability: A Survey of the Mathematical Theory, *Second Edition*

LANGE, RYAN, BILLARD, BRILLINGER, CONQUEST, and GREENHOUSE · Case Studies in Biometry

LARSON · Introduction to Probability Theory and Statistical Inference, *Third Edition*

LAWLESS · Statistical Models and Methods for Lifetime Data

LAWSON · Statistical Methods in Spatial Epidemiology

LE · Applied Categorical Data Analysis

LE · Applied Survival Analysis

LEE and WANG · Statistical Methods for Survival Data Analysis, *Third Edition*

LePAGE and BILLARD · Exploring the Limits of Bootstrap

LEYLAND and GOLDSTEIN (editors) · Multilevel Modelling of Health Statistics

LIAO · Statistical Group Comparison

LINDVALL · Lectures on the Coupling Method

LINHART and ZUCCHINI · Model Selection

LITTLE and RUBIN · Statistical Analysis with Missing Data

LLOYD · The Statistical Analysis of Categorical Data

MAGNUS and NEUDECKER · Matrix Differential Calculus with Applications in Statistics and Econometrics, *Revised Edition*

MALLER and ZHOU · Survival Analysis with Long Term Survivors

MALLOWS · Design, Data, and Analysis by Some Friends of Cuthbert Daniel

MANN, SCHAFER, and SINGPURWALLA · Methods for Statistical Analysis of Reliability and Life Data

MANTON, WOODBURY, and TOLLEY · Statistical Applications Using Fuzzy Sets

MARDIA and JUPP · Directional Statistics

MASON, GUNST, and HESS · Statistical Design and Analysis of Experiments with Applications to Engineering and Science

McCULLOCH and SEARLE · Generalized, Linear, and Mixed Models

McFADDEN · Management of Data in Clinical Trials

McLACHLAN · Discriminant Analysis and Statistical Pattern Recognition

McLACHLAN and KRISHNAN · The EM Algorithm and Extensions

McLACHLAN and PEEL · Finite Mixture Models

McNEIL · Epidemiological Research Methods

MEEKER and ESCOBAR · Statistical Methods for Reliability Data

MEERSCHAERT and SCHEFFLER · Limit Distributions for Sums of Independent Random Vectors: Heavy Tails in Theory and Practice

*MILLER · Survival Analysis, *Second Edition*

MONTGOMERY, PECK, and VINING · Introduction to Linear Regression Analysis, *Third Edition*

MORGENTHALER and TUKEY · Configural Polysampling: A Route to Practical Robustness

MUIRHEAD · Aspects of Multivariate Statistical Theory

MURRAY · X-STAT 2.0 Statistical Experimentation, Design Data Analysis, and Nonlinear Optimization

*Now available in a lower priced paperback edition in the Wiley Classics Library.

MYERS and MONTGOMERY · Response Surface Methodology: Process and Product Optimization Using Designed Experiments, *Second Edition*

MYERS, MONTGOMERY, and VINING · Generalized Linear Models. With Applications in Engineering and the Sciences

NELSON · Accelerated Testing, Statistical Models, Test Plans, and Data Analyses

NELSON · Applied Life Data Analysis

NEWMAN · Biostatistical Methods in Epidemiology

OCHI · Applied Probability and Stochastic Processes in Engineering and Physical Sciences

OKABE, BOOTS, SUGIHARA, and CHIU · Spatial Tesselations: Concepts and Applications of Voronoi Diagrams, *Second Edition*

OLIVER and SMITH · Influence Diagrams, Belief Nets and Decision Analysis

PANKRATZ · Forecasting with Dynamic Regression Models

PANKRATZ · Forecasting with Univariate Box-Jenkins Models: Concepts and Cases

*PARZEN · Modern Probability Theory and Its Applications

PEÑA, TIAO, and TSAY · A Course in Time Series Analysis

PIANTADOSI · Clinical Trials: A Methodologic Perspective

PORT · Theoretical Probability for Applications

POURAHMADI · Foundations of Time Series Analysis and Prediction Theory

PRESS · Bayesian Statistics: Principles, Models, and Applications

PRESS and TANUR · The Subjectivity of Scientists and the Bayesian Approach

PUKELSHEIM · Optimal Experimental Design

PURI, VILAPLANA, and WERTZ · New Perspectives in Theoretical and Applied Statistics

PUTERMAN · Markov Decision Processes: Discrete Stochastic Dynamic Programming

*RAO · Linear Statistical Inference and Its Applications, *Second Edition*

RENCHER · Linear Models in Statistics

RENCHER · Methods of Multivariate Analysis, *Second Edition*

RENCHER · Multivariate Statistical Inference with Applications

RIPLEY · Spatial Statistics

RIPLEY · Stochastic Simulation

ROBINSON · Practical Strategies for Experimenting

ROHATGI and SALEH · An Introduction to Probability and Statistics, *Second Edition*

ROLSKI, SCHMIDLI, SCHMIDT, and TEUGELS · Stochastic Processes for Insurance and Finance

ROSENBERGER and LACHIN · Randomization in Clinical Trials: Theory and Practice

ROSS · Introduction to Probability and Statistics for Engineers and Scientists

ROUSSEEUW and LEROY · Robust Regression and Outlier Detection

RUBIN · Multiple Imputation for Nonresponse in Surveys

RUBINSTEIN · Simulation and the Monte Carlo Method

RUBINSTEIN and MELAMED · Modern Simulation and Modeling

RYAN · Modern Regression Methods

RYAN · Statistical Methods for Quality Improvement, *Second Edition*

SALTELLI, CHAN, and SCOTT (editors) · Sensitivity Analysis

*SCHEFFE · The Analysis of Variance

SCHIMEK · Smoothing and Regression: Approaches, Computation, and Application

SCHOTT · Matrix Analysis for Statistics

SCHUSS · Theory and Applications of Stochastic Differential Equations

SCOTT · Multivariate Density Estimation: Theory, Practice, and Visualization

*SEARLE · Linear Models

SEARLE · Linear Models for Unbalanced Data

SEARLE · Matrix Algebra Useful for Statistics

SEARLE, CASELLA, and McCULLOCH · Variance Components

SEARLE and WILLETT · Matrix Algebra for Applied Economics

SEBER · Linear Regression Analysis

*Now available in a lower priced paperback edition in the Wiley Classics Library.

*Now available in a lower priced paperback edition in the Wiley Classics Library.